AF675500

Protecting the polar marine environment

How can we best protect the polar marine environment against pollution? Leading scholars on environmental law, the law of the sea, and Arctic and Antarctic affairs examine this important question. To what extent do existing global instruments of environmental protection apply to the Arctic Ocean and the Southern Ocean? Can the arrangements adopted at regional, sub-regional and national levels provide adequate protection? This book examines and compares various levels of regulation in protecting the marine environment of the Arctic and Antarctic, with specific attention to land-based activities, radioactive waste dumping, and shipping in ice-covered waters. Recent developments since the establishment of the Arctic Council in 1996 and the entry into force of the Protocol on Environmental Protection to the Antarctic Treaty in 1998 are also discussed. This is a volume that will appeal to polar specialists and to all those interested in environmental law and policy.

DAVOR VIDAS is Director of the Polar Programme at the Fridtjof Nansen Institute in Norway, and a Senior Research Fellow at the Institute. His recent books include *Governing the Antarctic: The Effectiveness and Legitimacy of the Antarctic Treaty System* (with O. S. Stokke, 1996), *Arctic Development and Environmental Challenges* (1997), *Order for the Oceans at the Turn of the Century* (with W. Østreng, 1999) and *Implementing the Environmental Protection Regime for the Antarctic* (2000).

Protecting the polar marine environment

LAW AND POLICY FOR POLLUTION PREVENTION

Edited by
DAVOR VIDAS

CAMBRIDGE
UNIVERSITY PRESS

CAMBRIDGE UNIVERSITY PRESS
Cambridge, New York, Melbourne, Madrid, Cape Town, Singapore, São Paulo

Cambridge University Press
The Edinburgh Building, Cambridge CB2 2RU, UK

Published in the United States of America by Cambridge University Press, New York

www.cambridge.org
Information on this title: www.cambridge.org/9780521663113

© The Fridtjof Nansen Institute 2000

This publication is in copyright. Subject to statutory exception
and to the provisions of relevant collective licensing agreements,
no reproduction of any part may take place without
the written permission of Cambridge University Press.

First published 2000
This digitally printed first paperback version 2006

A catalogue record for this publication is available from the British Library

Library of Congress Cataloguing in Publication data

Protecting the polar marine environment: law and policy for pollution prevention /
edited by Davor Vidas
p. cm.
Includes index.
ISBN 0 521 66311 3
1. Marine pollution – Law and legislation – Polar regions. 2. Marine resources
conservation – Law and legislation – Polar regions. I. Vidas, Davor.
K3592.72.P76 2000
344′.046343′0911 – dc21 00-023021

ISBN-13 978-0-521-66311-3 hardback
ISBN-10 0-521-66311-3 hardback

ISBN-13 978-0-521-03296-4 paperback
ISBN-10 0-521-03296-2 paperback

MOJOJ DRAGOJ LIVICI ZA PETI ROĐENDAN,

TATA

Contents

List of figures

List of tables

Notes on the contributors

Alan Boyle is Professor of Public International Law at the University of Edinburgh. He is General Editor of the *International and Comparative Law Quarterly* and co-author, with P. Birnie, of *International Law and the Environment* (1993). Among his many publications are the recent edited volumes, *The International Law Commission and the Future of International Law* (with M. Anderson, A. V. Lowe and C. Wickremasinghe, 1998) and *International Law and Sustainable Development* (with D. Freestone, 1999).

Lawson Brigham is a researcher at the Scott Polar Research Institute, University of Cambridge. A US Coast Guard officer 1970–95, he has sailed on all oceans and was captain of the icebreaker *Polar Sea* on Arctic and Antarctic expeditions, including a 1994 scientific crossing of the Arctic Ocean. Captain Brigham has published widely on polar science, ice navigation and the Northern Sea Route, including editorship of the book, *The Soviet Maritime Arctic* (1991).

Douglas Brubaker is Senior Research Fellow at the Fridtjof Nansen Institute, where he was coordinator of the project on Legal, Political and Security Aspects within the International Northern Sea Route Programme (INSROP, 1993–8). He is the author of *Marine Pollution and International Law: Principles and Practice* (1993) and is currently preparing a monograph on Russian Arctic waters in international law.

Christopher Joyner is Professor of Government and International Law at Georgetown University, Washington, DC. He is the author or editor of numerous publications addressing various issues of international affairs and international law. An expert on Antarctic politics and law, his recent books include *Governing the Frozen Commons: The Antarctic Regime and Environmental Protection* (1998) and *Eagle Over the Ice: US Foreign Policy and Antarctica* (co-authored with E. Theis, 1997).

Donald Rothwell is Associate Professor at the University of Sydney Faculty of Law. He is also acting director of the Australian Centre for Environmental Law, and has published widely on environmental law, as well as on Arctic and Antarctic law, including *Polar Regions in International Law* (1996) and, with R. Davis, *Antarctic Environmental Protection: A Collection of Australian and International Instruments* (1997).

Olav Schram Stokke is Research Director at the Fridtjof Nansen Institute. He has written extensively on international institutions and regime theory, especially on the effectiveness and implementation of regional cooperation in resource management and environmental protection. He is co-editor of *Governing the Antarctic: The Effectiveness and Legitimacy of the Antarctic Treaty System* (with D. Vidas, 1996) and *The Barents Region: Cooperation in Arctic Europe* (with O. Tunander, 1994).

David VanderZwaag is Professor of Law at Dalhousie University, Nova Scotia, and former co-director of that university's Marine Affairs Program. An Associate with the Oceans Institute of Canada, he has published widely on ocean and environmental law and policy, including *Canada and Marine Environmental Protection* (1995) and *Oceans Law and Policy in the Post-UNCED Era: Australian and Canadian Perspectives* (co-editor, 1996).

Davor Vidas is Director of the Polar Programme at the Fridtjof Nansen Institute. He has been member/adviser to the Norwegian delegation to Antarctic Treaty meetings since 1992. His books address both Antarctic and Arctic affairs, law of the sea and environmental law, including most recently *Order for the Oceans at the Turn of the Century* (co-edited with W. Østreng, 1999) and *Implementing the Environmental Protection Regime for the Antarctic* (2000).

Budislav Vukas is Professor of Public International Law at the University of Zagreb Faculty of Law, and Judge at the International Tribunal for the Law of the Sea. A member of the *Institut de Droit international*, Professor Vukas has made substantial contributions to various fields of international law, including the law of the sea, environmental law and human rights. Among his many publication activities is the editorship of the five-volume series *Essays on the New Law of the Sea* (1985–96).

Preface and acknowledgements

This book represents part of the final outcome of a three-year international research project, 'Polar Oceans and the Law of the Sea' (POLOS, 1996–8).[1] The POLOS project, which was directed by the editor of this book, sought to analyse global and regional solutions in the law of the sea and ocean policy as these relate to the Arctic Ocean and the Southern Ocean.

In addition to the Fridtjof Nansen Institute – which initiated the POLOS project in 1995 and coordinated it until its finalisation in December 1998 – experts from four continents participated in this project. Here it should be stated that, for those officials who took part in the project, their participation and the views expressed are in their personal capacity only.

This book has been written by a group of nine contributors, all prominent experts in their respective fields, from scholarly institutions located in six countries: Australia, Canada, Croatia, Norway, the United Kingdom and the United States.

In the course of the POLOS project, the research team met once each year for project workshops, where various drafts for the book's chapters were discussed. Along with acknowledging the contributions made by participants in the project, a special acknowledgement is also due to the institutions which provided venues for the workshops: in 1998, the Inter-University Centre in Dubrovnik, Croatia; in 1997, the Fridtjof Nansen Institute at Polhøgda near Oslo, Norway; and in 1996, the Antarctic Cooperative Research Centre at the University of Tasmania in Hobart, Australia. In connection with these arrangements, I especially wish to extend my gratitude to several persons who provided hospitality and organisational help: for the 1998 workshop, *hvala* to Berta Dragičević, Srećko Kržić and Dubravka Kapetanić; for the 1997 workshop, *takk* to Jan Magne Markussen, and to Willy and Sonja Østreng; and for the 1996 workshop, *thanks* to Bruce and Rosalie Davis, and to Richard and Twila Herr. Moreover, on the occasion of the 1996 workshop held in

[1] Another book, titled *Governing High Seas Fisheries: The Interplay of Global and Regional Regimes* and edited by Olav Schram Stokke, is forthcoming by the Oxford University Press.

Hobart, His Excellency the Governor of Tasmania Sir Guy Green hosted a seminar on the polar oceans and law of the sea, attended by the project research team and several leading Australian/Tasmanian policy-makers and experts on Antarctic affairs. In connection with this arrangement, Sir Guy honoured the project team with his kind hospitality at Government House; and so did, jointly, the Minister and Shadow Minister for Antarctic Affairs of Tasmania, the Honourable Peter Hodgman and the Honourable John White at Parliament House.

A great many experts, scholars and decision-makers alike, have generously offered comments and advice on various parts of the manuscript. While it is not possible to single out all of them here, I wish in particular to thank William Bush, Robin Churchill, Edgar Gold, Franklyn Griffiths, Alex Oude Elferink and Ashley Roach, as well as two anonymous reviewers of Cambridge University Press who provided very useful comments on the draft manuscript submitted in 1998.

Appreciation and thanks are due also to the publisher, especially to Finola O'Sullivan, Commissioning Editor for Law at Cambridge University Press, for her professional assistance in various phases which led towards the production of this book, and not least for her patience and understanding in the final phase of manuscript completion.

Similarly, I wish to express my gratitude to Egil Nansen, grandson of Fridtjof Nansen, for permission to use the drawing by Fridtjof Nansen on the cover of the book, showing the polar vessel *Fram*, which was used in some of the most famous expeditions in both the Arctic and Antarctic.

My special thanks also go to Susan Høivik, who undertook the language editing of the manuscript in her superb way, as always.

I wish to acknowledge the invaluable assistance and support received from my colleagues at the Fridtjof Nansen Institute, in particular Olav Schram Stokke, for his never-ending energy and enthusiasm in reading and commenting on draft after draft of various chapters through all the project years, and for his unfailing ability to provide lucid comments; Anne Berteig, for her careful and conscientious copy-editing of the manuscript; Kari Lorentzen, for invaluable bibliographical assistance; Claes Ragner Lykke, for cartographic assistance; Maryanne Rygg and Ivar Liseter, for various formatting assistance; and Snorre Fjelstad, for technical assistance in organising the 1997 project workshop.

And, finally, I wish to extend sincere thanks and appreciation to my dear friend Willy Østreng, for many years the director of the Fridtjof Nansen Institute, without whose vision and support the initiation of the POLOS project and, ultimately, this book would hardly have been possible.

Funding for the POLOS project and for the editing of this book was provided by the Research Council of Norway and by the Fridtjof Nansen Institute, whose support is gratefully acknowledged.

Let me add that none of the above-mentioned individuals or organisations is responsible for any shortcomings of this book. Although the individual authors bear the responsibility for own chapters, the editor alone is answerable for the book as such. If the reader should find satisfaction and enrichment in the pages that follow, then of course all those concerned deserve a measure of credit.

The book is up to date as of 8 June 1999.

DAVOR VIDAS
Moščenička Draga, Croatia, August 1999

List of abbreviations

AAT	Australian Antarctic Territory
ACOPS	Advisory Committee on Protection of the Sea
AEPS	Arctic Environmental Protection Strategy (1991)
AFZ	Australian Fishing Zone
AMAP	Arctic Monitoring and Assessment Programme
AMEC	Declaration on Arctic Military Environmental Cooperation (1996)
ARPA	(United States) Arctic Research Policy Act (1984)
ASMA	Antarctic Specially Managed Area
ASOC	Antarctic and Southern Ocean Coalition
ASPA	Antarctic Specially Protected Area
ASPPR	(Canadian) Arctic Shipping Pollution Prevention Regulations (1978, 1991)
ATCM	Antarctic Treaty Consultative Meeting
ATS	Antarctic Treaty System
AWPPA	(Canadian) Arctic Waters Pollution Prevention Act (1970)
AWPPR	(Canadian) Arctic Waters Pollution Prevention Regulations (1978)
Basel Convention	Convention on the Control of Transboundary Movements of Hazardous Wastes and Their Disposal (1989)
BAT	best available techniques
BEAR	Barents Euro–Arctic Region
BEP	best environmental practices
CAGIO	Circumpolar Advisory Group on Ice Operations
CCAMLR	Convention on the Conservation of Antarctic Marine Living Resources (1980)
CCAS	Convention for the Conservation of Antarctic Seals (1972)
CEE	Comprehensive Environmental Evaluation
CEP	Committee for Environmental Protection
CFR	(United States) Code of Federal Regulations

CITES	Convention on International Trade in Endangered Species of Wild Fauna and Flora (1973)
CLC	International Convention on Civil Liability for Oil Pollution Damage (1969)
COMNAP	Council of Managers of National Antarctic Programmes
CRAMRA	Convention on the Regulation of Antarctic Mineral Resource Activities (1988)
CZMA	(United States) Coastal Zone Management Act (1972)
EEZ	exclusive economic zone
EIA	environmental impact assessment
Environmental Protocol	Protocol on Environmental Protection to the Antarctic Treaty (1991)
EU	European Union
FAO	Food and Agriculture Organisation
Fund Convention	International Convention on the Establishment of an International Fund for Compensation for Oil Pollution Damage (1971)
GPA	Global Programme of Action for the Protection of the Marine Environment from Land-Based Activities (1995)
HNS Convention	International Convention on Liability and Compensation for Damage in Connection with the Carriage of Hazardous and Noxious Substances by Sea (1996)
IACS	International Association of Classification Societies
IAEA	International Atomic Energy Agency
IAPG	Interagency Arctic Policy Group
ICJ	International Court of Justice
ICJ Reports	International Court of Justice, *Reports of Judgments, Advisory Opinions and Orders*
IEE	initial environmental evaluation
IGPRAD	Inter-Governmental Panel of Experts on Radioactive Waste Disposal at Sea
ILM	*International Legal Materials*
IMO	International Maritime Organisation
INSROP	International Northern Sea Route Programme
ITLOS	International Tribunal for the Law of the Sea
London [Dumping] Convention	Convention for the Prevention of Marine Pollution by Dumping of Wastes and Other Matter (1972)
LOS Convention	United Nations Convention on the Law of the Sea (1982)
LRTAP	Convention on Long-Range Transboundary Air Pollution (1979)
MARPOL 73/78	International Convention for the Prevention of Pollution from Ships (1973), as modified by the Protocol of 1978

MEPC	Marine Environmental Protection Committee (of IMO)
MMS	(United States) Minerals Management Service
NATO	North Atlantic Treaty Organisation
NSF	National Science Foundation
NSR	Northern Sea Route
NSRA	Northern Sea Route Administration
NSR Guide	(Russian) Guide to Navigating Through the Northern Sea Route (1995)
NSR Regulations	Regulations for Navigation on the Seaways of the Northern Sea Route (1990)
OCS	outer continental shelf
OCSLA	(United States) Outer Continental Shelf Lands Act (1953)
OECD	Organisation for Economic Cooperation and Development
OILPOL	International Convention for the Prevention of Pollution of the Sea by Oil (1954)
OPA	(United States) Oil Pollution Act (1990)
OPRC	International Convention on Oil Pollution Preparedness, Response and Cooperation (1990)
OSPAR Convention	Convention for the Protection of the Marine Environment of the North-East Atlantic (1992)
OTA	(United States) Office of Technology Assessment
OWG	Outside Working Group of technical experts (under IMO)
PAME	Protection of the Arctic Marine Environment Working Group
PCB	polychlorinated biphenyl
Polar Code	(Draft) International Code of Safety for Ships in Polar Waters
POP	persistent organic pollutant
RPA	Regional Programme of Action for the Protection of the Arctic Marine Environment from Land-Based Activities (1998)
RSNT	Revised Single Negotiating Text
SOLAS	International Convention for the Safety of Life at Sea (1974)
STCW	International Convention on Standards of Training, Certification and Watchkeeping for Seafarers (1978)
St.meld	*Stortingsmelding* (Norwegian Government Report to the *Storting*)
TOVALOP	Tanker Owners' Voluntary Agreement Concerning Liability for Oil Pollution (1969)
UN	United Nations
UNCED	United Nations Conference on Environment and Development (1992)
UNCLOS III	Third United Nations Conference on the Law of the Sea (1973–82)
UN/ECE	United Nations Economic Commission for Europe
UNEP	United Nations Environment Programme

UNIDO	United Nations Industrial Development Organisation
UNTS	*United Nations Treaty Series*
USAP	United States Antarctic Research Program
USC	United States Code
USCA	United States Code Annotated

Introductory overview

Protecting the polar marine environment: interplay of regulatory frameworks

DAVOR VIDAS

Recent years have witnessed important developments that affect the polar regions of our globe, as well as their marine environments. In 1998, the Protocol on Environmental Protection to the Antarctic Treaty came into force, and entered the phase of implementation.[1] As to the Arctic, the post-Cold War decade of regional collaboration has resulted in various outcomes as well: in particular the 1997–8 publication by the Arctic Monitoring and Assessment Programme (AMAP) of the two *Arctic Pollution Issues* reports,[2] and the current development of an Arctic Council Action Plan to Eliminate Pollution of the Arctic – though with some still-pending options for a follow-up on the level of regional policy. At the global level, vital developments have been the entry into force of the United Nations Convention on the Law of the Sea, with increasingly universal participation of states,[3] as well as the emergence of other global instruments and arrangements relevant to the polar marine environment.

This book has been prompted largely by those developments. We wish to examine various approaches to protecting the polar marine environment – at the global, regional, sub-regional and domestic levels – and their actual application in selected issue-areas of marine pollution in polar oceans. Let us begin by posing some basic questions.

1. In respect of the various global instruments of environmental protection: to what extent are they *applicable* to the Arctic Ocean and the Southern Ocean?
2. In respect of the more specific arrangements worked out at the regional, sub-regional or national level: are they *sufficient*?

[1] The Protocol was signed in Madrid, on 4 October 1991, and entered into force on 14 January 1998. Text reprinted ILM, Vol. 30, 1991, pp. 1,416ff. On issues involved in implementation of the Protocol see D. Vidas (ed.), *Implementing the Environmental Protection Regime for the Antarctic* (Dordrecht: Kluwer Academic Publishers, 2000).

[2] *AMAP Assessment Report: Arctic Pollution Issues* (Oslo: Arctic Monitoring and Assessment Programme, 1998); and *Arctic Pollution Issues: A State of the Arctic Environment Report* (Oslo: Arctic Monitoring and Assessment Programme, 1997).

[3] UN doc. A/CONF.62/122, 10 December 1982; text reprinted in ILM, Vol. 21, 1982, pp. 1,261ff. As at 8 June 1999, there were 130 parties to the Convention.

Why focus specifically on the polar regions? And why approach protection of their marine environments in a comparative fashion? As to the latter question, despite the abundance of studies on marine environmental protection, comparative studies of different regional marine environment protection regimes are still rare.[4] As we leave the 1990s, it is timely to assess marine environmental protection regimes applicable to the two polar regions which have witnessed such vital developments in the course of the decade. Moreover, the realities of the Arctic and Antarctic pose additional, unique challenges. Both polar oceans have distinctive features that render them special 'polar cases' in many respects, where solutions agreed for warmer seas may not be sufficient nor readily applicable. On the other hand, there is also the mutual polarisation of the Arctic and Antarctic, due to differences in their socio-economic and political settings.

AREAS DEALT WITH IN THIS BOOK

In the context of the law and policy of marine environmental protection, and pollution prevention in particular, we will be interested in *extremes* in global proportions. Our focus will be on a very special part of the global environment – the vast polar ocean areas, largely frozen on the surface but teeming with life beneath their cold covers. The Arctic Ocean and the Southern Ocean are special in the geographical and geophysical sense – situated at the 'ends of the earth', with their extreme conditions, in contrast to all the other, more temperate seas of our world. And they are special in the political and legal sense, not least since the problems of the polar oceans often seem to remain equally remote in the context of global instruments for marine environmental protection – which in turn may diminish any truly global application of their provisions.

Disregarding for a moment both their unique features and the impact of the Arctic and Southern Oceans on the global environment, their sheer size deserves closer notice. The combined surface of the two polar oceans would cover an area approximately five times the size of Europe. *Approximately*, since it is difficult to reach consensus on how to define the Arctic or the Antarctic regions, and, accordingly, to delimit precisely their maritime area; estimates vary by millions of square kilometres, with the criteria depending on the specific context.[5]

[4] See E. Franckx and M. Pallemaerts, 'Conference on "Toxic Reductions Programmes in the North Sea and Baltic Sea: A Comparative Perspective" – Introduction', *International Journal of Marine and Coastal Law*, Vol. 13, 1998, pp. 300–1, and the literature referred to therein.

[5] For example, while an area of approximately 14 million km^2 is most often referred to as the size of the Arctic Ocean, there are considerable variations. The esteemed *Encyclopaedia Britannica*, for instance, varies by almost 2 million km^2: compare 'The Arctic', in *The New Encyclopaedia Britannica, Macropaedia*, Vol. 14, 15th edn (Chicago: Encyclopaedia Britannica, 1986), p. 6, where the surface of the Arctic Ocean is estimated at 12,257 million km^2, with 'Oceans' in *ibid.*, Vol. 25, p. 125, which sets it at 14,090 million km^2. Another assessment adds a further 1 million km^2 to the latter figure, thus yielding a total figure of about 15 million km^2; see Working Group on the Protection of the Arctic Marine Environment, *Report to the Third Ministerial Conference on the Protection of the Arctic Environment, 20–21 March 1996, Inuvik, Canada* (Oslo: Norwegian Ministry of the Environment, 1996), p. 21. For AMAP, marine areas assessed cover approximately 20 million km^2; see *A State of the Arctic Environment Report*, p. 10.

As to the Arctic, that criterion may be based on, *inter alia*, climatic (10°C July isotherm), biological (the tree-line) or geographical (the Arctic Circle, i.e. 66°32′ North latitude) circumstances – to mention only those most often put forward. Thus, the geographer will disagree with the biologist, and both will disagree with the physicist; and this difficulty is further multiplied if we seek a definition of the Arctic relevant for *all* areas of science, including social science and international law. Moreover, policy-makers will often disagree with everyone else, as well as among themselves. Consequently, each of the eight Arctic countries – Canada, Denmark/Greenland, Finland, Iceland, Norway, Russia, Sweden and the United States – has adopted its own definition of the 'Arctic'.[6]

As to the extent of the Antarctic region, the question is complex as well, although made somewhat easier by the isolation of the continent of Antarctica from other landmasses. Moreover, there is the phenomenon of the Antarctic Convergence, which is significant as both the oceanographic and ecosystemic boundary.[7] However, the political and legal context of the Antarctic does not always permit its spatial extension to this Convergence.[8]

This all means that we will have to supplement any exclusively *spatial* determination of either the Arctic or the Southern Ocean with a *functional* criterion, concentrating on the patterns of use, as well as a *political* criterion, based on actual cooperation between states in respect of a certain area thus agreed as referring to a 'region'.[9] Here we must bear in mind the close natural interaction between the marine and terrestrial areas within the polar regions, all the while seeing the two polar oceans as integral parts of the Arctic and Antarctic regions in terms of their socio-economic and political settings.

It may make sense to use the notion of 'polar oceans' as a generic term when contrasting them to other, warmer oceans, but one question demands clarification at the outset: to what extent are the two polar regions comparable at all? And is there any benefit to be gained from treating them jointly? Let us begin by reviewing the basic differences and similarities of the two polar regions.

THE ARCTIC AND ANTARCTIC REGIONS: SIMILARITIES AND CONTRASTS

Do the polar conditions of both the Arctic and Antarctic make these two regions not only *special* but also *similar* cases, in terms of the international

6 In calling those *eight* countries the 'Arctic countries', the criterion used by the Arctic Environmental Protection Strategy and the Arctic Council, respectively, has been followed.

7 On the Southern Ocean in general, see Sir George Deacon, *The Antarctic Circumpolar Ocean* (Cambridge University Press, 1984). On the phenomenon of the Antarctic Convergence in particular, see *ibid.*, pp. 114–19; and on its significance as the natural boundary of the Antarctic ecosystems see M. W. Holdgate, 'The Use and Abuse of Polar Environmental Resources', *Polar Record*, Vol. 22, 1984, p. 28.

8 See the discussion by Vidas, Chapter 4 in this book. See also Boyle, Chapter 1 in this book.

9 For further discussion on understanding the scope of a 'region' as applied to the polar regions and their maritime space, see especially Boyle, Chapter 1 in this book. See also Vukas, Chapter 2; and Stokke, Chapter 6 in this book.

regulation needed for their environmental protection – from which some appropriate 'polar approaches' should be required? Or do their many different socio-economic and political features make the two regions as diametrically opposed as they are in terms of their geographical location and the resulting semantics behind their names: Arctic and *Anti*-Arctic?

Contrasting features

Chiefly as a consequence of major differences in the social, strategic and economic conditions of the two polar regions, they do differ considerably in legal and political terms. When the 1996 Antarctic Treaty Consultative Meeting reviewed the possible mutual relevance of developments in the Arctic and the Antarctic, the emphasis was on:

> the need to bear in mind that, as far as co-ordination was concerned, the political and legal context governing activities in the Arctic and the Antarctic differ considerably.[10]

Indeed, the Arctic still lacks any counterpart to the Antarctic Treaty System, governing the whole spectrum of human activities in the Antarctic with an increasing reliance on 'hard' law.[11] Cooperation among the Arctic Eight has emerged only since the late 1980s, and formally since 1991 within the framework of the Arctic Environmental Protection Strategy.[12] This has been a process based on declarations, i.e. on 'soft' law. Even the Arctic Council has been established, not by an international treaty, but by a declaration.[13] Clearly, these cooperative fora are placed in contrasting social, strategic and economic settings, and here several important differences between the two polar regions emerge.

First, there are *indigenous peoples* inhabiting the Arctic coasts, whereas Antarctica has no native human inhabitants.[14] This very absence of a native population in the Antarctic was, at the time when the Antarctic Treaty was being negotiated,[15] seen as a major factor favouring the founding of what later became the Antarctic Treaty System. A passage from the 1960 US Senate hearings on the ratification of the Antarctic Treaty may serve to illustrate this point:

[10] See paras. 33–7 of the *Final Report of the Twentieth Antarctic Treaty Consultative Meeting, Utrecht, the Netherlands, 29 April–10 May 1996* (The Hague: Netherlands Ministry of Foreign Affairs, 1997).

[11] For a comprehensive review see O. S. Stokke and D. Vidas (eds.), *Governing the Antarctic: The Effectiveness and Legitimacy of the Antarctic Treaty System* (Cambridge University Press, 1996).

[12] Arctic Environmental Protection Strategy, with the Action Plan, was adopted at the First Ministerial Conference on the Protection of the Arctic Environment, at Rovaniemi, Finland, on 14 June 1991. Text reprinted in ILM, Vol. 30, 1991, pp. 1,624ff.

[13] The Arctic Council was established as a 'high level forum' by the Declaration on the Establishment of the Arctic Council, signed by the Arctic Eight in Ottawa, Canada, on 19 September 1996; text reprinted in ILM, Vol. 35, 1996, pp. 1,387ff.

[14] Compare 'Peoples of the North', in *A State of the Arctic Environment Report*, pp. 51–69, with J. C. M. Beltramino, *The Structure and Dynamics of Antarctic Population* (New York: Vantage Press, 1993). Actually, there is some 'native population' even in Antarctica – a dozen or so babies born in Argentine and Chilean scientific bases there.

[15] The Antarctic Treaty was signed in Washington, DC, on 1 December 1959, and entered into force on 23 June 1961; published in UNTS, Vol. 402, pp. 71ff.

SENATOR LAUSCHE: Do you visualize this as an area which, under present conditions, lends itself most favorably to international administration?

MR JESSUP: It clearly is my opinion, Senator, that it has that quality more than any other place on earth, partly because one does not need to deal here with an indigenous population.[16]

In the Arctic context, on the contrary, the presence and demands of the indigenous population may be seen as a factor which prompted the establishment of the Arctic Council, not least linked with domestic policy concerns, especially in Canada and Denmark/Greenland. Nevertheless, while the Antarctic Treaty System is a true form of international administration, the Arctic Council is still largely confined to international consultation.

Secondly, the *strategic importance* of the Arctic, although in military terms significantly diminished in the post-Cold War period,[17] is still far greater than that of the Antarctic. True, this aspect now represents a considerably less striking difference between the two polar regions than only a decade or so earlier. For instance, in 1994 the US administration made an inter-agency review of its Arctic policy, listing environmental protection at the top and thus (at least nominally) 'downgrading' national security and defence considerations.[18] On the other hand, freedom of navigation has traditionally been the strategic military interest of the US Navy, globally as well as Arctic-regionally; and in the latter context particularly when it comes to submarine operations.[19] These concerns are largely distinct from environmental considerations.

This difference is clearly reflected in the constitutive documents of the two regional cooperative processes. While demilitarisation of the Antarctic figures among the basic principles of the Antarctic Treaty, which prohibits any measure of a military nature in the Antarctic,[20] the Arctic Council Declaration expressly states that the Council is *not* to deal with matters related to military security.[21] Instead, environmental protection related to military activity in the Arctic is, on the international level, relegated to separate arrangements among individual states, such as the trilateral Declaration on Arctic Military Environmental Cooperation signed between Russia, the United States and Norway in September 1996.[22]

[16] *Hearings Before the Committee on Foreign Relations, United States Senate, 86th Cong., 2nd Sess., 14 June 1960* (Washington, DC: US Government Printing Office, 1960), p. 55.

[17] See W. Østreng, 'The Post-Cold War Arctic: Challenges and Transitions During the 1990s', in D. Vidas (ed.), *Arctic Development and Environmental Challenges* (Copenhagen: Scandinavian Seminar College, 1997), pp. 33–49.

[18] See 'United States Announces New Policy for the Arctic Region', Press Release of the US Department of State, 29 September 1994. See comments by D. Scrivener, 'Environmental Cooperation in the Arctic: From Strategy to Council', *Security Policy Library*, No. 1 (Oslo: Norwegian Atlantic Committee, 1996), p. 22; and F. Griffiths, 'Environment in the US Discourse on Security: The Case of the Missing Arctic Waters', in W. Østreng (ed.), *National Security and International Environmental Cooperation in the Arctic – The Case of the Northern Sea Route* (Dordrecht: Kluwer Academic Publishers, 1999), pp. 179–203.

[19] See Griffiths, 'Missing Arctic Waters', pp. 197–8. See also Brubaker, Chapter 10 in this book.

[20] Preamble to and Art. I(1) of the Antarctic Treaty.

[21] See explanatory note to para. 1(a) of the Arctic Council Declaration.

[22] Text available at www.denix.osd.mil/denix/Public/Intl/AMEC/declar.html.

Thirdly, various *economic uses*, some of them quite extensive, are present in the Arctic, including the Arctic Ocean. A direct consequence of the differing nature and scope of economic uses of the two polar regions and their oceans – a difference highly relevant to the themes of this book – concerns the type and scale of *sources* of marine pollution situated *within* the polar regions. Of prime importance in this respect, the presence and intensity of land-based sources (by far the largest single source of marine pollution in global terms) in the two polar regions are quite different. Large urban settlements, ports and harbours and other coastal developments, and not least centres of heavy industry – all present in parts of the Arctic, the Russian Arctic in particular – are either absent or negligible in the Antarctic.[23] While some 3.8 million people live in the Arctic region (as assessed by AMAP[24]) approximately 15,000 tourists visit Antarctica annually for shorter periods, and few scientists and station personnel, barely exceeding 1,000 in total, stay year-round.[25] The largest – indeed the only – Antarctic 'town', Villa las Estrellas on King George Island off the Antarctic Peninsula, numbers 50 inhabitants; there can of course be no comparison with Arctic centres such as Murmansk, with its population of close to half a million (and over 1 million in Murmansk *Oblast*). The industrial complexes in the Norilsk area and on the Kola Peninsula, comprising the world's largest nickel-copper smelter, Severonickel, as well as Pechenganickel, are of quite a different order from even the most densely concentrated Antarctic land-based sources of marine pollution: less than two dozen scientific stations and bases, with their related facilities, scattered on King George Island and on the tip of the Antarctic Peninsula.

In contrast to the Arctic, which according to some estimates 'may contain some of the world's largest petroleum reserves . . . located both on land and on the continental shelves',[26] the Antarctic coastal areas have yielded only some indications of mineral resources but no real discoveries. A study made by the US Congress Office of Technology Assessment confirmed that, in the Antarctic, 'there are no known mineral deposits of commercial interest'; it concluded that it 'does not expect that either an oil deposit or metal mine would be developed in Antarctica sooner than about three decades, if ever'.[27] For this reason and others, to be discussed in greater detail in Chapter 4 of this book, all mining in the Antarctic has been prohibited.[28]

There is one other prohibition in force in the Antarctic – that related to the disposal of radioactive waste material.[29] The Arctic, by contrast, is characterised by

[23] See VanderZwaag, Chapter 8 in this book.

[24] *AMAP Assessment Report*, p. 142; on the AMAP area see *ibid.*, pp. 9–10.

[25] See Beltramino, *The Structure and Dynamics of Antarctic Population*. For up-to-date Antarctic tourism statistics, see the website of the International Association of Antarctica Tour Operators, www.iaato.org.

[26] *A State of the Arctic Environment Report*, p. 146.

[27] US Congress, Office of Technology Assessment, *Polar Prospects: A Minerals Treaty for Antarctica* (Washington DC: US Government Printing Office, 1989), pp. 3 and 17.

[28] For the law of the sea implications of this prohibition see D. Vidas, 'Southern Ocean Seabed: Arena for Conflicting Regimes?', in D. Vidas and W. Østreng (eds.), *Order for the Oceans at the Turn of the Century* (The Hague: Kluwer Law International, 1999), pp. 291–314.

[29] See Art. V(1) of the Antarctic Treaty.

the high density of nuclear sources. Questions concerning the storage of spent nuclear fuel, nuclear reactors on land as well as on submarines and icebreakers, and the special issue of decommissioned nuclear submarines – all these represent problems for the Arctic environment, and that of the Russian Arctic especially. The problem of disposal of radioactive waste, and its dumping in the Arctic Ocean in particular, has attracted considerable attention in recent years.[30]

On the other hand, some uses of the two polar oceans are similar in nature, shipping being the most important among these; ice conditions prevail in both the Arctic and the Southern Oceans.[31] However, as a consequence of different geography and human activities, shipping patterns differ considerably. Shipping routes in the Arctic are long circular ones, often passing close to coastlines and through waters under the sovereignty of Arctic coastal states, where different domestic legislation on various ice navigation regimes applies.[32] By contrast, in approaching the Antarctic, shipping has a north–south orientation, traffic volumes are considerably smaller, and there are no domestic navigation regimes in force.

Also several other uses of the polar oceans may be similar in nature, for instance harvesting of marine living resources, scientific research, and tourism. As is the case with shipping, they too will often differ considerably in intensity and patterns.

It should be borne in mind that the sources of pollution affecting the polar oceans do not originate solely within the respective polar regions. *Extra-regional* sources of pollution, often remote from the polar areas themselves, may exert a significant impact on the polar marine environment. Indeed, sources situated at one pole may affect the environment of the other pole. Camplin and Hill have described a typical journey for a nuclide dumped in the cold Arctic water, travelling through the Atlantic, finally reaching the bottom waters of the Southern Ocean, and surfacing in Antarctica, in waters mixed vertically by surface cooling.[33] Recent reports indicate the presence of persistent organic pollutants of extra-regional origin in both polar regions,[34] although it is in the Arctic that this type of environmental contamination may exert significant effects on the indigenous population, for whom local foods remain important dietary and cultural resources.[35]

[30] See Stokke, Chapter 9 in this book. See also S. G. Sawhill, 'Cleaning-Up the Arctic's Cold War Legacy: Nuclear Waste and Arctic Military Environmental Cooperation', *Cooperation and Conflict*, Vol. 35, 2000, pp. 5–36. For an overview of radioactivity in the Arctic, see especially P. Strand, 'Radioactivity', in *AMAP Assessment Report*, pp. 525–620.

[31] On these features, and on the background for the development of an International Code of Safety for Ships in Polar Waters (Polar Code), see Brigham, Chapter 11 in this book.

[32] For Russian regulation of navigation in the Northern Sea Route see Brubaker, Chapter 10 in this book.

[33] W. C. Camplin and M. D. Hill, 'Sea Dumping of Solid Radioactive Waste: A New Assessment', *Radioactive Waste Management and the Nuclear Fuel Cycle*, Vol. 7, 1986, p. 242.

[34] See *Global Environment Outlook 2000: UNEP's Millennium Report on the Environment* (London: Earthscan Publications, 1999), pp. 177–96; and 'Persistent Organic Pollutants', in *AMAP Assessment Report*, pp. 183–371.

[35] See 'Peoples of the Arctic: Characteristics of Human Populations Relevant to Pollution Issues', in *AMAP Assessment Report*, pp. 141–82.

And, finally, although there are *sovereignty disputes* in both polar oceans, they too differ in nature. In the Antarctic such disputes relate to the uncertain status of sovereignty claims; in the Arctic they concern maritime delimitation. In the first half of the twentieth century, seven states – Argentina, Australia, Chile, France, New Zealand, Norway and the United Kingdom – put forward territorial claims to parts of the Antarctic. None of these claims has been expressly recognised by any other country apart from fellow claimants,[36] and even then only partially. Eventually, the seven claimant countries and other parties to the 1959 Antarctic Treaty agreed to put aside their competing positions on territorial claims in the Treaty area, and achieved an 'agreement to disagree' on the sovereignty issue (Article IV), for the sake of establishing a unique form of international governance for the Antarctic.[37] By contrast, in the Arctic there are several generally recognised sovereign coastal states, but several maritime boundaries and jurisdictional zones are disputed among them.

Sharing polar conditions

Notwithstanding all their differences, the Arctic Ocean and the Southern Ocean are both unquestionably characterised by polar conditions. Conventional wisdom tends to see this as the major feature that makes the two similar and thereby different from the rest of the world's oceans. The two polar oceans do share several important characteristics – despite their sharply different placement within the respective polar regions.[38]

Their *oceanographic boundaries* are specific to the polar oceans. The Southern Ocean, while not encircled by any landmass, is bound entirely by the Antarctic Convergence. This is a zone situated mainly between the 50° and 60°S (though extending towards 45°S in the meeting area with the Western Indian Ocean), which separates the cold, ascending and extremely productive waters of the Antarctic from the warmer, biologically less significant waters of the Pacific, Atlantic and Indian Oceans.[39] As to the Arctic, it too has a convergence as a marine boundary, in the zone where cold and diluted water meets warmer and saltier water from the south.[40] The Arctic Ocean is, in addition, almost completely encircled by the landmasses of North America, Eurasia and Greenland. Of the four openings, only one of these is a deep channel that connects the Arctic Ocean to the world

[36] The single possible exception being South Africa's implicit recognition of the Norwegian claim in 1959, in relation to use of an old Norwegian base in Queen Maud Land; see W. M. Bush, *Antarctica and International Law: A Collection of Inter-State and National Documents*, Vol. III (London: Oceana Publications, 1988), pp. 171 and 195.

[37] See an overview in D. Vidas, 'The Antarctic Treaty System in the International Community: An Overview', in Stokke and Vidas (eds.), *Governing the Antarctic*, pp. 35–60.

[38] While the Southern Ocean entirely surrounds the isolated continent of Antarctica, the Arctic Ocean is – quite the contrary – placed in the centre of the Arctic region, surrounded by continental landmasses.

[39] See *The Times Atlas of the Oceans* (London: Times Books, 1983), p. 51.

[40] On this phenomenon, and the coordinates of this convergence in the Arctic, see *A State of the Arctic Environment Report*, p. 7.

ocean space: this is the passage between the islands of Svalbard and Greenland (the Fram Strait), through which the major circulation of waters into and from the Arctic Basin occurs.[41]

Ecosystems found within these oceanographic and biological boundaries possess specific characteristics. While the living resources of the Arctic Ocean and those of the Southern Ocean differ greatly in various respects, they nevertheless share some important features, as a consequence of polar conditions.[42] First, the food chains of species inhabiting polar waters are characteristically short and simple, with a low number of species but large populations.[43] Secondly, their polar setting means short, site-specific breeding seasons. This in turn makes these living resources more exposed to environmental risks, an easy prey to any degradation of the polar environment. Those two specific polar features make the marine living resources of the polar oceans, while otherwise not necessarily fragile, extremely vulnerable to impacts of human activities. At the same time, the polar marine ecosystems are among the most productive in the world, as cold water is rich in the nutrients essential to marine life.

The presence of *ice*, and sea ice in particular, is one other notable feature common to both polar oceans. Most of the Arctic Ocean surface is covered by sea ice: while the perennial pack ice covers about 8 million km^2, the extent of sea ice is almost double between March and May, when it can cover as much as 15 million km^2.[44] Also, much of the Southern Ocean is covered with ice, though with considerably greater seasonal fluctuation than in the Arctic Ocean.[45] Ice formations do, it is true, differ considerably in the Arctic and the Antarctic. Basically, however, the presence of large areas of ice-infested waters, with the concomitant significantly higher exposure to environmental risks when compared with most areas of the warmer seas, makes the Arctic and the Southern Oceans similar as well as unique in global terms. The persistence of oil as a pollutant in ice-covered sea areas, in combination with the harsh climatic conditions which diminish the efficacy of available oil-slick clean-up methods and equipment, especially makes both polar oceans equally in need of special rules for safety of navigation. Oil spilled on or under ice cannot be cleaned by technologies used in warmer waters, and response capability is very limited. In such special conditions, the emphasis will have to be on the prevention of pollution, instead of remedial measures.

[41] *Ibid.*, pp. 10–11 and 31–2. See also *The Times Atlas of the Oceans*, pp. 24, 50–2 and 62–3; and 'The Arctic', in *Encyclopaedia Britannica*, pp. 6–7.

[42] See K. Sherman, 'Large Marine Ecosystems', in *Encyclopedia of Earth System Science*, Vol. 2 (New York: Academic Press, 1992), pp. 653–7, 661–2.

[43] For the European Arctic see J. R. Hansen, R. Hansson and N. Norris (eds.), *The State of the European Arctic Environment* (Copenhagen: European Environment Agency, 1996), p. 33; for the Antarctic see G. A. Knox, 'The Living Resources of the Southern Ocean: A Scientific Overview', in F. Orrego Vicuña (ed.), *Antarctic Resources Policy: Scientific, Legal and Political Issues* (Cambridge University Press, 1983), pp. 21–60.

[44] *A State of the Arctic Environment Report*, p. 12. See also Brigham, Chapter 11 and Figure 11.1, in this book.

[45] See Deacon, *The Antarctic Circumpolar Ocean*, p. 121; and Brigham, Chapter 11 and Figure 11.1, in this book.

It is the marine environment which not only comprises the major part of these areas, but also provides shipping routes to and through the polar regions. Common to both polar oceans in this respect is the fact that these are areas of *high risk* – significantly higher than in most other waters. Not only are climatic conditions extremely harsh, support facilities from the coast are either sparse or virtually non-existent.

Closely linked to these shared features of the polar oceans is yet another: their impact on *global climate*, and thus their potential role in global warming, which might be influenced by changes reducing the solar-radiation reflecting capability of the sea ice surface.[46] Such a scenario could lead to a vicious circle: the ensuing warmer temperatures could gradually result in the melting of sea ice as well as the Antarctic pack ice, in turn causing a significant rise in water levels of the world ocean.[47]

Here, however, not enough is known about the polar regions and the processes occurring there to project the effect of human activities on their terrestrial and marine ecosystems – and this *insufficiency in scientific knowledge* is yet another common characteristic of the polar oceans. Scientists themselves admit that their knowledge remains extremely tenuous.[48] Many basic questions, largely common to both polar regions, remain as yet unanswered: What constitutes a significant impact in environments that are relatively undisturbed by humans? What methods will minimise the environmental, health, and safety-related risks of living and working in polar regions? Could the polar regions, due to their pristine nature and remoteness, serve as early warning indicators of global climate change or global pollution?[49]

Crucial to such questions is the role of monitoring in assessments of environmental impacts of human activities in the polar oceans, and the possibility of their *comparison*. With their many unique features, it is hardly surprising that the marine components of the Arctic and Antarctic regions should attract particular attention as to the need for their protection.

It is due to those shared polar conditions that the recent Antarctic Treaty Consultative Meetings have not only pointed to differences in legal-political structures for the two polar regions, but also emphasised their similarities in other respects. Thus, the 1998 Meeting:

[46] For example, the covering pack ice of the Arctic Ocean reduces the exchange of energy between ocean and atmosphere by a factor of approximately 100; see 'The Arctic', *Encyclopaedia Britannica*, p. 7.

[47] See J. H. Zumberge, 'Potential Mineral Resource Availability and Possible Environmental Problems in Antarctica', in J. I. Charney (ed.), *The New Nationalism and the Use of Common Spaces: Issues in Marine Pollution and the Exploitation of Antarctica* (Totowa, NJ: Allanheld, Osmun, 1982), pp. 142–3; on various aspects connected with this problematique in respect of Antarctica, see C. Harris and B. Stonehouse (eds.), *Antarctica and Global Climatic Change* (London: Belhaven Press, 1991).

[48] M. A. Champ, D. A. Flemmer, D. H. Landers, C. Ribic and T. DeLaca, 'The Roles of Monitoring and Research in Polar Environments: A Perspective', *Marine Pollution Bulletin* – Thematic Issue on 'Environmental Awareness in Antarctica: History, Problems, and Future Solutions', Vol. 25, 1992, p. 220.

[49] *Ibid.* See also H. Cattle, 'Global Climate Models and Antarctic Climatic Change', in Harris and Stonehouse (eds.), *Antarctica and Global Climatic Change*, p. 22.

> echoed the view that there were several important points of convergence between the two polar areas, not the least with regard to the question of environmental protection.[50]

This has been supplemented by the view shared by some of the Arctic countries at the 1999 Antarctic Treaty Consultative Meeting, according to which 'bipolar approaches could provide an understanding of common environmental aspects' – for which reason it was considered 'desirable to strengthen cooperation in scientific research between the two regions'.[51]

STRUCTURE OF THIS BOOK: INTERPLAY OF REGULATORY FRAMEWORKS

Against the backdrop sketched out above, this book examines and compares various *levels of regulation* in protecting the polar marine environment against pollution – at the global, regional, sub-regional and domestic levels (Part I of the book); and then inquires into the modalities of their actual application in *selected issue-areas* of marine pollution in polar oceans (Part II).

We will pursue two basic questions. First, to what extent are the various global instruments of environmental protection applicable to, or relevant for, the Arctic Ocean and the Southern Ocean? And, secondly, are the more specific arrangements – worked out at the regional, sub-regional or national level – sufficient or adequate in all cases for protecting the polar marine environment?

Chapter 1 addresses the interrelationship of global and regional approaches to marine environmental protection, contrasting the advantages and disadvantages of the regional approach. Regionalism as a model for protecting the marine environment will be either restrictive (where the function of the regional rules is limited) or more liberal, thus allowing for greater action at the regional level – all depending on the source of pollution in question. *Vessel-source* pollution, but also *dumping* at sea in respect of a minimum standard, will tend to belong to the former; pollution from *land-based* activities, to the latter model. We revert to this framework in Part II. The limits of regionalism, its advantages and disadvantages are explored in Chapter 1, to provide a basis for further inquiry in Part I.

Chapters 2 and 3 examine the extent of application of global rules for marine environmental protection to specific conditions of the polar oceans and pollution issues there. These chapters show that existing global instruments contain two types of rules: some apply *equally* to the polar marine environment and the rest of the world ocean, whereas other rules, more limited in number, contain solutions *exclusively* applicable to the polar regions. In both respects, difficulties and contrasts become apparent. Global rules of general applicability are sometimes difficult to apply to special situations of the Arctic and the Southern

[50] Para. 99 of the *Final Report of the Twenty-Second Antarctic Treaty Consultative Meeting, Tromsø, Norway, 25 May–5 June 1998* (Oslo: Norwegian Ministry of Foreign Affairs, 1998).

[51] Para. 113 of the *Final Report of the Twenty-Third Antarctic Treaty Consultative Meeting, Lima, Peru, 24 May–4 June 1999*, available at www.rree.gob.pe/conaan/meeting1.htm.

Oceans, due to the unique ice conditions as well as the unique status of the Antarctic in terms of unresolved sovereignty questions. More specific yet global-level 'polar' rules can be found in various instruments: prominent examples include Article 234 of the LOS Convention ('Ice-covered areas'); the status of Special Area under Annexes I, II and V of MARPOL 73/78; the special provision contained in Article 4(6) of the 1989 Basel Convention; and the emerging Polar Code of Navigation. While these provisions have been included in the global instruments to apply in polar oceans only, they share an additional important feature: none of them is currently applied to *both* the Arctic and the Southern Ocean.

According to Article 197 of the LOS Convention, in protecting the marine environment, 'States shall co-operate on a global basis and, as appropriate, on a regional basis . . . taking into account characteristic regional features'. Chapters 4 to 6 of this book explore how this broad provision has been implemented in protecting the *polar* marine environment.

In Chapter 4, a brief account of the regional cooperative processes in the Arctic and the Antarctic, and their addressing the polar marine environment, is followed by a comparison of approaches actually taken in protecting the marine environment, and their current outcomes. It is here that we see the significant limitations of any common 'polar' approaches – not only in the differences between the respective regional cooperation processes as such, but also in their placement in relation to other levels of regulation. The *composition* of regulatory inventory applicable in the Arctic and Antarctic differs sharply; and this is well illustrated in the subsequent two chapters.

In the Antarctic, a comprehensive *regional* regime for (marine) environmental protection – the 1991 Environmental Protocol – is in force. Chapter 5 looks into the provisions of the Protocol and its annexes as relating to protection of the Antarctic environment against marine pollution. We see how much regulation in the Antarctic relies on regionally centralised arrangements.

In contrast, in the Arctic there is no such comprehensive, legally binding regional regime of marine environmental protection available. The international regulatory picture there is a diffuse one, with more than global–regional interplay. In this it may resemble various other regions of the world – but certainly not the Antarctic. Chapter 6 examines a level of regulation of Arctic marine environmental protection not found in the Antarctic: *sub-regional* cooperation, within the framework of the Barents Euro–Arctic Region and the bilateral Russo–Norwegian arrangements.

The picture is not yet complete, as much is left to action and regulation by individual states at their *domestic* level. Accordingly, the final chapter in Part I (Chapter 7) focuses on domestic perspectives and regulations adopted for the protection of the polar marine environment by several countries. The two-fold role of domestic law and policy, present also in marine environmental protection – that of providing new initiatives and of implementing the resultant international agreements – is here indicated. Three polar states have been selected for this analysis:

Australia, Canada and the United States. They were selected on the basis of several criteria: leadership or importance in recent initiatives for protecting the polar marine environment, elaboration of relevant domestic implementing legislation, and geographic placement. There is Australia, with its interests and legislation linked primarily to the Antarctic, and Canada for the Arctic. Then there is the 'bi-polar' case of the United States. One other important polar state, Russia, is more thoroughly addressed in terms of the various sources of pollution in the Arctic. This is largely the theme of Part II.

The first three chapters in Part II all focus on the major *regional* sources of pollution of the polar marine environment. They thus cut across various layers of regulation – global, regional, sub-regional and national – to arrive at patterns of implementation for each specific case. As to implementation of the framework for models of regionalism set out in Chapter 1, several major sources of regional marine pollution have been selected: land-based activities; the dumping of radio-active waste; and vessel-source pollution and related issues of safety of navigation in polar seas. The choice of these sources of pollution reveals at the outset that the Arctic Ocean is to a far larger extent affected by various human activities within the region than is its southern counterpart.

How do the restrictive and the liberal models of regionalism fare when applied to sources of marine pollution in the polar regions? Chapter 8 examines an array of instruments addressing *land-based* pollution – from the global to the extra-regional, regional and domestic levels. Chapter 9 addresses the *dumping* of radio-active waste in parts of the Arctic Ocean, and examines the direct interplay of the global regime – the London [Dumping] Convention – and Russian activities in the Barents and Kara Seas. Similarly, Chapter 10 looks at the application and interpretation of Article 234 of the LOS Convention in Russian regulation of navigation and *vessel–source* pollution in the Northern Sea Route, as well as relevant practice of some other Arctic states.

Chapter 11 highlights a new trend in regulation, an instrument developed at the global level yet with a potential for bi-polar application: the emerging Polar Code of Navigation. This final chapter seems to provide a suitable epilogue. Soon after the draft Polar Code reached the Antarctic *policy* scene, this to-be-bipolar instrument was in 1999 refocused back to what appears to have been the original intent – of Arctic origin, and thus solely for application to shipping in *Arctic* waters.

I Levels of regulation in the protection of the polar marine environment

1 Globalism and regionalism in the protection of the marine environment

ALAN BOYLE

This chapter is not concerned with what makes the polar regions different, or with the details of the legal and political regimes and institutions which govern them.[1] Our concern here is the relationship between regional regimes and the broader global context of the law of marine environmental protection. No study of the international law relating to protection of the marine environment can fail to note the interplay of global, regional, sub-regional and national rules and institutions, or the variety of interrelated and sometimes overlapping treaties which deal with the marine environment at these various levels.[2] This phenomenon has been likened to a 'Russian doll effect': as one layer of international regulation is peeled away, other layers appear beneath, until eventually the purely national layer is reached.[3]

This portrayal may oversimplify the position of the polar regions, and especially that of the Antarctic.[4] Partly because of the contested legal status of the Antarctic, and partly because of the ambiguities of the 1959 Antarctic Treaty,[5] the relationship between the Antarctic Treaty System and the law of the sea is a complex and uncertain one. Whether these two bodies of law conflict or co-exist is beyond the scope of this chapter, but the question is important to an understanding of the law relating to the protection of the marine environment in polar regions.[6]

[1] For discussion of these matters see in particular the Introductory overview and Chapters 4–6 in this book. For recent comprehensive studies see O. S. Stokke and D. Vidas (eds.), *Governing the Antarctic: The Effectiveness and Legitimacy of the Antarctic Treaty System* (Cambridge University Press, 1996); and D. R. Rothwell, *The Polar Regions and the Development of International Law* (Cambridge University Press, 1996).

[2] For a recent study see H. Ringbom (ed.), *Competing Norms in the Law of Marine Environmental Protection – Focus on Ship Safety and Pollution Prevention* (London: Kluwer Law International, 1997).

[3] S. Sadowski, 'Protection of the Marine Environment of the North Sea: The "Russian Doll" Effect', in Ringbom (ed.), *Competing Norms*, p. 109.

[4] See the discussion by Vidas, Chapter 4 in this book.

[5] UNTS, Vol. 402, pp. 71ff.

[6] See Vukas, Chapter 2 in this book. For comprehensive studies see F. Orrego Vicuña, 'The Law of the Sea and the Antarctic Treaty System: New Approaches to Offshore Jurisdiction', in C. C. Joyner and S. K. Chopra (eds.), *The Antarctic Legal Regime* (Dordrecht: Martinus Nijhoff, 1988), pp. 97–127; C. C. Joyner, 'The Antarctic Treaty System and the Law of the Sea – Competing Regimes in the

Whether regional regimes are part of or separate from a global framework of regulation is but one aspect of the relationship between global, regional and sub-regional approaches to protection of the marine environment. More important for policy-makers is to have an understanding of the comparative advantages and disadvantages of global or regional approaches when deciding whether to regulate and how to do so. Both the Arctic and Antarctic illustrate well the sometimes difficult choices which may have to be made between these different levels of international protection. Decision-makers must deal not only with the question whether to initiate action at a regional or sub-regional level, rather than at a global level: they must also consider what constitutes a 'region' or 'sub-region'. The variety of answers to this basic question reflects both the diversity of state practice, and the complexity of international legal and political responses to the problems of protecting and preserving the marine environment. That is the theme which this chapter will address.

GLOBALISM AND REGIONALISM IN THE LAW OF THE SEA

Regionalism in the pre-UNCLOS III law of the sea

The law of the sea is inherently global. The International Law Commission assumed as much in its codification of the subject in the 1950s; and the words 'region' and 'regional' appear only twice in the four Geneva Conventions of 1958.[7] Nor has there been any suggestion in the case law of the International Court of Justice that it is applying local or regional customary law when adjudicating law of the sea disputes. While the Court's decisions do take account of special circumstances, such as geography or dependence on fisheries,[8] and naturally pay particular attention to the practice of the parties in dispute, the Court has always been careful to articulate its conclusions in terms of a general law of the sea applicable to all states. The Court's general approach suggests that, while there may be, for example, a Latin American perspective on the law of the sea, or Latin American

Footnote 6 (*cont.*)
Southern Ocean?', *International Journal of Marine and Coastal Law*, Vol. 10, 1995, pp. 301–31; D. Vidas, 'The Antarctic Treaty System and the Law of the Sea: A New Dimension Introduced by the Protocol', in Stokke and Vidas (eds.), *Governing the Antarctic*, pp. 61–90; and T. Scovazzi, 'The Antarctic Treaty System and the New Law of the Sea: Selected Questions', in F. Francioni and T. Scovazzi (eds.), *International Law for Antarctica*, 2nd edn (The Hague: Kluwer Law International, 1996), pp. 377–94.

[7] See Art. 4(4) of the Convention on the Territorial Sea and the Contiguous Zone (straight baselines) and Art. 12(2) of the Convention on the High Seas (search and rescue); these conventions are published in UNTS, Vol. 516, pp. 205ff and UNTS, Vol. 450, pp. 82ff, respectively. See J. Crawford, 'Universalism and Regionalism from the Perspective of the Work of the International Law Commission', in *International Law on the Eve of the Twenty First Century: Views from the International Law Commission* (New York: United Nations, 1997), p. 99.

[8] See, e.g., *Fisheries* (*United Kingdom* v. *Norway*), ICJ Reports 1951, p. 116; *Fisheries Jurisdiction* (*United Kingdom* v. *Iceland*; *Federal Republic of Germany* v. *Iceland*), ICJ Reports 1974, pp. 3 and 175; *North Sea Continental Shelf* (*Federal Republic of Germany/Denmark*; *Federal Republic of Germany/Netherlands*), ICJ Reports 1969, p. 3.

practice contributing to the development of the law of the sea,[9] there can be no Latin American law of the sea distinct from what prevails elsewhere.

Regionalism in the LOS Convention

The 1982 United Nations Convention on the Law of the Sea[10] presents a more complex picture, however. On the one hand its explicit purpose is to articulate a comprehensive, uniform and global legal order for the world's oceans, and it seeks to sustain that legal order in several ways. Article 309 prohibits reservations and thus compels states to make an 'all or nothing' choice when deciding whether to become a party to the Convention. Article 311 gives the Convention pre-eminence over other agreements; it specifically limits the freedom of parties to create new agreements which are incompatible with the effective execution of the object and purpose of the Convention or which affect either the application of 'the basic principles embodied herein' or the rights and obligations of other parties. This article thus provides a significant constraint on the making of regional agreements by parties to the LOS Convention. At the same time, Article 237 specifically preserves the freedom of states to make further agreements relating to the protection and preservation of the marine environment, provided these are 'concluded in furtherance of the general principles and objectives of this Convention'. The same article also preserves obligations under existing agreements on the marine environment, but requires them to be 'carried out in a manner consistent with the general principles and objectives' of the LOS Convention.

Moreover, Part XV of the Convention subjects disputes concerning the interpretation or application of the Convention to compulsory, binding dispute settlement. Although there are certain exceptions to this principle, disputes concerning the Convention's articles on protection of the marine environment will generally fall within the requirement of compulsory settlement.[11] Regional agreements which derogate from the Convention in violation of Articles 237 or 311 would therefore be open to unilateral challenge by other states parties in one or other of the various fora on which the Convention confers jurisdiction.

The Convention is thus equipped with strong and sophisticated mechanisms intended to preserve its integrity and universality. On the other hand, while recognising that the problems of ocean space are 'closely interrelated' and 'need to

[9] See F. C. Garcia-Amador, 'Latin America and the Law of the Sea', in L. M. Alexander (ed.), *The Law of the Sea: A New Geneva Conference. Proceedings of the 6th Annual Conference of the Law of the Sea Institute, Kingston, Rhode Island, 21-24 June 1971* (Kingston, RI: University of Rhode Island, Law of the Sea Institute, 1972); A. Szekely, *Latin America and the Development of the Law of the Sea*, 2 vols. (Dobbs Ferry, NY: Oceana, 1976); F. Orrego Vicuña (ed.), *The Exclusive Economic Zone: A Latin American Perspective* (Boulder, CO: Westview Press, 1984).

[10] Text reprinted in ILM, Vol. 21, 1982, pp. 1,261ff.

[11] See A. E. Boyle, 'UNCLOS, the Marine Environment and the Settlement of Disputes', in Ringbom (ed.), *Competing Norms*, pp. 241–56. Vukas, Chapter 2 in this book, takes a more cautious view of the extent to which environmental disputes fall within compulsory jurisdiction. On this question, as on others, Art. 297 of the LOS Convention is far from clear.

be considered as a whole',[12] the Convention is replete with references to regional rules, regional programmes, regional cooperation and so on. It makes specific provision for regional cooperation in the case of enclosed and semi-enclosed seas.[13] Moreover, in the case of fisheries management, regional cooperation and regulation are required if the provisions of the Convention[14] and the 1995 Implementing Agreement on Straddling and Highly Migratory Fish Stocks[15] are to be implemented effectively. Part XII of the Convention, dealing with protection of the marine environment, also makes significant reference to regional rules and standards in various contexts.

It is clear therefore that a global law of the sea *can* accommodate regional approaches to certain problems, including protection of the marine environment. There will be no necessary incompatibility with the LOS Convention, provided any regional arrangements are consistent with the object and purpose of the Convention as set out in Articles 237 and 311, and provided they comply with the framework for regulation of the marine environment established by Part XII.

Regionalism in Part XII of the LOS Convention

The interplay between globalism and regionalism in the law of the sea is at its most evident and most complex in Part XII of the LOS Convention. There is no doubt that the fundamental elements of the law of the marine environment – both conventional and customary – are found in these articles of the Convention. They not only build on pre-existing law, including prior regional agreements such as the Baltic and Mediterranean Conventions of 1974 and 1976, respectively,[16] but have provided the basis for subsequent developments, whether at global, regional or national level. There are important linkages between this part of the Convention and other, sectoral, treaties dealing with the marine environment, including the 1972 Convention on the Prevention of Marine Pollution by Dumping of Wastes and Other Matter (London [Dumping] Convention) and its successor Protocol of 1996,[17] as well as the 1973 International Convention for the Prevention of Pollution from Ships, as modified by the Protocol of 1978 (MARPOL 73/78)[18] and other IMO conventions. Part XII also provided the framework for Chapter 17 of Agenda 21 of the Report of the 1992 Rio Conference on Environment and Development, and is specifically referred to in that report as representing the international law on the subject.[19] If that view is correct, then it is not merely regional arrangements

[12] See the Preamble to the LOS Convention.

[13] Arts. 122–123 of the LOS Convention. See the further discussion by Vukas, Chapter 2 in this book.

[14] Arts. 61–70 and 116–120.

[15] Text reprinted in ILM, Vol. 34, 1995, pp. 1,547ff.

[16] For those two conventions see below in this chapter.

[17] ILM, Vol. 11, pp. 1,291ff (Convention); ILM, Vol. 36, 1997, pp. 7ff (Protocol).

[18] ILM, Vol. 12, 1973, pp. 1,319ff (Convention); and ILM, Vol. 17, 1978, pp. 546ff (Protocol).

[19] *Report of the UN Conference on Environment and Development*, Annex II, Agenda 21, Chapter 17, para. 1, UN doc. A/CONF.151/26/Rev.1, Vol. 1; see U. Beyerlin, 'New Developments in the Protection of the Marine Environment: Potential Effects of the Rio Process', *Zeitschrift für ausländisches öffentliches Recht und Völkerrecht*, Vol. 55, 1995, p. 544; P. W. Birnie and A. E. Boyle,

between parties to the Convention which must comply with the constraints of Part XII: so must regional arrangements between non-parties, who will be bound as a matter of customary law.[20]

While setting out a global framework of rules and principles governing marine pollution and the protection of marine ecosystems, Part XII also reflects a pragmatic acceptance that, in certain instances, regional approaches will be necessary or more appropriate even within a broadly uniform and comprehensive global legal order. It does, however, treat different sources of pollution differently in this respect. Within the global framework, two contrasting models of regionalism can be noted – one restrictive, the other more liberal.

The restrictive model of regionalism

This model is exemplified by the provisions of the LOS Convention on dumping at sea and pollution from ships.[21] Here the function of regional rules or treaties is relatively limited: it is to reinforce enforcement and application of the global rules found in the LOS Convention itself and in the 1972 London Convention and MARPOL 73/78. These latter conventions are also global in scope; neither permits regional derogation or the separate adoption of lower regional standards. Their purpose is to provide international *minimum* standards, especially for flag states, and the LOS Convention articles largely serve to reinforce this objective.

At the same time, some elements of regionalism are permissible even here. Although dumping at sea is now globally almost entirely prohibited,[22] regional treaties had for some time been more stringent than was required by the 1972 London Convention in its original form.[23] Neither the LOS Convention nor the London Convention in any way limits the freedom exercised by states to impose additional controls on dumping in response to the environmental circumstances of certain regional seas, including those, such as the Baltic, that are shallow and semi-enclosed.

The scope for regionalism with regard to pollution from ships is necessarily more limited. In the interests of freedom of navigation, MARPOL 73/78 is not merely a minimum standard for flag states, it is also a maximum standard for exclusive economic zone regulation by coastal states.[24] There is some room for

International Law and the Environment (Oxford: Clarendon Press, 1992), pp. 251–99; and A. Yankov, 'The Law of the Sea Convention and Agenda 21: Marine Environmental Implications', in A. E. Boyle and D. Freestone (eds.), *International Law and Sustainable Development: Past Achievements and Future Challenges* (Oxford: Clarendon Press, 1999), pp. 271–96.

[20] *North Sea Continental Shelf*, p. 3. On Part XII of the LOS Convention and customary law see the discussion by Vukas, Chapter 2 in this book.

[21] Arts. 210 and 211 of the LOS Convention.

[22] See the 1996 Protocol to the 1972 London Convention.

[23] See the 1972 Oslo Convention for the Prevention of Marine Pollution by Dumping from Ships and Aircraft (ILM, Vol. 11, 1972, pp. 262ff); the 1974 Helsinki Convention on the Protection of the Marine Environment of the Baltic Sea Area (ILM, Vol. 13, 1974, pp. 546ff); the 1976 Barcelona Protocol for the Prevention of Pollution of the Mediterranean Sea by Dumping from Aircraft and Ships (ILM, Vol. 15, 1976, pp. 300ff); and the 1986 Noumea Protocol for the Prevention of Pollution by Dumping (ILM, Vol. 26, 1987, pp. 38ff).

[24] LOS Convention, Art. 211(5).

additional regional action, however. MARPOL 73/78 itself provides for stricter discharge rules in designated special areas,[25] while the LOS Convention does not prevent coastal states from exercising some control over navigation in environmentally sensitive areas,[26] or the exercise of port state control over compliance with international rules and standards.[27] Article 234 of the LOS Convention also permits additional measures to be taken nationally or regionally to control pollution from ships in ice-covered areas, while Article 211(6) allows for other special areas to be designated by IMO. Under this article IMO has a special responsibility for ensuring that regional or national action affecting navigation falls within the narrow boundaries of acceptability under the LOS Convention and its own conventions. It is really only under Article 234 that there is a significant autonomous discretion conferred on coastal states. The full implications of this article are further considered below, in several other chapters of this book.[28]

The liberal model of regionalism

The more liberal approach is found in the LOS Convention's articles on land-based (including airborne) sources of pollution, and in the practice of states on these. Here, although the negotiation of global rules and standards is encouraged by its Articles 207 and 212, no attempt is made in the LOS Convention either to impose a uniform global standard comparable to that for ships, or even a minimum standard comparable to that for dumping at sea. Indeed, no such global standards exist for land-based or airborne pollution, nor are they likely to be agreed, given the great diversity of sources and the widely differing socio-economic priorities of states when asked to control pollution originating in industrial and agricultural activities. Instead, states are free to set their own standards of regulation, provided only that these meet the more general requirements of Article 194 of the LOS Convention. Briefly, this article requires states to take 'all measures consistent with this Convention that are necessary to prevent, reduce and control pollution of the marine environment from any source, using for this purpose the best practicable means at their disposal'. These measures must minimise to the fullest extent the release of toxic, harmful or noxious substances. States are free to take such measures nationally or jointly, including regionally, as they deem appropriate.

In practice, international action to tackle these sources of pollution remains almost entirely regional. Prior to the 1992 Rio Conference, no agreement could be reached on a stronger global approach to land-based marine pollution.

[25] Annex I, Regulations 9 and 10.

[26] See generally *International Journal of Marine and Coastal Law*, Special Issue on Particularly Sensitive Sea Areas, Vol. 9, 1994.

[27] See LOS Convention, Art. 218, which provides for port state jurisdiction over pollution offences at sea. See also below in this chapter.

[28] See especially Vukas, Chapter 2; Rothwell and Joyner, Chapter 7; and Brubaker, Chapter 10 in this book.

Since Rio, there has been the adoption in 1995 of the non-binding Washington Declaration and the Global Plan of Action for the Protection of the Marine Environment from Land-Based Activities,[29] but this neither sets global standards of pollution control nor does it limit or preclude regional action.[30] Precisely because it does so little, it does not alter the liberal attitude of the LOS Convention towards regionalism in the control of these sources of pollution.

The limits of regionalism: conclusions

What we see when we look at the international law of the marine environment is that rules on pollution from ships are essentially uniform and international at the global level; rules on dumping at sea are given a minimum standard internationally, but have been supplemented and strengthened by a number of regional agreements or by national legislation; and rules on land-based and airborne sources of marine pollution are primarily regional, sub-regional or national in character, with little or no attempt to deal with this problem globally.

How far the LOS Convention constrains regional action thus depends principally on the source of the pollution, and in particular on whether freedom of navigation at sea will be affected. Regional action is least appropriate in this latter case. It is most appropriate in the case of industrial pollution affecting enclosed or semi-enclosed seas. This is where the states in question will share a common interest in taking measures to protect the marine environment, but they will also inevitably want a wide measure of autonomous discretion in deciding when and how far they should act. On other matters, such as pollution emergencies, environmental impact assessment (EIA) or environmental monitoring, the LOS Convention has very little to say beyond a general requirement for states to take action or cooperate.[31] In these cases, regional cooperation is both sensible and permissible. Indeed, looking beyond the marine environment, it is evident that most international action on emergencies, environmental impact assessment and monitoring has been at a regional rather than a global level. There is, for example, no global treaty on EIA, but there is an important UN/ECE treaty covering potentially all of Europe and North America,[32] as well as various other regional and sub-regional agreements.

The LOS Convention both encourages and constrains regionalism with regard to the marine environment. What it does not do is specify what a 'region' is, in any context.

[29] UNEP (OCA)/LBA/IG.2/L.4; reprinted in *Yearbook of International Environmental Law*, Vol. 6, 1995, pp. 883–6. See further T. A. Mensah, 'The International Legal Regime for the Protection and Preservation of the Marine Environment from Land-Based Sources of Pollution', in Boyle and Freestone (eds.), *International Law and Sustainable Development*, pp. 297–324.

[30] For further discussion, see VanderZwaag, Chapter 8 in this book.

[31] Arts. 199, 200, 204 and 206 of the LOS Convention.

[32] The 1991 Espoo Convention on Environmental Impact Assessment in a Transboundary Context; text reprinted in ILM, Vol. 30, 1992, pp. 800ff.

WHAT IS A 'REGION'?

Attempts at definition

The terms 'region' or 'regional', both of which appear in the LOS Convention, are not defined by that convention nor by any other relevant instrument, including Agenda 21. This omission has not hindered reliance on the concept of regionalism, but it gives it an amorphous and open-textured character which makes any attempt at definition essentially descriptive rather than prescriptive.

Literature on the subject distinguishes two or possibly three senses in which the term 'region' has been used in a maritime context: the formal, the functional and the political.[33] A *formal* definition of a marine region would focus on its physical and geographical character, such as the fact that it is an enclosed or semi-enclosed sea. A *functional* definition would concentrate on patterns of use – resource exploitation, navigation, fisheries, defence and so on. A *political* region is essentially defined by little more than the decision of a group of states to cooperate, although some element of geographical propinquity may be implicit even here;[34] for example, an agreement among members of the British Commonwealth should probably not be described as 'regional' in any sense.

These descriptions are probably of more use in understanding *how* a particular region comes to be composed than in telling us what a region *is*. Not surprisingly, after considering use of the term 'regional' in the LOS Convention, one author concludes that 'any kind of co-operation developed by states in a given part of the ocean is regional'.[35] There is no reason to doubt the accuracy of this view. The records of the Third UN Conference on the Law of the Sea disclose no discussion of the term. There is some attempt to define one category of region – the enclosed or semi-enclosed sea – in Article 122.[36] This is presented as an essentially formal concept determined by reference to the geography of the surrounding landmass. On the other hand, such regions also require special treatment for functional reasons – i.e. because they are especially vulnerable to certain environmentally harmful uses.

[33] See also Stokke, Chapter 6 in this book. See generally L. M. Alexander, 'Regional Arrangements in the Oceans', *American Journal of International Law*, Vol. 71, 1977, p. 84; L. M. Alexander, 'New Trends in Marine Regionalism', *Ocean Yearbook*, Vol. 11, 1994, pp. 1–8; A. Vallega, 'The Regional Scale of Ocean Management and Marine Regional Building', *Ocean and Coastal Management*, Vol. 24, 1994, pp. 17–38; and B. A. Boczeck, 'Global and Regional Approaches to the Protection and Preservation of the Marine Environment', *Case Western Reserve Journal of International Law*, Vol. 16, 1984, p. 39.

[34] On cooperation within the Antarctic Treaty System characterised as being 'regional', see also Vidas, Chapter 4 in this book.

[35] Vallega, 'The Regional Scale of Ocean Management'.

[36] See further Vukas, Chapter 2 in this book. See also L. M. Alexander, 'Regionalism and the Law of the Sea: The Case of Semi-Enclosed Seas', *Ocean Development and International Law*, Vol. 2, 1974, p. 151; and Vallega, 'The Regional Scale of Ocean Management'.

UNEP established a 'regional seas' programme in the mid-1970s. Its first regional seas treaty was the 1976 Barcelona Convention for the Protection of the Mediterranean Sea Against Pollution,[37] a sea which meets all the requirements of Article 122 of the LOS Convention. Subsequent UNEP Regional Seas have, however, been wholly eclectic in composition. Some are enclosed or semi-enclosed, such as the Mediterranean and Black Seas; some are oceanic, such as West and East Africa; some are based on island groupings, such as the Caribbean. Some involve ecosystem management or coastal zone management, whereas others do not. No consistent pattern or definition of what constitutes a region is apparent here, beyond a shifting mixture of formal, functional and political elements, whose balance varies from case to case.[38]

Agenda 21 similarly lacks any definition of a marine region. It does, however introduce the idea of integrating the protection of the marine and coastal environment, requiring states to manage the marine environment and adjacent land areas as a single entity.[39] This approach is reflected in the 1995 revision of the Barcelona Mediterranean Convention,[40] and it is a significant innovation. It means that a state may be considered to be in a marine region even if it has no sea coast, provided its adjacent land area falls within the ambit of integrated coastal zone management. There are obvious implications here for the status of Finland and Sweden as 'Arctic' states, as these two countries have no coastline on the Arctic Ocean.

From all of this we can see not only that it is impossible and probably pointless to try to define a region in the law of the sea, but that it is also impossible to draw a clear dividing line between the marine environment and the land environment. This is scarcely surprising, given that the greatest impact on the marine environment comes not from the use of the sea but from the use of the land.

Defining a region thus resolves itself largely into a question of policy: what is the most sensible geographical and political area within which to address the interrelated problems of marine and terrestrial environmental protection? As one author correctly points out:

> development of the basic regional concept has not been stimulated by scientific thought but by the decision-making context and practice of the UN system.[41]

[37] Text reprinted in ILM, Vol. 15, 1976, pp. 290ff.

[38] Vallega, 'The Regional Scale of Ocean Management'; P. M. Haas, 'Save the Seas: UNEP's Regional Seas Programme and the Coordination of Regional Pollution Control Efforts', *Ocean Yearbook*, Vol. 9, 1991, pp. 188–212.

[39] See Agenda 21, Chapter 17; see Alexander, 'Regionalism and the Law of the Sea', but contrast Vallega, 'The Regional Scale of Ocean Management'.

[40] On which see A. Vallega, 'Geographical Coverage and Effectiveness of the UNEP Convention on the Mediterranean', *Ocean and Coastal Management*, Vol. 31, 1996, pp. 199–218; and A. Vallega, 'Regional Level Implementation of Chapter 17: The UNEP Approach to the Mediterranean', *Ocean and Coastal Management*, Vol. 29, 1995, pp. 251–328.

[41] Vallega, 'The Regional Scale of Ocean Management'.

From this perspective it does not matter how a 'region' is defined, so long as it works. What does seem to be important is that there should be close correspondence between the 'political' region and the 'geographical' region: and that is undoubtedly one of the central lessons of UNEP's regional seas programme.[42]

The Antarctic as a marine region

In what sense is the Antarctic marine environment a region? As a polar continent, Antarctica itself is of course a unique region, for various physical, geographical and political reasons. Our concern, however, is to see how the legal regime which now governs the Antarctic defines its marine environment. That legal regime is constituted principally by three main treaties, all interlinked, which belong to the Antarctic Treaty System: the 1959 Antarctic Treaty, the 1980 Convention on the Conservation of Antarctic Marine Living Resources[43] (CCAMLR) and the 1991 Protocol to the Antarctic Treaty on Environmental Protection.[44]

The Antarctic Treaty applies to the area south of 60° South latitude, including all ice shelves.[45] It thus covers both land and sea, including high seas,[46] within what is known as the Antarctic Treaty area. CCAMLR has a broader territorial scope, applying not only to the Antarctic Treaty area, but also to living resources within the 'Antarctic Convergence which form part of the Antarctic marine ecosystem'.[47] The Antarctic Convergence is a distinct mass of cold Antarctic water which constitutes a largely self-contained marine ecosystem with a seasonally fluctuating boundary.

Two points of special importance emerge from these definitions. First, the Antarctic Convergence appears to create the boundary of the biggest example of a 'large marine ecosystem' being adopted as a region for regulatory purposes. Large marine ecosystems have been defined by one author as:

> relatively large regions of the world ocean . . . characterized by unique bathymetry, hydrography, and productivity within which marine populations have adapted reproductive, growth, and feeding strategies.[48]

[42] *Ibid.* [43] Text reprinted in ILM, Vol. 19, 1980, pp. 837ff.

[44] Text of the Environmental Protocol reprinted in ILM, Vol. 30, 1991, pp. 1,416ff. In addition to the Antarctic Treaty, its Protocol and CCAMLR, there is the 1972 Convention for the Conservation of Antarctic Seals (CCAS; reprinted in ILM, Vol. 11, 1972, pp. 251ff). Art. 1(e) of the Protocol provides a legal definition of the ATS, by listing its main components. [45] Art. VI of the Antarctic Treaty.

[46] The application of the Treaty to the high seas was a controversial question during the negotiations, and Art. VI is without prejudice to the rights of states under international law in the high seas area. On the drafting of Art. VI, see A. Van der Essen, 'The Application of the Law of the Sea to the Antarctic Continent', in F. Orrego Vicuña (ed.), *Antarctic Resources Policy: Scientific, Legal and Political Issues* (Cambridge University Press, 1983), pp. 232–3.

[47] Art. I(1) of CCAMLR. See C. Redgwell, 'Protection of Ecosystems under International Law: Lessons from Antarctica', in Boyle and Freestone (eds.), *International Law and Sustainable Development*, pp. 205–24.

[48] See K. Sherman, 'Biomass Yields of Large Marine Ecosystems', *Ocean Yearbook*, Vol. 8, 1989, p. 117 and the literature referred to therein. See also L. M. Alexander, 'Large Marine Ecosystems', *Marine Policy*, Vol. 17, 1993, p. 186.

The region within the Antarctic Convergence certainly fits this description, although scientists have identified some 50 large marine ecosystems in all. Some of these are shallow areas with vertical mixing of nutrients and high productivity; some are current-driven systems, such as the Gulf Stream; others are enclosed or semi-enclosed seas, including some of those now covered by UNEP's regional seas agreements. Unlike any of the UNEP treaties, the 1980 CCAMLR defines the Antarctic marine environment in these terms for the purposes of conservation of living resources only, rather than for protection of the environment as such. However, the 1991 Environmental Protocol also adopts a modified variant of this ecosystem approach. On the one hand, Articles 3, 6 and 8 regulate activities only in the narrower Antarctic Treaty area, rather than the Antarctic Convergence. On the other hand, Article 2 commits the parties to 'the comprehensive protection of the Antarctic environment and dependent and associated ecosystems', while Article 3(1) refers to the 'protection of the Antarctic environment and dependent and associated ecosystems and the intrinsic value of Antarctica' as 'fundamental considerations in the planning and conduct of all activities in the Antarctic Treaty area'. The Convergence is most probably for this purpose a 'dependent and associated ecosystem', which to that extent is covered by the Protocol.[49]

Secondly, the Antarctic is one of the few examples of a region where protection of the terrestrial and marine environments has been significantly integrated in the manner called for by Chapter 17 of Agenda 21. The 1991 Protocol does include a specific annex on prevention of marine pollution, but otherwise its provisions on environmental protection, environmental monitoring and impact assessment, the ban on mineral activities and so on appear to apply equally to the whole land and sea area within the Antarctic Treaty area.[50] The Protocol is indeed the sole example of a single international environmental regime covering an entire continent and its surrounding ocean. Thus, it is probably inaccurate to speak of the Antarctic marine environment as a 'region' in itself: rather, it is simply part of a much larger 'macro-region' of land *and* sea to which the Antarctic Treaty System applies throughout. In this sense, it is once again unique.

The Arctic as a marine region

Although, unlike Antarctica, the Arctic has an indigenous population whose interests need to be accommodated, it is far from being integrated socially or economically. Indeed from this point of view the Antarctic would appear more

[49] See also the discussion by Vidas, Chapter 4 in this book.

[50] The accuracy of this proposition depends on the interpretation of Art. VI of the Antarctic Treaty. Views differ in particular on whether the ban on mining under the 1991 Protocol applies to the Antarctic deep seabed; compare J. I. Charney, 'The Antarctic System and Customary International Law' and L. Migliorino, 'The New Law of the Sea and the Deep Seabed of the Antarctic Region', in Francioni and Scovazzi (eds.), *International Law for Antarctica*, pp. 59–61 and 400–5, respectively. For a recent analysis see D. Vidas, 'Southern Ocean Seabed: Arena for Conflicting Regimes?', in D. Vidas and W. Østreng (eds.), *Order for the Oceans at the Turn of the Century* (The Hague: Kluwer Law International, 1999), pp. 291–314.

closely integrated than the Arctic. Nevertheless, the Arctic is arguably a marine region in several senses. First, it is geographically a large semi-enclosed sea mostly covered by ice.[51]

Secondly, it is functionally a distinct region with its own unique or special environmental characteristics and problems arising from the prevalence of ice and the extremes of climate. Navigation, protection of the environment, and resource management all present special problems, some of these similar to those in Antarctica.[52] For all these reasons it merits coherent treatment as a marine region in its own right.

Thirdly, the Arctic Ocean can be seen as an ecosystem. Like the Antarctic, protection of the terrestrial and marine environments is intimately linked and requires integrated treatment. Unlike the Antarctic, however, it is far from clear what the boundaries of the Arctic ecosystem should be. The tree line? One of the temperature isotherms? Latitude? All are possibilities; none is uniquely compelling in the same way that the Antarctic Convergence represents an obvious ecosystem boundary. Where the Arctic begins and ends is more diffuse, and the answer may be that it should be defined differently for different purposes. Thus, the 1973 International Agreement on the Conservation of Polar Bears and Their Habitats,[53] Article 234 of the 1982 LOS Convention, and the 1991 Arctic Environmental Protection Strategy[54] (AEPS) all apply to different geographical areas, and none defines the Arctic in any definitive sense. Indeed, the AEPS leaves it to each Arctic state to define the geographical scope of the Arctic as a matter of national choice. At the same time, despite these uncertainties, the Arctic Ocean and surrounding landmass is certainly a political region, with evidence of long-standing patterns of cooperation even during the Cold War. This perhaps illustrates once more the essential eclecticism or relativity of the notion of a marine region. What ultimately makes a region cohere as a usable analytical tool is the political and institutional will to see that cooperation is effective within whatever boundaries are chosen.[55]

ADVANTAGES AND DISADVANTAGES OF REGIONALISM

Political uses of regionalism

The growing importance of regional management of the marine environment is evident in various ways. Probably the most notable examples, and certainly the most extensive ones, are to be found in UNEP's regional seas programme. This

[51] Whether the Arctic Ocean meets the criteria of a semi-enclosed sea as defined in Art. 122 of the LOS Convention is more questionable. See the discussion by Vukas, Chapter 2 in this book.

[52] See the Introductory overview to this book. See also an overview in D. R. Rothwell, 'The Arctic Environmental Protection Strategy and International Environmental Cooperation in the Far North', *Yearbook of International Environmental Law*, Vol. 6, 1995, pp. 65–105. For detailed reviews see *AMAP Assessment Report: Arctic Pollution Issues* (Oslo: Arctic Monitoring and Assessment Programme, 1998); and *Arctic Pollution Issues: A State of the Arctic Environment Report* (Oslo: Arctic Monitoring and Assessment Programme, 1997).

[53] Text reprinted in ILM, Vol. 13, 1974, pp. 13ff.

[54] Text reprinted in ILM, Vol. 30, 1991, pp. 1,624ff.

[55] Vallega, 'The Regional Scale of Ocean Management'.

now comprises some twelve separate regions and involves some 160 countries, some in several regions.[56] A few regions have agreed only to adopt action plans (e.g., East and South Asian Seas) but the majority have evolved into a complex network of treaties, protocols and action plans. In most cases there are also institutional arrangements and trust funds, some of which have helped foster significant levels of political and technical cooperation. Of these, the Mediterranean and East Africa are generally thought to be the most successful; the Red Sea and the Gulf are probably the least effective, largely because they lack adequate political and institutional support.[57] The polar regions, the North Sea[58] and the Baltic[59] fall outside UNEP's programme, but here too we find evidence of effective and developed regional cooperation and regulation to protect the marine environment, as other chapters in this book will show.[60]

Another important example of the uses of regionalism can be observed in the arrangements for port state control of shipping. The oldest scheme of this kind involves European states cooperating under the 1982 Paris Memorandum of Understanding on Port State Control[61] to ensure that vessels entering and leaving European ports meet international standards of seaworthiness and pollution control. This particular scheme has undoubtedly helped to deter sub-standard vessels from using European ports, and has incidentally reduced some of the competitive advantages of lower standards enjoyed by some non-European flag of convenience vessels. Comparable regional schemes have thus far (as of 8 June 1999) been adopted in Latin America, Asia-Pacific, the Caribbean, the Mediterranean, the Indian Ocean and West and Central Africa.[62]

Advantages of regionalism

The most important argument for a regional approach to protection of the marine environment is that in many cases it works better than a global

[56] Haas, 'Save the Seas'.

[57] *Ibid.*; P. A. Verlaan and A. S. Khan, 'Paying to Protect the Commons: Lessons from the Regional Seas Programme', *Ocean and Coastal Management*, Vol. 31, 1996, pp. 83–104.

[58] See the papers collected in *International Journal of Estuarine and Coastal Law*, Vol. 5, 1990; Sadowski, 'Protection of the Marine Environment'; S. Saetevik, *Environmental Co-operation between North Sea States* (London: Belhaven, 1988); T. IJlstra, 'Regional Co-operation in the North Sea: An Inquiry', *International Journal of Estuarine and Coastal Law*, Vol. 3, 1988, pp. 181–207; and M. Pallemaerts, 'The North Sea Ministerial Declarations from Bremen to the Hague: Does the Process Generate any Substance?', *International Journal of Estuarine and Coastal Law*, Vol. 7, 1992, pp. 1–26.

[59] See M. Fitzmaurice, *International Legal Problems of the Environmental Protection of the Baltic Sea* (Dordrecht: Martinus Nijhoff, 1992).

[60] See Vidas, Chapter 4; Joyner, Chapter 5; and Stokke, Chapter 6 in this book.

[61] Text reprinted in ILM, Vol. 21, 1982, pp. 1ff.

[62] See G. C. Kasoulides, *Port State Control and Jurisdiction. Evolution of the Port State Regime* (Dordrecht: Martinus Nijhoff, 1993); R. W. J. Schiferli, 'Regional Concepts of Port State Control: A Regional Effort with Global Effects', *Ocean Yearbook*, Vol. 11, 1994, pp. 202–17; D. Anderson, 'Port States and Environmental Protection', in Boyle and Freestone (eds.), *International Law and Sustainable Development*, pp. 325–44; and M. Valenzuela, 'Enforcing Rules against Vessel-Source Degradation of the Marine Environment: Coastal, Flag and Port State Jurisdiction', in Vidas and Østreng (eds.), *Order for the Oceans*, pp. 496–501.

solution.[63] Regional approaches eliminate the disadvantages of unilateralism while enabling states to agree on commitments for common action that may be more feasible to implement than under a more broadly based global scheme. Regional schemes are more likely to respond to common interests in dealing with a common problem.[64] This point is true in many cases for fisheries, dumping of waste, port state control of shipping, pollution emergencies, and probably also for land-based sources of marine pollution. Within an overall global framework, largely provided by the LOS Convention, these problems all appear potentially better handled at the regional level.

Regional approaches also tend to produce institutions that have more cohesion and may be more effective for that. The South Pacific Forum and the Indian Ocean Marine Affairs Commission are perhaps good examples of this argument.[65] On the other hand, some regional institutions undoubtedly fail. Here the Red Sea and the Gulf again show that regionalism does not inevitably work.[66]

A third argument is that regional cooperation may be easier to organise and may prove more effective on technical matters such as monitoring of pollution, environmental impact assessment, scientific research and the dissemination of information and expertise. Again, however, this is not an inevitable outcome.

A fourth benefit is that regional approaches may have an emerging role as a good way of giving effect to Chapter 17 of Agenda 21 and meeting the goals of sustainability and integrated ecosystem management.[67] This is certainly the aim of new treaties adopted under regional seas programmes in the Mediterranean and the Baltic. These do show something of a shift away from the older focus on pollution prevention in favour of ecosystem management and sustainable development.

Finally, regional agreements have been a significant means of implementing the framework provisions of Part XII of the LOS Convention, even before its entry into force in 1994. The state practice evident in these agreements is one reason why Part XII has so quickly come to be regarded as largely a codification of customary law. At the same time, by facilitating some flexibility in implementation, regional arrangements do help accommodate the special needs and varying circumstances of a range of seas with diverse oceanographic and ecological characteristics within a global international law of the sea. Much the same is true of regional regulation of fisheries.

Disadvantages of regionalism

Taken too far, regionalism may weaken the consensus on a genuinely global law of the sea. Fragmentation is an inherent risk in any system of law built

[63] R. W. Knecht, 'A Commentary on the Institutional and Political Aspects of Regional Ocean Governance', *Ocean and Coastal Management*, Vol. 24, 1994, pp. 39–50.
[64] Schiferli, 'Regional Concepts of Port State Control'.
[65] Alexander, 'New Trends in Marine Regionalism'. [66] Haas, 'Save the Seas'.
[67] Alexander, 'New Trends in Marine Regionalism'.

on the consent of states; in a universal medium such as the oceans it carries special risks. There is, however, no real evidence that this has been the effect of regional environmental cooperation. On the contrary, as we have seen, it has arguably strengthened the LOS Convention.

A more significant objection to regional cooperation is that it may fragment the possibilities for, and the effectiveness of international supervision of compliance with environmental standards. The lack of any global oversight has been a real problem with regard to land-based sources of marine pollution. Not only have some of the regional bodies responsible for controlling this source of pollution failed to do an effective job, in some regions there simply are no such institutions.[68] Without an overarching global scheme comparable to the London Convention, there is in these cases no alternative supervisory mechanism and no accountability.[69] This is not *per se* an argument against regionalism, but it is a reminder of the need to integrate both regional and global approaches into an effective whole.

Finally, a problem which remains is that regional agreements dealing with common spaces may create conflict with third parties. This is a potential risk in Antarctica, where non-treaty parties are, in principle, not formally bound by the rules of the Antarctic Treaty System.[70] It is less of a problem in other maritime regions, where the overarching effect of the LOS Convention will give parties rights and dispute settlement options which they can use in the event of any regional-level interference with their rights.

CONCLUSIONS

First, there is no inherent reason why interested states should not or cannot cooperate to produce regional regimes for protection of the marine environment in either the Arctic or the Antarctic.

Secondly, there is nothing in the 1982 LOS Convention or in general international law to inhibit the making of such regional arrangements, provided they do not contravene the objectives of the LOS Convention or the rights of third states.

Thirdly, it is self-evidently essential to define the area of application of any new legal regime, but there can be different definitions for different purposes within the same basic region. Neither 'the Arctic' nor 'the Antarctic' needs to be given a single all-purpose definition – nor have states done so.

And, finally, the real test of regional arrangements is the existence of institutions with the political will and scientific input to make them work effectively. Rules alone cannot solve any of the problems.

[68] See Birnie and Boyle, *International Law and the Environment*, pp. 304–19.

[69] The 1995 Washington Declaration on Protection of the Marine Environment from Land-Based Activities does not create such an institution, but it does seek to strengthen institutional cooperation; see also VanderZwaag, Chapter 8 in this book.

[70] But see Art. X of the 1959 Antarctic Treaty. For the argument that the main principles of the ATS may have acquired customary status *vis-à-vis* non-parties, see Charney, 'The Antarctic System and Customary International Law'.

2 United Nations Convention on the Law of the Sea and the polar marine environment

BUDISLAV VUKAS

The 1982 United Nations Convention on the Law of the Sea (LOS Convention) was conceived as a framework convention regulating the relations of states in respect of all ocean space: it had to regulate all the different legal regimes at sea and all human activities on the seas and oceans.[1] In addition to many other subjects, the Convention deals with the marine environment: it contains a system of rules on the protection and preservation of the marine environment. The application of those general rules to particular parts of the ocean space has often been examined. This chapter will scrutinise the environmental provisions of the LOS Convention with a view to their applicability to the polar oceans.

A very valid reason for such a study can be found in the Arctic Environmental Protection Strategy (AEPS), adopted at the First Ministerial Conference on the Protection of the Arctic Environment in Rovaniemi, Finland, on 14 June 1991, where eight Arctic countries expressed their opinion on the relevance of the LOS Convention also for the implementation of the Strategy, as the Convention reflects customary international law:

> The implementation of the Strategy will be carried out through national legislation and in accordance with international law, including customary international law as reflected in the 1982 United Nations Convention on the Law of the Sea.[2]

[1] The LOS Convention was negotiated through eleven sessions of the Third United Nations Conference on the Law of the Sea (UNCLOS III), in the period 1973–82. It was opened for signature on 10 December 1982, and entered into force on 16 November 1994. On 28 July 1994, the Agreement Relating to the Implementation of Part XI of the Convention was adopted by United Nations General Assembly Resolution 48/263 (the Agreement, which itself entered into force on 28 July 1996, is to be interpreted and applied together with Part XI as a single instrument). As of 8 June 1999, there were 130 parties to the Convention (i.e., 129 states and the European Community). Among them there are twenty-four of the total of twenty-seven Consultative Parties to the Antarctic Treaty; of the eight Arctic countries, Iceland, Finland, Norway, Russia and Sweden are parties to the LOS Convention. Texts of the Convention and the Agreement are reproduced in UN Pub. Sales No. E.97.V.10 (New York: United Nations, 1997).

[2] AEPS, Chapter 1; text reprinted in ILM, Vol. 30, 1991, pp. 1,624ff. On the AEPS see Vidas, Chapter 4 in this book.

As a consequence of a belief in the importance of the LOS Convention, the ministers of the Arctic countries concluded in the AEPS that the preventive measures they take will be 'consistent in particular with the 1982 United Nations Convention on the Law of the Sea',[3] and they agreed to apply 'the principles concerning the protection and preservation of the Marine Environment as reflected in the 1982 United Nations Convention on the Law of the Sea'.[4]

It is interesting to note that in another instrument relevant to the polar oceans and adopted almost simultaneously with the AEPS – the 1991 Protocol on Environmental Protection to the Antarctic Treaty – no reference whatsoever is made to the LOS Convention.[5] Neither the Protocol nor its Annex IV, dealing specifically with the prevention of marine pollution in the Antarctic Treaty area, contain any reference to the LOS Convention, which is supposed to regulate all ocean space.

The general, simplified statement that the LOS Convention reflected customary international law was not quite correct – even in respect of the environmental provisions – at the time of the adoption of the LOS Convention in 1982 or at the time of the adoption of the AEPS in 1991. Currently (as of 8 June 1999), with 130 parties to the Convention, and its solutions being applied to many other treaties as well as to national legislation, the conclusion concerning the customary character of the LOS Convention could be correct in respect of more provisions than at the end of UNCLOS III, or before the entry of the LOS Convention into force. Yet, any particular provision deserves scrutiny before being considered customary law.

The relation between the LOS Convention and customary law remains a subject of considerable interest. Notwithstanding 130 ratifications/accessions, a large number of states are not yet bound by the Convention. Among them are three Consultative Parties to the Antarctic Treaty (Ecuador, Peru and the USA) as well as some other important maritime states (including Canada, Denmark, Iran, Israel and Liberia). However, customary law is of great interest for all states in respect of its rules which have not been codified in the LOS Convention, for example the rules on internal waters. On the other hand, there are customary rules which are being developed independently of the solutions adopted in the LOS Convention. Naturally, while touching upon these complex issues within the context of its main theme, this chapter cannot deal with all those aspects of the relations between treaty and customary law of the sea.

APPLICABILITY OF THE LOS CONVENTION TO THE POLAR OCEANS

Due to the specific geographical, climatic, historical and political circumstances in the polar oceans, and the fact that the LOS Convention does not

[3] *Ibid.*, Chapter 7. [4] *Ibid.*, Chapter 7(i).

[5] Text of the Protocol with Annexes I–IV, adopted in Madrid, Spain, on 4 October 1991; reprinted in ILM, Vol. 30, 1991, pp. 1,461ff. For a discussion of marine pollution prevention under the Protocol, see Joyner, Chapter 5 in this book. See also Vidas, Chapter 4 in this book.

indicate any sea or ocean to which it is or is not applicable, it is often asked whether and to what extent the Convention applies to the polar oceans.

There is much to indicate that the states participating in UNCLOS III intended to draft a 'Charter of the Oceans' – a basic framework convention that would deal with all the major issues of the *entire* ocean space. This intention is revealed in the first preambular paragraph of the LOS Convention, where Conference participants stated that they were prompted 'by the desire to settle . . . all issues relating to the law of the sea'. Furthermore, they expressed their awareness 'that the problems of ocean space are closely interrelated and need to be considered as a whole' (third preambular paragraph). Following this philosophy, 'pollution of the marine environment' has been defined in general terms, in Article 1(1)(4) of the LOS Convention, as:

> the introduction by man, directly or indirectly, of substances or energy into the marine environment . . . which results or is likely to result in such deleterious effects as harm to living resources and marine life, hazards to human health, hindrance to marine activities, including fishing and other legitimate uses of the sea, impairment of quality for use of sea water and reduction of amenities.

The general applicability of the LOS Convention is confirmed also by another characteristic of its contents: it takes into account the specific features of some categories of seas. A special Part is dedicated to enclosed or semi-enclosed seas (Part IX) and another to archipelagic states (Part IV).

At first glance it could seem that Article 234 of the Convention, which provides a specific provision concerning the prevention, reduction and control of marine pollution from vessels in ice-covered areas within the exclusive economic zone, could be a major argument in favour of the global application of the LOS Convention. This provision belongs to Part XII of the Convention, which deals with the protection and preservation of the marine environment, and it aims at resolving the particular problems of some specific seas – the ice-covered areas. Taking into account the drafting history of Article 234, Nordquist, Rosenne and Yankov explain the value of Article 234 as follows:

> The inclusion of article 234 in the Convention as Part XII, section 8, notwithstanding its geographical scope – limited in reality to ice-covered polar regions, *principally the Northern Hemisphere* – emphasises the global character of the whole convention, which applies to all the seas and oceans of the world.[6]

The above quotation discloses the hidden side of Article 234. It was negotiated at UNCLOS III between Canada, the Soviet Union and the United States, and is 'sometimes called the "Arctic" article'.[7] Thus, in negotiating and adopting Article 234, states participating in UNCLOS III did *not* have in mind its application to

[6] M. H. Nordquist (editor-in-chief) with S. Rosenne and A. Yankov (eds.), *United Nations Convention on the Law of the Sea 1982, A Commentary*, Vol. IV (Dordrecht: Martinus Nijhoff, 1991), p. 393 (emphasis added). [7] *Ibid.*

ice-covered sea areas of the Antarctic.[8] This is in line with the dominant opinion at the Conference. Its President, Hamilton Shirley Amerasinghe (speaking as representative of Sri Lanka), formulated this opinion when he indicated in 1975 at the 30th Session of the UN General Assembly one limitation of the scope of UNCLOS III:

> I should make it clear that the question of the status of Antarctica is in no way linked with the issues before the United Nations Conference on the Law of the Sea and, therefore, this question should not delay agreement on a new Convention on the Law of the Sea.[9]

However, this statement by the first President of UNCLOS III should not be understood as generally excluding the legal issues of the Southern Ocean from the scope of the Conference and the Convention it adopted. Amerasinghe only wanted to exclude any linkage of the problems discussed at UNCLOS III with the 'status of Antarctica'. Thus, all law of the *sea* issues, that do not impinge on the unresolved problem of the status of Antarctica (e.g., the regime of the high seas, the main principles on the protection of the marine environment, and the dispute settlement system relating to law of the sea issues) are beyond doubt applicable also to marine areas of the Southern Ocean.

It is not always easy to draw the line between law of the sea rules that do or do not concern the 'status of Antarctica'. However, it is clear that the application of Article 234 is contrary to the approach suggested by President Amerasinghe, namely that this provision is based on the existence of a 'coastal State' to which special rights are given to protect the ice-covered areas within the exclusive economic zone. It is a concept that should not be applied to the waters off Antarctica – where, according to the dominant opinion, there are no generally recognised coastal states and, consequently, there should be no exclusive economic zones.[10]

Notwithstanding the limited scope of this study, many provisions or Parts of the LOS Convention are indirectly linked and relevant to the topic of our concern. They include not only those dealing directly with marine pollution, but also rules on navigation, the establishment of artificial islands, and the exploration of non-living resources, etc. In the following, however, we will focus more closely on three Parts of the Convention that do have major relevance for our topic: Part IX (enclosed or semi-enclosed seas), Part XII (protection and preservation of the marine environment) and Part XV (settlement of disputes). We begin by indicating some of the provisions from other Parts of the Convention that deal directly with protection of the marine environment; most of these relate to navigation.

[8] Alfred van der Essen is cautious: although he accepts its 'general nature', in his view 'Article 234 is principally applicable to the Arctic, where the coastal States are not disputed and the geographical complexity is exceptional'; the realities of the Antarctic 'do not make strict application of it very probable'. See A. van der Essen, 'The Arctic and Antarctic Regions', in R. J. Dupuy and D. Vignes (eds.), *A Handbook on the New Law of the Sea*, Vol. 1 (Dordrecht: Martinus Nijhoff, 1991), pp. 527–8.

[9] See *30th General Assembly Official Records*, 2380th meeting, 1975, para. 36.

[10] See, however, Australian legislation on the EEZ, as discussed by Rothwell and Joyner, Chapter 7 in this book.

NAVIGATION AND THE PROTECTION OF THE MARINE ENVIRONMENT

According to Part II of the LOS Convention, passage of a foreign ship through the territorial sea 'shall be considered to be prejudicial to the peace, good order or security of the coastal State' if it engages in 'any act of wilful and serious pollution contrary to this Convention' (Article 19(2)(h)). The coastal state may adopt laws and regulations in conformity with the Convention and other rules of international law, relating to innocent passage through the territorial sea, in respect of 'the preservation of the environment of the coastal State and the prevention, reduction and control of pollution thereof' (Article 21(1)(f)). When the coastal state designates or prescribes sea lanes and traffic separation schemes in its territorial sea, it may particularly require tankers, nuclear-powered ships and ships carrying nuclear or other inherently dangerous or noxious substances or materials to confine their passage to such sea lanes (Article 22). Such ships, when exercising their right to innocent passage, are to 'carry documents and observe special precautionary measures established for such ships by international agreements' (Article 23).

All these rules on the protection of the marine environment in respect of ships enjoying the right of innocent passage are applicable also to straits used for international navigation (Article 45) and to archipelagic waters (Article 52) when the regime of innocent passage is applied in these areas.

Special rules on the marine environment are contained also in the new regime agreed upon at UNCLOS III for straits used for international navigation – the transit passage regime. Ships in transit passage are required to 'comply with generally accepted international regulations, procedures and practices for the prevention, reduction and control of pollution from ships' (Article 39(2)(b)). States bordering straits may adopt laws and regulations relating to transit passage through straits in respect of 'the prevention, reduction and control of pollution, by giving effect to applicable international regulations regarding the discharge of oil, oily wastes and other noxious substances in the strait' (Article 42(1)(b)).

In the specific legal regime of the exclusive economic zone, the coastal state has the jurisdiction as provided for in the relevant provisions of the Convention with regard to 'the protection and preservation of the marine environment', as will be further elaborated below in this chapter.

The following provisions, although contained in Part VII on the high seas, concern a general duty of the flag state. Every state shall take measures for ships flying its flag to ensure safety at sea with regard to 'the construction, equipment and seaworthiness of ships'; such measures shall include those necessary to ensure 'that the master, officers and, to the extent appropriate, the crew are fully conversant with and required to observe the applicable international regulations concerning . . . the prevention, reduction and control of marine pollution' (Article 94(3)(a) and (4)(c)).

Finally, there is yet another area we should address, although the activity endangering the marine environment is not navigation. In Part XI of the Convention (entitled 'The Area') special consideration is given to the protection of the marine environment in the international seabed area. The duty to take necessary measures to ensure effective protection of the marine environment from harmful effects which may arise from the activities of exploration and exploration of the Area is proclaimed in Article 145 of the Convention.[11] However, specific duties are given to the organs of the International Seabed Authority, in particular the Council (Article 162(2)(x)) and the Legal and Technical Commission (Article 165(2)).

ENCLOSED OR SEMI-ENCLOSED SEAS: PART IX

The topic of 'enclosed or semi-enclosed seas' was listed among the issues that UNCLOS III was convened to resolve (Item 17 of the agenda); at the Conference it was allocated to the Second Committee.[12] Although there has never been a clear-cut definition or an adopted list of such regional seas, some fifteen to twenty states bordering on smaller seas (primarily the Gulf, the Mediterranean and the Baltic) negotiated throughout the Conference the contents of a Part of the Convention dedicated to such seas. Two major fields of disagreement emerged among them: the contents and the legal nature of the provisions on enclosed or semi-enclosed seas.

The most zealous states in these negotiations (Algeria, Iran, Iraq, Turkey, Yugoslavia) insisted on having specific rules for such seas concerning the extension and/or delimitation of coastal marine areas, as well as the regulation of navigation. However, unanimity was absent among states bordering such seas, and other states were unwilling to accept rules for enclosed or semi-enclosed seas that deviated from general rules applicable to all other seas. Such specific rules on navigation, protection of the marine environment etc., intended to regulate the relations among coastal states in such seas, could also affect the rights and duties of third states, as some of these rules would have restricted the freedoms traditionally enjoyed by ships flying all flags.

Thanks to its 'innocent' contents and nature, Part IX eventually survived the controversies among states keen to include provisions on enclosed or semi-enclosed seas, as well as the general opposition of third states. In the final version, Part IX deals only with the living resources of the sea, scientific research and the protection of the marine environment. And in respect of all these issues, merely 'cooperation' has been suggested.

[11] See, however, in relation to the Antarctic Treaty area, D. Vidas, 'Southern Ocean Seabed: Arena for Conflicting Regimes?', in D. Vidas and W. Østreng (eds.), *Order for the Oceans at the Turn of the Century* (The Hague: Kluwer Law International, 1999), pp. 291–314.

[12] See UN doc. A/CONF.62/28, 20 June 1974; and UN doc. A/CONF.62/29, 2 July 1974; *Third United Nations Conference on the Law of the Sea, Official Records*, Vol. III, UN Pub. Sales No. E.75.V.5 (New York: United Nations, 1975), pp. 57 and 59.

The reason for dealing with enclosed or semi-enclosed seas in this chapter is that cooperation in 'the implementation of their rights and duties with respect to the protection and preservation of the marine environment' (Article 123(b)) has been indicated as one field of cooperation for states bordering on an enclosed or semi-enclosed sea. As already mentioned, there is no official list of enclosed or semi-enclosed seas, and commentators differ as to which particular seas are included in this category; often contrasting opinions are expressed with respect to the Mediterranean. From a technical point of view, the definition of enclosed or semi-enclosed seas (Article 122) leaves much to be desired:

> For the purposes of this Convention, 'enclosed or semi-enclosed sea' means a gulf, basin or sea surrounded by two or more States and connected to another sea or the ocean by a narrow outlet or consisting entirely or primarily of the territorial seas and exclusive economic zones of two or more coastal States.

A broad interpretation of that definition would permit the inclusion of almost *all* seas in this category. The phrase permitting consideration of an enclosed or semi-enclosed sea to be every gulf, basin or sea 'consisting entirely or primarily of the territorial seas and exclusive economic zones of two or more coastal States' excludes only the three major oceans and the Southern Ocean from the application of the definition.

However, a flexible interpretation of that definition, that does not account for whether coastal states have actually proclaimed exclusive economic zones, accords with the original reasons for including the question of enclosed and semi-enclosed seas on the agenda of UNCLOS III, and for the insertion of special rules for such seas in the LOS Convention. The following characteristics of such seas were essential for demonstrating the need to adopt special rules for enclosed or semi-enclosed seas:

1. the complexity of navigation in these seas due to their small surface and poor connection with other seas;
2. the growing danger from all types of pollution because of their small size and poor interchange of their waters with adjacent seas; and
3. the necessity of taking specific precautionary measures in relation to the management, conservation and exploitation of the living resources of such seas, as they are endangered by their natural characteristics and pollution.

Taking into account the motives for establishing the category of enclosed or semi-enclosed seas, and the adopted definition, the characteristics of the Arctic Ocean would seem to justify its being considered as an enclosed or semi-enclosed sea.[13]

As already mentioned, the cooperation of coastal states with respect to

[13] Without explaining the reasons for his position, Alfred van der Essen claims that the provisions 'concerning enclosed or semi-enclosed seas (Art. 122), are not applicable to these [i.e., polar] regions, by the very reason of the definition of the said seas': see van der Essen, 'The Arctic and Antarctic Regions', p. 525.

the marine environment is envisaged in Part IX. Yet, it is important to analyse, first, what the scope of that cooperation should be; secondly, who the potential subjects of the cooperation are; and, thirdly, what the probability of their engagement is.

The scope of cooperation of states bordering an enclosed or semi-enclosed sea is 'to coordinate the implementation of their rights and duties with respect to the protection and preservation of the marine environment' (Article 123(b)). Closely related to marine environment protection are the two other fields of cooperation: coordination of the management, conservation, exploration and exploitation of the living resources of the sea; and the coordination of scientific research policies, as well as undertaking of joint programmes of scientific research in the area (Article 123(a) and (c)).

Secondly, Article 123 invites 'States bordering an enclosed or semi-enclosed sea' to cooperate (chapeau of Article 123). The way in which they cooperate among themselves is 'directly or through an appropriate regional organization'. However, in addition to their mutual cooperation, coastal states are requested 'to invite, as appropriate, other interested States or international organizations to cooperate with them' (Article 123(d)).

Thirdly, Article 123 is drafted in such a manner that the legal nature of the commitments of states concerning cooperation is far from clear, yet the drafting history of that provision testifies that the Conference did not want to impose a strict legal obligation to cooperate for states bordering such seas. An early draft of the Convention, the so-called 'Informal Single Negotiating Text' (ISNT, 1975) imposed the duty to cooperate ('shall cooperate'). However, the subsequent draft, the so-called 'Revised Single Negotiating Text' (RSNT, 1976), reverted to the present formula of the chapeau of Article 123. Thus, the final text of the first sentence of Article 123 reads:

> States bordering an enclosed or semi-enclosed sea *should* cooperate with each other in the exercise of their rights and in the performance of their duties under this Convention. (emphasis added)

The flexibility of the first sentence has to a certain extent been corrected by the second sentence, which requires of states that 'they *shall endeavour*, directly or through an appropriate regional organization' (emphasis added) to coordinate their activities in the three abovementioned fields.

Scholars have a hard time explaining the nature of the outcome of such a clumsy compromise. Thus, Nandan and Rosenne concluded that Article 123 'emphasises the need and desirability of cooperation between States bordering an enclosed or semi-enclosed sea'; later, they add that Article 123 'encourages States to initiate attempts to coordinate the functions, activities and policies' mentioned in that Article.[14]

[14] See M. H. Nordquist (editor-in-chief) with S. N. Nandan and S. Rosenne (eds.), *United Nations Convention on the Law of the Sea 1982, A Commentary*, Vol. III (Dordrecht: Martinus Nijhoff, 1995), pp. 356 and 366.

It cannot be said that the commitment of states under Article 123 is entirely devoid of legal force. Although states are not obliged to coordinate their activities, it can be claimed that acts systematically rejecting any negotiations on the protection and preservation of the marine environment of an enclosed or semi-enclosed sea would represent a contravention of the Convention. Thus, there is a *sui generis* legal obligation relative to the establishment of the cooperation concerning the living resources, the marine environment and marine scientific research in enclosed or semi-enclosed seas. However, there is also another aspect of such an obligation: once the cooperation is established – for example, the 1991 Arctic Environmental Protection Strategy and various activities and programmes such as the Arctic Monitoring and Assessment Program (AMAP) – states must cooperate *bona fide*.[15] This means that they must, *inter alia*, provide correct information, permit access to their territory and engage with all available resources in the common endeavours.

PROTECTION AND PRESERVATION OF THE MARINE ENVIRONMENT: PART XII

Part XII of the LOS Convention deals with the protection and preservation of the marine environment. It applies to the entire marine environment, the polar oceans included. Among these rules of general application are provisions of particular relevance for the polar oceans. For example, Article 194(5) deals with vulnerable seas:

> The measures taken in accordance with this Part shall include those necessary to protect and preserve rare or fragile ecosystems as well as the habitat of depleted, threatened or endangered species and other forms of marine life.

It is interesting to note Joyner's remark in respect of Article 194 in general:

> It is in Article 194 of the LOS Convention that legal clout is given to the duty not to pollute ocean space, inclusive of Antarctic seas.[16]

Another provision very important for polar oceans is Article 197 ('Cooperation on a global or regional basis'):

> States shall cooperate on a global basis and, as appropriate, on a regional basis, directly or through competent international organizations, in formulating and elaborating international rules, standards and recommended practices and procedures consistent with this Convention, for the protection and preservation of the marine environment, taking into account characteristic regional features.

[15] See T. Scovazzi, 'Implications of the New Law of the Sea for the Mediterranean', *Marine Policy*, Vol. 5, 1981, p. 307.

[16] See C. C. Joyner, 'The Antarctic Treaty System and the Law of the Sea – Competing Regimes in the Southern Ocean', *International Journal of Marine and Coastal Law*, Vol. 10, 1995, p. 314.

Cooperation on a global basis under Article 197 is unconditionally mandatory ('States shall cooperate'), while cooperation on a regional basis depends upon the circumstances of each particular region ('as appropriate'). The hortatory character of the regional legislative cooperation is even more clear in the French text of the Convention, where it is said that states cooperate on a regional basis *'le cas échéant'*. However, on both levels, global as well as regional, 'characteristic regional features' have to be taken into account in drafting international legislation for the protection and preservation of the marine environment.

The way in which regional cooperation is envisaged in Article 197 approaches the manner in which regional cooperation in a specific type of region – enclosed or semi-enclosed seas – has been provided for in Article 123(b). Yet there are several differences between the two provisions. Article 197 quite clearly indicates the field of cooperation on a regional basis: the formulation and elaboration of international rules, standards and recommended practices and procedures. Article 123(b) is not so explicit: states bordering enclosed or semi-enclosed seas 'coordinate the implementation of their rights and duties with respect to the protection and preservation of the marine environment'. It is not clear whether that formulation was intended to eliminate the right of the states bordering such seas to adopt international rules on marine protection, and to require and permit them only to coordinate implementation of international rules adopted elsewhere. States bordering an enclosed or semi-enclosed sea should be entitled to adopt national and international (regional, sub-regional) rules as much as coastal states in all other seas; they have the right to act in accordance with Part XII of the LOS Convention, which envisages extensive legislative activities of states on an international as well as a national level.

Having mentioned so often the notions of 'enclosed or semi-enclosed seas' and of 'regions', we should look into their meaning, mutual relations and relevance for the polar oceans.

As noted above, a poorly drafted definition exists in the LOS Convention for an 'enclosed or semi-enclosed sea' (Article 123). On the other hand, there is no definition of the widely used term 'region', either in the LOS Convention or in any other relevant international instrument. However, it is obvious that the term is used to describe a variety of seas with different geographic characteristics – such as the Mediterranean, the Gulf area, and the Caribbean – and even for parts of the oceans (e.g., the North East Atlantic). The general use of the term 'region', including in the UNEP's Regional Seas Programme, suggests that the only limit to the term 'region' would be the entire 'ocean space' – 'regional' as being all that is not 'global'. 'Enclosed or semi-enclosed seas', as defined in the LOS Convention, are but a specific type of 'region'.[17]

Considering all the engagements of states in different parts of ocean

[17] For a discussion of the notion of 'region', see Boyle, Chapter 1 in this book; see also Stokke, Chapter 6 in this book.

space, we can conclude that 'regions' are determined by taking into account geographic realities of the seas and oceans, but also the decision of states to deal specifically with maritime activities in one part of the ocean space. That is why Hugo Caminos and Vicente Marotta Rangel were tempted to speak about 'functional regionalism' and 'geographic regionalism'.[18] But these are only slightly different motives for dealing with maritime issues on a 'regional' and not 'global' (universal) basis. Therefore, notwithstanding their differences, both the Arctic and the Antarctic could be considered as 'regions' in the context of the contemporary law of the sea and the actual cooperation of states as to the demanding tasks of exploring, exploiting and protecting the seas and the natural environment in general.[19]

While mentioning the reference by the LOS Convention to other international rules on the protection and preservation of the marine environment, it is important to stress that the Convention does not provide any clear-cut answer on how its provisions relate to other international norms.

First, one of the Final Provisions (Part XVI) of the Convention deals generally with the relation of the Convention to 'other conventions and international agreements' (Article 311). In addition to general rules on that issue, Article 311 contains a provision on the precedence of the LOS Convention, as between states parties, over the 1958 Geneva Conventions on the Law of the Sea (paragraph 1), and on the inviolability of the 'basic principle relating to the common heritage of mankind set forth in article 136' (paragraph 6).

For present purposes, however, the most important provision is paragraph 5 of Article 311, which stipulates that 'This article does not affect international agreements expressly permitted or preserved by other articles of this Convention.' Any doubt concerning the application of that provision to global or regional treaties concluded for the protection of the marine environment has to be eliminated, as there is a *lex specialis* concerning previous and subsequent agreements in this very field. Article 237, the final provision in Part XII, is entitled 'Obligations under other conventions on the protection and preservation of the marine environment'. It stipulates:

> 1. The provisions of this Part are without prejudice to the specific obligations assumed by States under special conventions and agreements concluded previously which relate to the protection and preservation of the marine environment and to agreements which may be concluded in furtherance of the general principles set forth in this Convention.
>
> 2. Specific obligations assumed by States under special conventions, with respect to the protection and preservation of the marine environment, should be carried out in a manner consistent with the general principles and objectives of this Convention.

[18] H. Caminos and V. Marotta Rangel, 'Sources of the Law of the Sea', in Dupuy and Vignes (eds.), *A Handbook on the New Law of the Sea*, pp. 54–7.

[19] See also Boyle, Chapter 1 in this book.

Paragraph 1 of Article 237 is in accordance with the concept of the LOS Convention as an 'umbrella treaty' as regards its environmental provisions. The Convention contains only basic, general principles on the protection of the marine environment. Provisions dealing with particular sources of pollution, with the protection of different seas and with specific questions in relation to the protection of the seas should be embodied in special international instruments. The duty of states to cooperate in formulating international norms, as stipulated in Article 197, stems also from the realisation that the LOS Convention is insufficient as regards the regulation of marine environmental protection.

The concept of the 'umbrella treaty' also has another aspect: it sees the LOS Convention as a set of environmental provisions possessing a higher value than other international rules in the field. This results from the requirement that all future international rules, standards, recommended practices and procedures must be 'consistent with this Convention' (Article 197) and from the provision that 'Specific obligations assumed by States under special conventions . . . should be carried out in a manner consistent with the general principles and objectives of this Convention' (Article 237(2)).

This last provision is vague, as it seems inconsistent with the contents of paragraph 1 of the same Article, and it could affect obligations of states parties to the LOS Convention towards third states which concluded with the above-mentioned 'special conventions' – for whom the LOS Convention is *res inter alios acta*.

PART XII AND CUSTOMARY INTERNATIONAL LAW

The relations of treaty provisions with customary international law are two-fold. They crystallise existing and/or emerging customary law, or they influence the creation of new customary rules. The International Court of Justice has confirmed not only the first process – codification[20] – but also the passing of treaty provisions into the general *corpus* of international law.[21] To this second phenomenon to be possible, the provision concerned should 'be of a fundamentally norm-creating character such as could be regarded as forming the basis of a general rule of law'.[22] Additionally, the Court requires a demonstration of the wide acceptance of the new rule that can sometimes be expressed in only a short period of time. It seems that, in the Court's view, acceptance by the international community would be manifested either by 'a very widespread and representative participation in the convention . . . provided it included that of States whose interests were specially affected'[23] or when 'State practice, including that of States

[20] *Continental Shelf* (*Tunisia/Libyan Arab Jamahiriya*), Judgment, ICJ Reports 1982, p. 38, para. 24.

[21] *North Sea Continental Shelf* (*Federal Republic of Germany/Denmark*; *Federal Republic of Germany/Netherlands*), Judgment, ICJ Reports 1969, p. 41, para. 71.

[22] *Ibid.*, p. 42, para. 72.

[23] *Ibid.*, p. 42, para. 73.

whose interests are specially affected, should have been both extensive and virtually uniform in the sense of the provision invoked – and should moreover have occurred in such a way as to show a general recognition that a rule of law or legal obligation is involved'.[24]

Research in customary international law includes much subjective evaluation on the basis of vague criteria. All the same, given the widespread ratification of the LOS Convention and its abundant quotation in national legislation, for the majority of its provisions of a 'norm-creating character' it is not too risky to qualify them as customary law. The basic provisions on the protection and preservation of the marine environment have also been confirmed in many specific treaties, as well as at the 1972 UN Conference on the Human Environment and the 1992 UN Conference on Environment and Development (UNCED).

General provisions

The basic principle of Part XII of the LOS Convention, as well as customary law in the field (Principle 7 adopted at the 1972 Stockholm UN Conference on the Human Environment),[25] is expressed in Article 192:

> States have the obligation to protect and preserve the marine environment.

The Convention also affirms a state's sovereign right to exploit its natural resources. This is a right to be exercised in accordance with the state's environmental policies and its duty to protect and preserve the marine environment (Article 193; Stockholm Principle 21). This is reconfirmed in Principle 2 of the 1992 Rio Declaration on Environment and Development:

> States have, in accordance with the Charter of the United Nations and the principles of international law, the sovereign right to exploit their own resources pursuant to their own environmental and developmental policies, and the responsibility to ensure that activities within their jurisdiction or control do not cause damage to the environment of other States or of areas beyond the limits of national jurisdiction.[26]

States are to take all necessary measures to carry out their basic duty to protect and preserve the marine environment; to this end they shall use 'the best practicable means at their disposal', taking measures 'in accordance with their capabilities' (Article 194(1)). Here we should note how the drafters of the LOS Convention have taken into account the differences that exist between states. The main consequence to be derived from this provision is the possibility of differentiating between developed and developing states in relation to the interpretation and application of some provisions of the Convention, as well as in regard to future

[24] *Ibid.*, p. 43, para. 74.

[25] *Declaration of the United Nations Conference on the Human Environment*, UN doc. A/CONF.48/14/Rev.1.

[26] *Rio Declaration on Environment and Development*, UN doc. A/CONF.151/5/Rev.1.

national and international actions. Moreover, some of the environmental provisions in the Convention have already been stipulated, with due regard for the special needs of developing states. For example, scientific and technical assistance is to be provided for such countries; furthermore, states parties to the Convention are to promote programmes of scientific, educational, technical and other assistance to developing states for the protection of the marine environment (Article 202(a)). The duties to provide appropriate assistance for both the minimisation of the effects of major incidents and concerning the preparation of environmental assessments are obligatory upon all states parties, especially in relation to developing states (Article 202(b) and (c)). Furthermore, for the purpose of abating pollution, developing states are to be granted preference in the allocation of appropriate funds and technical assistance by international organisations and in the utilisation of their specialised services (Article 203).

All such provisions that take into account the specific situation of developing states represent the implementation of Stockholm Principles 11 and 23. These state that 'the environmental policies of all states should enhance and not adversely affect the present or future development potential of developing countries', that it is necessary to meet 'the possible national and international economic consequences resulting from the application of environmental measures' and that it will be essential to consider 'the extent of the applicability of standards which are valid for the most advanced countries but which may be inappropriate and of unwarranted social cost for the developing countries'.

The same philosophy of differentiating developed from developing countries can be found in UNCED Principle 11 dealing with environmental legislation:

> States shall enact effective environmental legislation. Environmental standards, management objectives and priorities should reflect the environmental and developmental context to which they apply. Standards applied by some countries may be inappropriate and of unwarranted economic and social cost to other countries, in particular developing countries.

This differentiated approach to developed and developing countries could nowadays be considered a general customary principle of international law. The rationale behind it lies not only in the difference in the achieved level of development which would determine a state's ability to contribute to the activities necessary to protect the environment, but also in the difference in the contribution (responsibility) of states to today's global environmental degradation. Thus, Principle 7 of the Rio Declaration reads:

> States shall cooperate in a spirit of global partnership to conserve, protect and restore the health and integrity of the Earth's ecosystem. In view of the different contributions to global environmental degradation, States have common but differentiated responsibilities. The developed countries acknowledge the responsibility that they bear in the international pursuit of sustainable development in view of the pressures their societies place on the

global environment and of the technologies and financial resources they command.

Apart from the principle stated in Article 192, some other general principles of environmental law have also been codified in the LOS Convention. Among them is the duty of states to take all necessary measures to ensure that pollution arising from incidents or activities under their jurisdiction or control does not spread beyond those areas and does not cause damage to other states and their environment (Article 194(2)). Furthermore, states are obliged not to transfer damage or hazards from one area to another, or to transform one type of pollution into another (Article 195).

They are also obliged to take all measures necessary to protect the marine environment from pollution resulting from the use of technologies under their jurisdiction or control, and from the introduction of species, alien or new, to a particular part of the marine environment to which these may cause significant and harmful changes (Article 196(1)).

Cooperation of states

Unlike Section 1 (General provisions), other Sections in Part XII of the LOS Convention do not contain many principles and provisions which could easily be qualified as customary law. As concerns Section 2 (Global and regional cooperation), however, this could be claimed in relation to the duty of states to cooperate in formulating and elaborating international rules and standards (Article 197), on the basis of Principle 21 of the Stockholm Declaration, and consolidated by the conclusion of a multitude of bilateral, sub-regional, regional and global treaties on environmental law.

Another provision contained in Section 2 which has apparently also acquired the nature of customary international law is the obligation of each state to notify other states of imminent or actual damage if it deems them likely to be affected by such damage (Article 198). Closely linked to Article 198 is the general obligation of states and competent international organisations to 'cooperate, to the extent possible, in eliminating the effects of pollution and preventing or minimizing the damage' (Article 199).

The remainder of Section 2, as well as Sections 3 (Technical assistance) and 4 (Monitoring and environmental assessment), is composed of provisions representing programmes of cooperation of states parties to the LOS Convention. However, some of these are derived from provisions which do constitute customary international law.

Sources of pollution

Three Sections of Part XII deal with specific sources of pollution: Section 5 (International rules and national legislation to prevent, reduce and control pollution of the marine environment), Section 6 (Enforcement) and Section 7 (Safeguards).

Section 5 represents a detailed elaboration of the principle of Article 197, requiring states to cooperate, directly or through competent international organisations, in adopting international rules and standards. This duty is dealt with specifically in relation to each source of marine pollution.

The Convention envisages that the protection and preservation of the marine environment is to be regulated not only by international law, but by national legislation as well. The right of states to adopt laws and regulations has been provided for in relation to all sources of pollution, but the relation of national to international law is determined in different ways.

With some sources of pollution, national legislation must not be 'less effective' than international rules (e.g., in relation to dumping – Article 210(6)). Concerning some other sources of pollution, national laws and regulations shall be adopted 'taking into account internationally agreed rules, standards and recommended practices and procedures' (as in the case of pollution from or through the atmosphere – Article 212(1)).

The duty of states to adopt laws and regulations to prevent, reduce and control pollution of the marine environment is an important element of the general obligation of states to protect and preserve the marine environment (Article 192). The specific details concerning the relation of domestic with international law are, as yet, conventional rules needing further acceptance by states in order to acquire the status of customary law.

The provisions on enforcement (of national and international rules) can be considered as the application of the general rules relating to the competence of states in respect to flag state jurisdictions and maritime zones under their jurisdiction. However, these general rules are in Part XII amplified by some additional new elements. A special study and extreme caution would be necessary to distinguish customary rules from new provisions concerning enforcement with respect to pollution from ships. However, it is clear that more innovations have been adopted in respect to enforcement by port states (Article 218) than with regard to enforcement by flag states (Article 217) and coastal states (Article 220).

The foundation of one provision in customary law has been expressly mentioned: it has been stressed that the rights of states to take and enforce measures beyond the territorial sea in order to avoid pollution arising from maritime casualties are based on customary and conventional international law (Article 221). In making such an assertion, the drafters of the LOS Convention were in fact not stating anything new. Not only are these rights of coastal states contained in the 1969 International Convention Relating to Intervention on the High Seas in Cases of Oil Pollution Casualties,[27] they are also confirmed as customary law by a resolution of the *Institut de Droit international* adopted at its Edinburgh Session in 1969.[28]

[27] See text in A. C. Kiss (ed.), *Selected Multilateral Treaties in the Field of the Environment*, UNEP Reference Series 3 (Nairobi: United Nations Environment Programme, 1983), pp. 230ff.

[28] Institut de Droit international, *Tableau des Résolutions adoptées (1957–1991)* (Paris: A. Pedone, 1992), p. 75.

Concerning safeguards to be applied in the exercise of the powers of enforcement against foreign ships (Section 7), only a few principles can be pointed out as belonging to customary international law: the duty to avoid adverse consequences either for foreign vessels or for the marine environment in the exercise of the powers of enforcement (Article 225), the duty not to discriminate in form or in fact against vessels of any state (Article 227) and the provisions on liability for damage or loss arising from unlawful or disproportionate enforcement measures (Article 232).

Other issues

Of the remaining four Sections (8–11), only Section 9 (Responsibility and liability) and Section 10 (Sovereign immunity) contain principles of customary law.

Two of the principles contained in Article 235 represent customary law. States are responsible for the fulfilment of their international obligations concerning marine protection; they are to be liable for non-compliance with such obligations in accordance with international law. Beside the duty to implement the existing law, they shall further develop international law on responsibility and liability. In practice, however, states try to avoid implementation of this duty, on a global as well as a regional basis.[29]

The widespread non-application of the LOS Convention's provisions regarding the protection and preservation of the marine environment to warships and other vessels or aircraft owned or operated by a state and used only on government non-commercial service (Article 236) reflects the general principle of the sovereign immunity enjoyed by such ships and aircraft. This is a serious shortcoming of the Convention, particularly in respect of the polar oceans, where many such ships are involved in various activities. Unfortunately, the legal position of these categories of ships has in the LOS Convention generally been blurred by Article 58(1), which does not spell out clearly which are the 'lawful uses of the sea related to these freedoms [i.e., freedom of navigation, etc.], such as those associated with the operation of ships' to which they are entitled in the exclusive economic zone. That is why there is no clear answer to the question whether military and similar ships enjoying freedom of navigation in the exclusive economic zone are restricted at least in some military uses of the sea.[30] Be this as it may, on the basis of Article 58(2), the principle that the sea is to be reserved for peaceful purposes (Article 88) applies to both the high seas and the exclusive economic zone.

[29] For a brief account, see Joyner, Chapter 5 in this book. For a detailed, well-informed review, see M. Skåre, 'A Liability Annex or Annexes to the Environmental Protocol: A Review of the Process within the Antarctic Treaty System', in D. Vidas (ed.), *Implementing the Environmental Protection Regime for the Antarctic* (Dordrecht: Kluwer Academic Publishers, 2000).

[30] See B. Vukas, 'L'utilisation pacifique de la mer, dénucléarisation et désarmement', in R. J. Dupuy and D. Vignes (eds.), *Traité du Nouveau Droit de la Mer* (Paris and Brussels: Economica and Bruylant, 1985), pp. 1,055–6.

PART XII AND THE POLAR OCEANS

In addition to the specific significance that some of the general provisions of Part XII may have for the environment of the polar oceans, Part XII of the LOS Convention contains specific features relating solely to the polar marine environment. Thus, application of Part XII to the Arctic Ocean is subject to specific geographic and climatic conditions, with a view to which a special provision (Article 234) has been adopted.

However, the two polar oceans are not in the same position as far as the application of the provisions of Part XII is concerned. Many of these provisions are based on sovereignty, sovereign rights and jurisdiction in the coastal waters. As there are no generally recognised coastal states in Antarctica, all provisions based on the existence of a coastal state and its coastal marine areas have to be interpreted and applied *mutatis mutandis*, in accordance with the realities of Antarctica.

Thus, one of the first, most basic general provisions, Article 193, has been adopted with the states' territories in mind:

> States have the sovereign right to exploit their natural resources pursuant to their environmental policies and in accordance with their duty to protect and preserve the marine environment.

The situation is similar with respect to many provisions on protection of the marine environment from pollution from different sources. Thus, for example, Article 207 stipulates that states shall adopt laws and regulations to prevent, reduce and control pollution of the marine environment from land-based sources, taking into account internationally agreed rules, standards and recommended practices and procedures (paragraph 1). Although it is not explicitly stated, this provision has in mind coastal states, i.e. states having sovereignty over a coastline. This is even more evident from the text of Article 208 ('Pollution from seabed activities subject to national jurisdiction') under which coastal states are required to adopt rules and regulations and take other measures in order to prevent, reduce and control marine pollution from that source. To the extent that seabed activities off the coast of Antarctica are permitted, the most appropriate solution at present would be to see the adoption of rules and other measures as a joint obligation of *all* states engaged in such activities.

A further example of a provision not easily applicable in the Antarctic is contained in Article 210(5), which mentions a system of *approval of the coastal state* of dumping within its territorial sea and its exclusive economic zone.

Concerning not only the adoption of rules and procedures for the protection of the marine environment, but also the enforcement in respect of the mentioned sources of pollution, the LOS Convention has based its approach on the activity of the coastal state: land-based pollution (Article 213), pollution from seabed activities (Article 214) and pollution by dumping (Article 216).

Due to the controversy concerning sovereignty over Antarctica, the two most important aspects of enforcement under the LOS Convention with respect to pollution from vessels – enforcement by port states (Article 218) and enforcement by coastal states (Article 220) – are unlikely to be applied in the Antarctic unless the necessary adjustments and compromises can be made. Only the least efficient solution – enforcement by flag states (Article 217) – remains uncontroversial in respect of this very important source of pollution.

However, the problem of the existence of 'coastal states' on Antarctica concerns not only the general rules contained in Part XII, but also the sole provision inserted especially for the polar oceans. This provision is contained in Article 234, which reads:

> Coastal States have the right to adopt and enforce non-discriminatory laws and regulations for the prevention, reduction and control of marine pollution from vessels in ice-covered areas within the limits of the exclusive economic zone, where particularly severe climatic conditions and the presence of ice covering such areas for most of the year create obstructions or exceptional hazards to navigation, and pollution of the marine environment could cause major harm to or irreversible disturbance of the ecological balance. Such laws and regulations shall have due regard to navigation and the protection and preservation of the marine environment based on the best available scientific evidence.

Due to the drafting history of that provision and the fact that it is based on the notions of 'coastal states' and the 'exclusive economic zone', the applicability of Article 234 to Antarctica is still a controversial issue. Nonetheless, it remains to be seen what innovation this provision actually adds to all the other general provisions on the protection of the marine environment from pollution from vessels.[31]

The jurisdiction of the coastal state in respect of 'the protection and preservation of the marine environment' in its exclusive economic zone is confirmed in the basic provisions on the rights, jurisdiction and duties of the coastal state in the exclusive economic zone (Article 56(1)(b)(iii)). Jurisdiction of the coastal state comprises the adoption of 'laws and regulations for the prevention, reduction and control of pollution from vessels' (Article 211(5)). However, such laws and regulations must conform with and give effect to 'generally accepted international rules and standards established through the competent international organization or general diplomatic conference'. All other states are required to comply with the laws and regulations adopted by the coastal state (Article 58(3)).

In addition to the general rule on the right of the coastal state to legislate concerning the protection of the marine environment of its exclusive economic zone, Article 211(6) contains provisions in respect of situations where general international rules are inadequate for especially vulnerable areas within the exclusive economic zone.

[31] On the application and interpretation of Art. 234 by some Arctic states, see Brubaker, Chapter 10 in this book.

If the coastal state convinces the competent international organisation, the International Maritime Organisation (IMO), of the existence of special oceanographical and ecological conditions, then, in accordance with Article 211(6)(a):

> the coastal State may, for that area, adopt laws and regulations for the prevention, reduction and control of pollution from vessels implementing such international rules and standards or navigational practices as are made applicable, through the organization, for special areas.

With the approval of the IMO, the coastal state may also adopt additional laws and regulations relating to 'discharges or navigational practices but shall not require foreign vessels to observe design, construction, manning or equipment standards other than generally accepted international rules and standards' (Article 211(6)(c)).

After this sketchy overview of Article 211(6), which deals with vulnerable areas inside exclusive economic zones in general, what remains to be seen are the additional, specific elements given in Article 234 concerning such areas which are also ice-covered. The main gain of the Arctic coastal countries which negotiated this provision was their right to adopt special laws and regulations without seeking the IMO's permission. These laws and regulations should satisfy certain conditions:

1. they must be non-discriminatory, which means that there should be no discrimination in their contents and enforcement in respect of ships flying different flags, including the flag of the coastal state;
2. they must 'have due regard for navigation', i.e. they must not unnecessarily hamper navigation; and
3. in protecting the marine environment the laws and regulations should be based on the highest scientific achievements ('best scientific evidence').

SETTLEMENT OF DISPUTES: PART XV

The LOS Convention includes rules on a compulsory, binding system of settlement of disputes between states parties concerning the interpretation or application of the Convention (Part XV of the Convention). If a dispute cannot be settled by an exchange of views between the parties to the dispute, by conciliation or any other means of their choice, each party is entitled to submit the dispute to the court or tribunal having jurisdiction under the Convention. According to Article 287, states may choose one or more of the following compulsory procedures entailing binding decisions: the International Tribunal for the Law of the Sea (ITLOS), the International Court of Justice (ICJ), an arbitral tribunal or a special arbitral tribunal. If the parties to a dispute do not accept the same procedure for settlement of the dispute, it may be submitted only to the arbitral tribunal.

Some provisions especially affecting disputes relating to the marine environment should be addressed here. First of all, we should note that not all disputes relating to the marine environment are expressly provided for as being subject to the procedures entailing binding decisions. According to Article 297(1)(c), disputes concerning the exercise by a coastal state of its sovereign rights or jurisdiction provided for in the LOS Convention shall be submitted to a court or tribunal 'when it is alleged that a coastal State has acted in contravention of specified international rules and standards for the protection and preservation of the marine environment which are applicable to the coastal State and which have been established by this Convention or through a competent international organization or diplomatic conference in accordance with this Convention'.

This rule, whose drafting leaves much to be desired – as does the whole of Article 297 ('Limitations on applicability of section 2') – seems to provide that disputes in which it would be alleged that a state other than the coastal state acted in contravention of international rules and standards for the protection and preservation of the marine environment of the exclusive economic zone are *not* to be subjected to the procedures before the courts and tribunals listed in the Convention.

Some remarks should also be added in respect of the courts and tribunals mentioned in the Convention. In drafting the provisions concerning special arbitral tribunals, provided for in Article 287(1)(d), special attention was accorded to disputes concerning the marine environment. Under Annex VIII to the Convention (Special arbitration), special arbitral tribunals have been provided for disputes relating to (1) fisheries; (2) protection and preservation of the marine environment; (3) marine scientific research; and (4) navigation, including pollution from vessels and by dumping (Annex VIII, Article 1). For such disputes a list of experts is to be established and maintained; the list of experts in the field of the protection and preservation of the marine environment shall be drawn up and maintained by the United Nations Environment Programme, and the list in the field of navigation, including pollution from vessels and by dumping, by the International Maritime Organisation (Annex VIII, Article 2(1) and (2)).

Two other jurisdictions mentioned in Article 287 have used the right to establish standing special chambers under their respective Statutes for disputes relating to environmental issues. In July 1993 the ICJ established a seven-member Chamber for Environmental Matters.[32] Very soon after ITLOS was established, in February 1997 it formed the Chamber for Marine Environment, also composed of seven members.[33] In both cases, the chambers are to deal only with disputes submitted to them by joint request of the parties to the dispute.

FINAL REMARKS

Some comment is due on the various criticisms levelled against the LOS Convention concerning its utility for the protection of the polar marine environment.

[32] ICJ, *Yearbook 1994–1995*, No. 49, p. 17.

[33] Doc. ITLOS/1997/Res.2, 28 April 1997.

A general criticism has been that there are issues which are either insufficiently addressed by the LOS Convention, or not addressed at all. Although this assertion is essentially correct, we should bear in mind the purpose and nature of the LOS Convention. Unlike the 1958 Geneva codification, the LOS Convention had to cover in one single instrument all substantive, organisational and procedural provisions regarding the law of the sea. In comparison with the first UN codification of the law of the sea, it had to include and develop new topics, such as the exploration and exploitation of the seabed beyond national jurisdiction, the protection and preservation of the marine environment, marine scientific research, and the development and transfer of marine technology.

With such an ambitious task, the LOS Convention had to be the 'Charter for the Oceans'. Thus, in respect of many issues it deals with, it had to remain merely an 'umbrella treaty' – a framework convention (on navigation, exploration and exploitation of the living resources, and protection and preservation of the marine environment). In all those fields, there are already in force many treaties providing an elaborate international regulation on the global and regional level. However, even with such an approach, the LOS Convention appears to be the most voluminous treaty ever concluded. One of the fields in which UNCLOS III engaged prematurely in drafting too detailed provisions – exploration and exploitation of the Area – almost proved fatal for the survival of the LOS Convention.

In respect of the provisions on the protection and preservation of the marine environment – an issue mentioned only marginally at Geneva in 1958 – the LOS Convention offers the codification and progressive development of many principles dispersed elsewhere in numerous treaties and other instruments. Some of these instruments, for example the Action Plans for individual regional seas, can afford to deal with details, and even go beyond the purely legal issues (the protection of historical sites, the sound planning of different economic activities and their coordination with the protection of the environment, etc.). By contrast, the LOS Convention had to be a treaty dedicated exclusively to the 'law of the sea'.

Having in mind its essential purpose – to establish the legal order for the oceans – the negotiators at UNCLOS III left unresolved some of the issues almost intrinsically linked to their work. The status of Antarctica and the Southern Ocean was not the only such issue. The military uses of the sea were even more important at the time of the UNCLOS III negotiations. Yet, the LOS Convention does not offer many clear answers concerning the use of the seas and oceans for military purposes.

The provisions of the LOS Convention on the protection and preservation of the marine environment, as well as all other rules it contains, will represent a useful contribution to contemporary international law only in so far as they can co-exist with other general and regional norms in this field. As indicated in this chapter, one very sensitive issue involves their relation to the already-developed treaty system for the Antarctic, and the initial cooperation of the Arctic countries. However, the high level of participation of the Antarctic Treaty Consultative Parties and the Arctic countries in the LOS Convention proves that the attitude of UNCLOS III was not wrong. The participation of these countries is a proof of their will to contribute to the co-existence of the Treaty system for the Antarctic and the regime for the seas and oceans established at UNCLOS III.

3 Global environmental protection instruments and the polar marine environment

DONALD R. ROTHWELL

The marine environment is subject to many legal regimes, some applying only within defined regions. In the cases of the Arctic and the Antarctic, significant regional initiatives include the regimes created by the 1959 Antarctic Treaty and its 1991 Environmental Protocol, as well as those adopted under the 1991 Arctic Environmental Protection Strategy and the 1996 Arctic Council.[1] Beyond such regional regimes, there exists a sizeable body of international law which applies globally: legal regimes which set out to impose obligations upon all states, in principle covering all parts of the earth.

Developments in international environmental law during the past three decades have seen the emergence of several core principles which provide a framework of customary environmental law. These principles include: the obligation of all states to conserve the environment and its natural resources; the obligation upon states to assess potential, and monitor actual environmental impact; the obligation upon states to conserve the environment both within and beyond areas of national jurisdiction; and sustainable development.[2]

This list is not exhaustive; it may well be possible to identify other principles which are in a state of development, or which have particular application for specific environmental problems. These principles of international environmental law, emerging from state practice as well as incorporated in international environmental instruments, provide the underlying framework for marine environmental protection globally, thus including the polar regions. As to the polar regions in particular, given the sensitivity of their marine environments and the unknown consequences that may result from environmental degradation, the standard of due diligence is of special importance – not least in the Arctic, in view of the development activity which has occurred there thus far and which may be

[1] On regional environmental protection regimes in the Arctic and the Antarctic, see the discussion by Vidas, Chapter 4 in this book.

[2] A. Kiss and D. Shelton, *International Environmental Law* (Ardsley-on-Hudson, NY and London: Transnational Publishers and Graham & Trotman, 1991), pp. 145–54.

expected to expand even further, along with the potential for transboundary marine pollution.[3]

Considerable impact in regulating the environmental protection of the polar regions is also exerted by various specific marine environmental agreements and conventions, perhaps the most prominent being the 1973 International Convention for the Prevention of Pollution from Ships and its 1978 Protocol[4] (MARPOL 73/78) and the 1982 UN Convention on the Law of the Sea[5] (LOS Convention). In addition, a vast array of other international environmental instruments also assist in protecting various aspects of the polar marine environment.[6]

The purpose of this chapter is to review and assess the relevance of global agreements for the polar marine environment, so as to develop a greater understanding of the application of 'global international environmental instruments' in the polar regions. In the process it will be possible to appreciate the growing extent of global environmental responsibility that exists for the polar marine environment, and how developments in international environmental law impose upon the polar states obligations of sustainable development and ecosystem management.

MARINE POLLUTION CONVENTIONS AND THE POLAR OCEANS

Like other oceans of the world, the Arctic and Southern Oceans are subject to the existing international legal regime dealing with marine pollution. The legal regime which has come into being has primarily done so without the benefit of the provisions of the LOS Convention being in place, as most of the relevant conventions were negotiated during the late 1960s and the 1970s in response to growing international concerns over marine environmental pollution, especially following several major maritime incidents. These conventions have primarily dealt with discrete types of pollutants or polluting activities, which has meant a focus on ship-sourced marine pollution, primarily from oil and other related substances. MARPOL 73/78 is the principal global convention dealing with these matters. In relation to the dumping of substances at sea, the 1972 Convention on the Prevention of Marine Pollution by Dumping of Wastes and Other Matter[7] (London Convention) is the principal global convention.

3 For a review of transboundary marine pollution issues in the Arctic, see the reports of the Working Group on Protection of the Arctic Marine Environment founded under the Arctic Environmental Protection Strategy (www.nrc.ca/arctic/pame) and in particular a report of 2 May 1997 entitled 'Protection of the Arctic Marine Environment: Status Report to Senior Arctic Officials' (www.girda.no/prog/polar/pame).

4 ILM, Vol. 12, 1973, pp. 1,319ff (Convention); and ILM, Vol. 17, 1978, pp. 546ff (Protocol).

5 For a detailed analysis of the LOS Convention as it relates to the polar marine environment see Vukas, Chapter 2 in this book.

6 During preparation of the Arctic Environmental Protection Strategy, altogether twenty-six 'global conventions' were identified as potentially having an impact upon the protection of the Arctic environment, many of which also provided protection for the *marine* environment; see *List of Major International Instruments and Policy Declarations Pertaining to the Arctic Environment*, final version, as presented at the Officials and Ministerial Meeting, 10–14 June 1991 at Rovaniemi, Finland, pp. 2–31.

7 UNTS, Vol. 1,046, pp. 120ff.

Even though it is considered to constitute approximately 70 per cent of marine pollution,[8] land-based pollution has been the least regulated form of marine pollution at the global level. There is no specific convention dealing with this problem, though the international community has begun to pay greater attention to this problem following the 1995 Washington Declaration and the Global Programme of Action for the Protection of the Marine Environment from Land-Based Activities.[9]

Global-level international regimes have also been developed to deal with maritime emergencies that can have polluting consequences, as well as liability resulting from marine pollution and other maritime incidents resulting in impact upon the marine environment.[10]

The result is a global body of law dealing with marine pollution and protection of the marine environment which is relatively sophisticated and certainly more advanced than any other area of international environmental law. These developments will now be reviewed from a sectoral perspective.

INTERNATIONAL REGULATION OF SHIP-SOURCED POLLUTION

International regulation of ship-sourced marine pollution has been the subject of global attention since the 1950s following the adoption of the 1954 International Convention for the Prevention of Pollution of the Sea by Oil[11] (OILPOL). However, the Convention suffered from difficulties in implementation and enforcement and eventually became outdated following the development of the so-called 'supertankers' and the consequent challenges these vessels posed for international regulators seeking to limit pollution by ships at sea. MARPOL 73/78 was eventually adopted as a replacement for OILPOL, but did not enter into force until 1983 following the adoption of an amending Protocol in 1978.

As MARPOL 73/78 has now been in place for a relatively longer period, it is possible to judge its impact at both the global level and also in the polar oceans. The Convention is designed to prevent pollution of the marine environment by prohibiting and limiting the discharge of harmful substances or effluents from ships. It has global application and contains general provisions for all oceans, but it also contains enhanced protection for nominated 'special areas' which can include oceans or parts of oceans. One way to measure the success of MARPOL 73/78 is to examine the extent to which ship-generated pollution has fallen: from

[8] Agenda 21, Chapter 17, para. 18.

[9] Reproduced in *Environmental Policy and Law*, Vol. 26, 1996, pp. 37ff. For an analysis see VanderZwaag, Chapter 8 in this book. On globalism and regionalism, as regards land-based sources of pollution, see Boyle, Chapter 1 in this book.

[10] See the 1969 International Convention on Civil Liability for Oil Pollution Damage (UNTS, Vol. 973, pp. 3ff); the 1971 International Convention on the Establishment of an International Fund for Compensation for Oil Pollution Damage (ILM, Vol. 11, 1972, pp. 284ff); and the 1996 International Convention on Liability and Compensation for Damage in Connection with the Carriage of Hazardous and Noxious Substances by Sea (ILM, Vol. 35, 1996, pp. 1,406ff); these conventions are not addressed in this chapter.

[11] UNTS, Vol. 327, pp. 3ff.

being estimated at approximately 35 per cent of global marine pollution sources in the early 1970s to approximately 10 per cent by the early 1990s[12] – and indeed, to examine the causal relationship between MARPOL 73/78 and this trend.

MARPOL 73/78 is a framework convention containing specific provisions that regulate certain types of pollution in attached annexes. The main body of the Convention deals with extent of application (Article 3), violation and enforcement (Articles 4–6) and other procedural issues. The six annexes that are attached to the Convention concern:

1. the prevention of pollution by oil (Annex I);
2. the control of pollution by noxious liquid substances in bulk (Annex II);
3. the prevention of pollution by harmful substances in packaged forms (Annex III);
4. the prevention of pollution by sewage from ships (Annex IV);
5. the prevention of pollution by garbage from ships (Annex V); and
6. the prevention of air pollution from ships (Annex VI).[13]

'Discharge' is widely defined to include any release, including 'any escape, disposal, spilling, leaking, pumping, emitting or emptying' (Article 2(3)(a)). Exceptions exist in the case of dumping under the 1972 London Convention, or the release of harmful substances arising from seabed mineral resource activities or scientific research (Article 2(3)(b)). MARPOL 73/78 applies to all ships flying the flag of contracting parties, or operating under the authority of a party (Article 3(1)) and operates through a combination of flag state enforcement and port state control, with parties encouraged to cooperate in the detection of violations and in enforcement (Article 6).

The prevention of pollution by oil

The most detailed operational provisions of MARPOL 73/78 are found in Annex I concerning the prevention of pollution by oil. This Annex prohibits the discharge into the sea of oil unless certain conditions are met (Annex I, Regulation 9). Within designated 'special areas', discharges are absolutely prohibited,[14] with the exception of ships of less then 400 gross tonnage (other than oil tankers) which may discharge under strict conditions (Annex I, Regulation 10). Exceptions also apply in cases where a discharge is necessary to secure the safety of the ship and save life at sea, or where the discharge has resulted from damage to the ship (Annex I,

[12] See the figures quoted in R. M. M'Gonigle and M. W. Zacher, *Pollution, Politics and International Law* (Berkeley, CA: University of California Press, 1979), p. 17; and in *Agenda 21: Programme of Action for Sustainable Development* (New York: United Nations, 1993), Chapter 17, para. 18.

[13] Annex VI was adopted in 1997 and has yet to enter into force; as of 8 June 1999 there had only been two ratifications (by Norway and Sweden).

[14] The 'Special Areas' that were originally designated under MARPOL cover the areas of the Mediterranean Sea, the Baltic Sea, the Black Sea, the Red Sea and the 'Gulfs Area'; see Annex I, Regulation 10.

Regulation 11). In an effort to ensure that ships have access to adequate oil reception facilities, Annex I details requirements for the provision of such facilities at various ports and for a variety of vessel types (Annex I, Regulation 12). Provisions also exist dealing with the use, cleaning and maintenance of ballast tanks, and the use of certain oil discharge monitoring equipment in addition to the need to complete an oil record book (Annex I, Regulations 13–20).

The Antarctic Treaty area was included as a 'Special Area' under Annex I by the amendments of 1990 (in force 1992), with the result that any discharge of oil is prohibited. This extension of the MARPOL 73/78 Special Area concept to the Antarctic is a welcome development that provides additional protection for the Southern Ocean, given the vast degree of international acceptance of MARPOL 73/78.[15] The same cannot be said for the Arctic, however. It has yet to be recognised as a Special Area, and at present there is insufficient support amongst Arctic states for such status being conferred.[16] A major problem with Annex I in the Arctic context is the lack of adequate and reasonably priced port reception facilities.[17] This is, however, an issue not only in the Arctic, where there are some significant ports, but also globally. With the vast distances involved in shipping between Arctic ports and those to the south this is a matter which needs to be addressed if commercial shipping in Arctic waters continues to increase.

Control of other forms of ship-sourced pollution

Annex II of MARPOL 73/78, on pollution by noxious liquid substances in bulk, complements Annex I and provides additional protection to the polar marine environment from the discharge of such substances. The Southern Ocean was listed as a Special Area for the purposes of Annex II by amendments adopted in 1992. At present, however, Annex II does not absolutely prohibit the discharge of ballast waters. This has recently become a matter of great concern to some coastal states;[18] given the commercial shipping and port traffic in the Arctic and the developing maritime traffic in the Southern Ocean, it will be most unfortunate if foreign marine organisms are introduced into polar waters through the discharge of ballast waters.

[15] As at 8 June 1999, the 1978 MARPOL Protocol and Annexes I and II had 108 parties, which represented some 94 per cent of the world tonnage; for updated status see the International Maritime Organisation website at www.imo.org.

[16] At neither the 1996 nor the 1997 Meeting of Ministers of Arctic Countries under the 1991 Arctic Environmental Protection Strategy was support forthcoming for having the Arctic declared a MARPOL 'Special Area'; see the 1996 Inuvik Declaration on Environmental Protection and Sustainable Development, available at the Arctic Council website at http://arctic-council.usgs.gov; and the 1997 Alta Declaration on the Arctic Environmental Protection Strategy, at *ibid.* See the discussion in D. VanderZwaag, *Canada and Marine Environmental Protection: Charting a Legal Course Towards Sustainable Development* (London: Kluwer Law International, 1995), p. 155; and D. VanderZwaag, 'International Law and Arctic Marine Conservation and Protection: A Slushy, Shifting Seascape', *Georgetown International Environmental Law Review*, Vol. 9, 1997, p. 322.

[17] D. Brubaker, *Marine Pollution and International Law* (London: Belhaven Press, 1993), p. 128.

[18] 'Report of the 37th Session of the Marine Environment Protection Committee', *Environmental Policy and Law*, Vol. 26, 1996, p. 18.

Annex III, dealing with pollution by harmful substances in packaged forms and containers, has important provisions which apply to vessels engaged in Antarctic resupply. However, sovereign immunity exceptions do limit its application. The regulation of sewage is dealt with by Annex IV, but this Annex has yet to gain sufficient support to enter into force.[19] It is particularly applicable for the polar regions, as all vessels navigating through those waters carry sewage on board which, if discharged, may have severe environmental consequences. Disposal of garbage from vessels is dealt with in Annex V. Regulations apply to the disposal of various types of garbage, including acceptable distances from land and the manner in which it may be disposed of. Vessel-source garbage in the polar regions has been the subject of some concern for environmental groups, especially with the increase in voyages by tourist vessels.

Assessment

The enforcement of MARPOL 73/78 in polar waters faces several difficulties. Especially significant is the fact that both the Arctic and Southern Oceans contain vast stretches of high seas areas beyond the reach of coastal state jurisdiction. As flag state jurisdiction prevails on the high seas, this results in inconsistent application of MARPOL 73/78 due to the varying standards that flag states adopt for enforcement. In polar waters within the limits of the EEZ, coastal states have enforcement jurisdiction; however, in polar conditions, enforcement capacity may be constrained by logistical and operational conditions. In the Southern Ocean there are additional constraints created by the Antarctic Treaty's limitation on the exercise of jurisdiction, plus the political reality that many states do not recognise the existence of coastal states around the Antarctic continent.

The MARPOL 73/78 Annexes are under constant review and have undergone continual revision and expansion over the years. All the same, the only specific consideration given to the polar regions in MARPOL 73/78 has been the listing of the Antarctic as a Special Area under Annexes I, II and V.[20] The IMO has considered a proposal that the Antarctic Special Area under MARPOL 73/78 be expanded to make it identical with the outer limits of the 1980 Convention on the Conservation of Antarctic Marine Living Resources, but no decision has been taken.[21] MARPOL 73/78 has additional limitations in its application to polar waters.[22] The growing number of flag-of-convenience vessels operating in polar waters will make it more difficult to ensure that MARPOL 73/78 standards are being adhered to. This is especially an issue for tourist vessels, which are visiting polar

[19] As at 8 June 1999, Annex IV had received seventy-five ratifications, representing only some 43 per cent of world tonnage; see updated status at the International Maritime Organisation website at www.imo.org.

[20] See the discussion in M. White, *Marine Pollution Laws of the Australasian Region* (Annandale: Federation Press, 1994), pp. 261–2.

[21] 'Report of the 37th Session of the Marine Environment Protection Committee', *Environmental Policy and Law*, Vol. 26, 1996, p. 17. See also the discussion by Vidas, Chapter 4 in this book.

[22] See the discussion in White, *Marine Pollution Laws in the Australasian Region*, p. 261.

waters in increasing numbers. Moreover, MARPOL 73/78 often establishes limitations on discharges from nearest land; due to the ice-fringed nature of polar coastlines, however, this can be difficult to determine.[23] In any event, with much of the polar oceans covered with ice of various forms even during the summer months,[24] any legitimate discharge of oil or other pollutants under MARPOL 73/78 standards in these waters has the potential to create serious environmental impact on the marine environment and the marine ecosystem. A more practical and environmentally sensitive definition of 'nearest land' in the polar regions would therefore be helpful. A further area where MARPOL 73/78 could be expanded is through greater recognition of the 'Particularly Sensitive Sea Area' concept, to allow for additional special measures to be taken to protect designated marine areas. This also would have obvious applications in polar waters and be a further step towards recognising the special environmental conditions of the polar oceans.

INTERNATIONAL REGULATION OF DUMPING AT SEA

The 1972 London Convention

The 1972 Convention on the Prevention of Marine Pollution by Dumping of Wastes and Other Matter[25] is another international instrument designed to prohibit and limit marine pollution. To that end, the London Convention supports the terms of the LOS Convention and is similar in its goals to MARPOL 73/78. The London Convention seeks to ensure that all practicable steps are taken to prevent pollution of the sea by the dumping of waste, or the dumping of other matter that is liable to create hazards to human health, or to harm living resources and marine life (Article I). Contracting parties are also to take measures individually, according to their capabilities, to prevent pollution caused by dumping (Article II). Enforcement is primarily through the flag state (Article VII). 'Dumping' includes any deliberate disposal at sea of wastes or other matter from vessels, aircraft, platforms or other man-made structures at sea (Article III). It also extends to any deliberate disposal at sea of vessels, aircraft, platforms or other man-made structures (Article III). However, dumping does not extend to the disposal at sea of wastes which are incidental to the normal operations of vessels, aircraft, platforms and other man-made structures (Article III). The Convention distinguishes between three types of waste: (1) wastes which are prohibited from being dumped; (2) wastes which require a special prior permit in order to be dumped; and (3) wastes which require a prior general permit in order to be dumped (Article IV).

[23] See MARPOL 73/78, Annex I, Regulation 1(9). With the exception of the Great Barrier Reef area along the northeastern coast of Australia, 'nearest land' is defined as being 'the baseline from which the territorial sea in question is established in accordance with international law'; MARPOL 73/78, Annex I, Regulation 1(9).

[24] See Brigham, Chapter 11 in this book, Figure 11.1.

[25] The Convention was originally known as the 'London Dumping Convention'; however, in 1992 it was agreed that the short name of the Convention should become the 'London Convention'.

Annexes to the Convention list the various substances which fall into each of these categories. Annex I substances are those which cannot be dumped, while Annex II substances cannot be dumped without a prior special permit. In cases where a permit is required for dumping, 'careful consideration of all the factors set forth' in Annex III is to be taken into account prior to a permit being issued (Article IV).[26] An exception to the above provisions exists in instances where dumping is necessary to secure the safety of human life or the safety of vessels and aircraft (Article V). Contracting parties are required to establish and nominate appropriate authorities which are to be responsible for the issuing of permits and maintaining records on dumping activities. (Article VI).

The London Convention has been subject to ongoing review, and amendments have been introduced prohibiting the dumping of radioactive waste, industrial waste, and the incineration at sea of industrial waste.[27] The most substantial amendments took place in 1996 following the adoption of a new Protocol.[28] The effect of the Protocol is to place considerable limitations on material that may be dumped at sea. Article 4 provides that only those materials listed in Annex I may be dumped,[29] with the sole exceptions being in case of *force majeure* or in any case which constitutes a danger to human life or a real threat to vessels.

Assessment

The London Convention has particular application to polar waters, given the potential that exists for some states, ship owners and waste disposers to exploit the polar regions as a potential dumping ground for hazardous wastes. In that regard, special attention has in recent years been given to the dumping of hazardous wastes in the Arctic, especially radioactive waste in Russian waters.[30] At the 1997 Arctic Environmental Protection Strategy Ministerial Meeting, express reference was made to the need for regional cooperation to 'enhance nuclear reactor safety and to increase and promote the safe management, storage and disposal of spent nuclear fuel and radioactive waste'.[31] The ministers recognised the importance of ongoing cooperation to provide for the early completion of facilities needed to implement the ban on the dumping of radioactive waste at sea adopted under the London Convention.

[26] Factors listed in Annex III relate to the characteristics and composition of the matter, and the characteristics of the dumping site and method of deposit.

[27] See 'Ban on Sea Dumping of Radioactive Wastes Takes Effect', *Marine Pollution Bulletin*, Vol. 28, 1994, p. 194. For background on the issue of radioactive waste dumping, see D. P. Calmet and J. M. Bewers, 'Radioactive Waste and Ocean Dumping: The Role of the IAEA', *Marine Policy*, Vol. 15, 1991, pp. 413–30.

[28] ILM, Vol. 36, 1997, pp. 1ff. See the discussion by Stokke, Chapter 9 in this book.

[29] The exceptions listed in Annex I include dredged material, sewage sludge, fish waste, vessels and platforms and man-made structures, inert inorganic material, organic material of natural origin and various listed bulky items.

[30] A detailed discussion is provided by Stokke, Chapter 9 in this book.

[31] Alta Declaration, para. 16.

The application of the London Convention in polar waters is enhanced by the obligations that it imposes upon the polar states, whether they be territorial claimants or not, to apply the Convention within the polar waters and to their own flagged vessels. In the Antarctic, it is difficult to apply the Convention by claimant states against third states who are not parties to the Antarctic Treaty, due to the effect of Articles IV and VIII of the Treaty and the uncertainty as to the status of Antarctic 'coastal waters'. One of the difficulties with the London Convention is the emphasis on flag state enforcement, thereby requiring vigilance on the part of flag states to ensure that their flagged vessels are not engaging in activities in breach of the Convention. While there is adequate scope under the London Convention for coastal state action to enforce the Convention's provisions in polar waters, it is especially in the Antarctic that states may experience difficulty in collecting sufficient evidence to prosecute, given the relative remoteness of the areas in question. The application of the London Convention in the Arctic has also been subject to criticism, where questions have arisen concerning the disposal of vessels, platforms and other man-made structures.[32] An exception applies for these objects provided they are not considered 'industrial waste'.[33] Many of the criticisms that have been levelled against the London Convention were addressed in the 1996 Protocol. While it remains uncertain as to when the Protocol can be expected to enter into force,[34] the emphasis it gives to enhanced marine environmental protection and sustainable development will represent significant advances in the regime controlling dumping at sea.

INTERNATIONAL REGULATION OF LAND-BASED POLLUTION

Notwithstanding the impact land-based polution has upon the marine environment, this remains the least regulated of all forms of marine pollution.[35] This is evident from Article 207 of the LOS Convention, which anticipates the future development of international regulation in this area. The task of reducing land-based marine pollution is more difficult than other marine pollutants: regulation goes to the heart of state sovereignty as well as economic development, because the point sources of land-based marine pollutants are primarily factories, industry and agriculture. Any regulation of such activities directly impacts upon the economic sovereignty of a state, and this explains why this area has not

[32] See VanderZwaag, 'International Law and Arctic Marine Conservation', p. 322.

[33] In 1995, the London Convention Consultative Meeting considered a proposal by Denmark that a moratorium be adopted on the disposal at sea of decommissioned offshore installations; however, the proposal did not secure the support of the Meeting. See E. J. Molenaar, 'Ocean Dumping', *Yearbook of International Environmental Law*, Vol. 6, 1995, p. 256.

[34] As at 8 June 1999, there were six ratifications; twenty-six ratifications are required before the Protocol enters into force.

[35] A. E. Boyle, 'Land-Based Sources of Marine Pollution: Current Legal Regime', *Marine Policy*, Vol. 16, 1992, p. 24; and, more generally, M. Qing-nan, *Land-Based Marine Pollution: International Law Development* (London: Graham & Trotman and Martinus Nijhoff, 1987).

been as extensively regulated as have other sources of marine pollution. An additional problem in the context of polar regions is that some land-based marine pollutants are sourced in states outside the region, states with scant interest in polar affairs.[36]

Since 1982 the United Nations Environment Programme (UNEP) has taken an active role in this area. It has sponsored a global programme of action designed primarily to assist states to prevent, reduce and control degradation of the marine environment from land-based sources. Part of this programme has reviewed the effectiveness of the Regional Seas Programme in dealing with land-based pollution, while there has also been considerable discussion of scientific and developmental problems associated with this issue. In November 1995, at the UNEP Intergovernmental Conference to Adopt a Global Programme of Action for the Protection of the Marine Environment from Land-Based Activities held in Washington DC, a Declaration on Protecting the Marine Environment from Land-Based Activities was adopted.[37] The Declaration seeks to enhance coordination at the national, regional and international levels and sets common goals of sustained and effective action to deal with land-based impacts upon the marine environment. Particular pollutants identified as in need of action are persistent organic pollutants, radioactive materials, heavy metals, oils (hydrocarbons), nutrients, sediment mobilisation and land litter.[38]

Assessment

Land-based marine pollution is a problem of considerable importance in the Arctic, and was identified as a major source of marine environmental pollution in the AEPS.[39] In subsequent years the AEPS has continued to give attention to addressing land-based marine pollution problems in the Arctic, especially through the Arctic Monitoring and Assessment Programme (AMAP) and the Protection of the Arctic Marine Environment Working Group (PAME). At the 1996 Inuvik and 1997 Alta Ministerial Meetings for the AEPS, continued emphasis was given to the need to develop a more comprehensive programme to deal with land-based marine pollution. In 1997 the ministers endorsed a proposal to complete and implement a Regional Programme of Action for the Protection of the Arctic Marine Environment from Land-Based Activities,[40] and at the 1998 Arctic Council Meeting at Iqaluit they gave formal endorsement to this proposal as well as to the development of an Arctic

[36] See the discussion by VanderZwaag, Chapter 8 in this book.

[37] Declaration on Protecting the Marine Environment from Land-Based Activities, reproduced in *Environmental Policy and Law*, Vol. 26, 1996, p. 37.

[38] For further details, see VanderZwaag, Chapter 8 in this book.

[39] See Arctic Environmental Protection Strategy, ILM, Vol. 30, 1991, pp. 1,633ff, listing noise, persistent organic pollutants, oil pollution, heavy metals, radioactive materials and acidification as major Arctic pollutants.

[40] Alta Declaration, para. 9. See E. Leighton, 'PAME Working Group Meets in Canada', *WWF Arctic Bulletin*, No. 4, 1997, p. 10. See the discussion by VanderZwaag, Chapter 8 in this book.

Council Action Plan to Eliminate Pollution of the Arctic.[41] The concurrent development of the Global Programme of Action should prove an important catalyst for the Arctic states, working within the AEPS/Arctic Council and global legal framework, to respond to the problem of land-based marine pollution. The difficulties in dealing with this issue at both the global and regional levels remain serious.[42] To date, land-based marine pollution has not been identified as a concern in the Southern Ocean. Part of the reason for this is the lack of industrial activities conducted in Antarctica as well as on the northern landmasses adjacent to the Southern Ocean. However, the possible effects of long-range transportation of pollutants should not be underestimated.

INTERNATIONAL REGULATION OF MARITIME EMERGENCIES

In addition to the international conventions discussed above dealing with specific marine pollution sources, there are several international conventions dealing with other maritime affairs that have an important role to play in protecting the polar marine environment. These conventions are essentially those which deal with maritime emergencies and the obligations upon states to respond, and the capacity of coastal states in particular to take action to deal with such emergencies swiftly and thereby limit the environmental impact.

The 1969 Intervention Convention

The 1969 Convention Relating to Intervention on the High Seas in Cases of Oil Pollution Casualties[43] (Intervention Convention) deals with the ability of a coastal state to take action on the high seas in order to protect the marine environment within its national jurisdiction. Contracting parties may take such measures on the high seas as are necessary to prevent, mitigate or eliminate grave and imminent danger to their coastline or related interests from pollution or the threat of pollution of the sea following a maritime casualty (Article I). The exception concerns cases of warships and other non-commercial government vessels. Intervention can occur where there has been a collision of ships, a stranding, or any other incident of navigation or occurrence which results in material damage or imminent threat of material damage to a ship or its cargo (Article II). The ability to intervene, however, is not unlimited; the coastal state is required to consult with the flag state, notify the ship owner of the proposed action, and consult with independent experts before any action can be taken, though an exception exists in cases of extreme urgency (Article III). Action taken under the Convention must be proportionate to the actual or threatened damage (Article V), and the state is responsible

[41] See P. Prokosch, 'Environment Continues to be Core of Circumpolar Cooperation', *WWF Arctic Bulletin*, No. 3, 1998, p. 4.

[42] For comment, see VanderZwaag, 'International Law and Arctic Marine Conservation', pp. 332–4; and VanderZwaag, Chapter 8 in this book.

[43] ILM, Vol. 9, 1970, pp. 25ff.

for acts which go beyond a proportionate response (Article VI). The Convention was adjusted by a 1973 Protocol allowing for intervention on the high seas in cases of pollution by substances other than oil.

Assessment

Given the hazards of polar navigation and the marine environmental impact that would result from a major pollution incident, the Intervention Convention provides polar states with a capacity to intervene on the high seas and potentially avert an environmental accident, or disaster. However, the practical application of the Convention in polar waters is limited, due to the lack of established emergency response crews and vessels, the vast distances likely to be involved in any such operation, and the problems caused by polar climate and sea ice. In the Southern Ocean there remains the vexed issue of the recognition of 'coastal states', while the limitations against taking action against government vessels engaged in non-commercial service is also a factor. The *Bahiá Paraiso* incident in 1989 pointed up the difficulties of mounting such a clean-up operation in the Antarctic, but it also clearly demonstrated the need for continual vigilance to respond to such maritime casualties.[44]

The 1990 Oil Pollution Preparedness, Response and Cooperation Convention

The 1990 International Convention on Oil Pollution Preparedness, Response and Cooperation[45] (OPRC Convention), which entered into force in 1995, seeks to complement the Intervention Convention but also gives effect to many international environmental law obligations concerning cooperation and mutual assistance in dealing with environmental problems.[46] Parties are, individually or jointly, to take all appropriate measures in accordance with the terms of the Convention to prepare for and respond to an oil pollution incident (Article 1). Ships which fly the flags of state parties are required to have an oil pollution emergency plan as adopted by the IMO (Article 3). Procedures are also established for dealing with oil pollution reporting in the event of a discharge or probable discharge of oil. Once notification is received of a discharge of oil, the recipient of the information is to inform without delay other states whose interests are affected or likely to be affected (Article 5). The Convention also seeks to enhance national capacity to respond to oil pollution incidents; it requires national authorities to be designated as being responsible for oil pollution preparedness and response and the receipt of

[44] See M. Barinaga and D. Lindley, 'Wrecked Ship Causes Damage to Antarctic Ecosystem', *Nature*, Vol. 337, 1989, p. 495; 'Argentine Ship Sinks Near Palmer Station', *Antarctic Journal of the United States*, Vol. 24, 1989, pp. 3–7.

[45] ILM, Vol. 30, 1991, pp. 733ff.

[46] See in particular LOS Convention, Arts. 197–199.

oil pollution reports. Each state must also create a national contingency plan for preparedness and response (Article 6).[47]

Assessment

The OPRC Convention provides yet a further basis for polar states to develop comprehensive strategies to respond to maritime incidents which have environmental consequences. By placing minimum obligations upon flag states to ensure that their vessels have an oil pollution emergency response plan, polar states can be reassured that vessels from OPRC Convention parties which operate in their waters have met this standard. The requirements for cooperation amongst states to deal with such incidents are also important, given the difficulties in responding to emergencies in polar waters. One issue which arises here is whether vessels operating in polar waters should be required to meet even higher standards of oil pollution preparedness and response than vessels operating in more temperate climates. To that end, the requirements imposed by Canada under the Arctic Waters Pollution Prevention Act for vessels operating in parts of the Canadian Arctic are an interesting example.[48]

International Code of Safety for Ships in Polar Waters[49]

Consistent with the Intervention Convention and the OPRC Convention, and also maritime laws and international instruments dealing with safety of life at sea and shipping standards,[50] there has been a recent initiative to develop an 'International Code for Polar Navigation' (Polar Code) by certain polar states with a view to final promulgation by the IMO. The Polar Code is currently in a draft form only and continues to be developed by working groups of the Maritime Safety Committee of the IMO. The basis for the development of the Polar Code is the recognition that ships operating in ice-infested waters in both the Arctic and Antarctic regions are exposed to unique risks, and that existing maritime standards dealing with navigation, safety of life at sea, and ship safety and pollution do not deal adequately with these circumstances. A principal focus of the Polar Code has been safety of navigation, as well as prevention of pollution from ship operations

[47] For discussion see W. A. O'Neil, 'The International Convention on Oil Pollution Preparedness, Response and Co-Operation', in C. M. De La Rue (ed.), *Liability for Damage to the Marine Environment* (London: Lloyd's of London Press, 1993), pp. 23–8.

[48] See VanderZwaag, *Canada and Marine Environmental Protection*, pp. 330–48; see also Rothwell and Joyner, Chapter 7 in this book. As to requirements of Russian legislation for vessels navigating the Northern Sea Route, see Brubaker, Chapter 10 in this book.

[49] A separate chapter in this book – Chapter 11 by Brigham – discusses the details of the draft Polar Code; therefore, only a very brief overview is provided here.

[50] See, e.g., the 1966 International Convention on Load Lines (UNTS, Vol. 640, pp. 133ff); the 1972 Convention on the International Regulations for Preventing Collisions at Sea (UNTS, Vol. 1,050, pp. 16ff; and UNTS, Vol. 1,143, pp. 346ff); and the 1974 International Convention for the Safety of Life at Sea (UNTS, Vol. 1,184, pp. 2ff).

in polar waters. The Code is not meant to override existing pollution prevention standards for ships operating in polar waters; rather, the intention is to create additional requirements to mitigate the additional risks imposed on shipping due to the harsh climatic conditions in the polar waters.

MARINE ENVIRONMENTAL PROTECTED AREAS

Marine environmental conservation has given increased emphasis to the importance of marine protected areas, especially during the 1990s. The development of this concept for maritime areas has been a natural follow-up from the declaration of terrestrial protected areas and also the protection granted under regimes such as the 1971 Convention on Wetlands of International Importance Especially as Waterfowl Habitat[51] (Ramsar Convention), and the 1979 Convention on the Conservation of Migratory Species of Wild Animals[52] (Bonn Convention). In addition, various global international environmental agreements have provided a basis for specific action by the polar states.

The 1972 Convention for the Protection of the World Cultural and Natural Heritage

With 156 states parties (as at 8 June 1999), the 1972 Convention for the Protection of the World Cultural and Natural Heritage[53] (World Heritage Convention) is perhaps one of the most successful international environmental instruments providing for protected areas. The focus of the Convention is the protection of cultural and natural heritage, both of which are widely defined (Articles 1 and 2). State parties are required to identify and delineate properties which meet the criteria for being part of the world heritage; certain properties which meet the eligibility criteria then become subject to being placed on a 'World Heritage List' (Articles 3 and 6). World Heritage List properties are subject to increased obligations by the states parties to protect, conserve, present and transmit to future generations the identified cultural and natural heritage (Article 4). As at 8 June 1999, there were inscribed on the World Heritage List 582 sites, located in 114 countries which are parties to the Convention.

Assessment

There is the potential for more expanded use to be made of the World Heritage Convention in the polar regions, especially as the Convention can apply to marine and terrestrial areas alike. From time to time there has been speculation as to whether Antarctica, or parts of it, should be placed on the World Heritage List. While a formal application has never been placed before the World Heritage

[51] UNTS, Vol. 996, pp. 245ff. [52] ILM, Vol. 19, 1980, pp. 15ff. [53] ILM, Vol. 11, 1972, pp. 1,358ff.

Committee (no doubt related to the concerns such an application would create regarding the recognition of sovereignty), Australia has succeeded in having some of its sub-Antarctic islands placed on the World Heritage List.[54] Arctic states have also successfully had several sites placed on the World Heritage List. While a contentious issue remains the lack of uniformly applied standards for the management of world heritage areas, the Convention provides for international recognition of the special status of certain areas for either their cultural or their natural significance. This recognition assists states in enhancing laws and regulations which provide for the protection and conservation of those areas.

The 1989 Basel Convention on the Transboundary Movements of Hazardous Wastes

Since its adoption, the 1989 Convention on the Control of Transboundary Movements of Hazardous Wastes and Their Disposal (Basel Convention)[55] has grown in global importance due to the concern over the export, import and shipment of hazardous wastes. The Convention establishes a regime by which exporting states are required to notify importing states of arrangements regarding the transboundary movement of the hazardous waste. In addition, states through which the wastes transit are also entitled to be notified (Articles 4 and 6). The transboundary movement of the waste may be halted in certain circumstances if permission is not forthcoming from exporting and transit states.

Assessment

The Basel Convention has particular application in the Arctic, because of the amount of industrial waste generated in areas such as Siberia and the Kola Peninsula, and the potential for such wastes to be transported through Arctic waters. Its relevance for the Antarctic is highlighted in the Convention's terms, which expressly prohibit 'the export of hazardous wastes or other wastes for disposal within the area south of 60° South latitude, whether or not such wastes are subject to transboundary movement' (Article 4(6)). The movement of all hazardous wastes within the Antarctic Treaty area is not prohibited, however; rather, the Antarctic Treaty area has been given a special protected status which completely prohibits the export or disposal of such wastes. This has particular application to the seven Antarctic claimants, which are prohibited from exporting wastes to their claimed territories as the limitation applies even if a transboundary movement has not taken place.[56]

[54] The islands include Macquarie Island (administered by the State of Tasmania) and Heard and McDonald Islands (external territories of Australia); see Australian Antarctic Division, *Heard Island Wilderness Reserve Management Plan* (Hobart: Australian Antarctic Division, 1995), p. 8.

[55] ILM, Vol. 28, 1989, pp. 649ff.

[56] This would be the case with the removal of waste from the territory of any claimant state to the Antarctic continent.

The 1992 Convention on Biological Diversity

The Convention on Biological Diversity was one of the most significant achievements of the UN Conference on Environment and Development. The importance attached by the international community to the Convention was later demonstrated by the speed with which it entered into force. The Convention brings together many of the fundamental conservation elements already existing in the 1973 Convention on International Trade in Endangered Species of Wild Fauna and Flora (CITES) and the 1979 Bonn Convention in order to create an overall protection, conservation and management regime for global biological diversity. To that end the Convention's objectives, as stated in its Article 1, are:

> the conservation of biological diversity, the sustainable use of its components and the fair and equitable sharing of the benefits arising out of the utilisation of genetic resources.

The ultimate goal of the Convention, the maintenance of biological diversity, is described in its Article 2 as:

> the variability among living organisms from all sources including, inter alia, terrestrial, marine and other aquatic ecosystems and the ecological complexities of which they are a part; this includes diversity within species, between species and of ecosystems.

Parties are required under the Convention to give priority to *in situ* and *ex situ* conservation (Articles 8 and 9). Of particular significance to dealing with marine environmental pollution are the following obligations regarding *in situ* conservation:

1. the establishment of a system of protected areas to conserve biological diversity;
2. the promotion of the protection of ecosystems, natural habitats and the maintenance of viable populations of species in natural surroundings;
3. the promotion of environmentally sound and sustainable development; and
4. the rehabilitation and restoration of degraded ecosystems and the promotion of the recovery of threatened species (Article 8).

The Convention's provisions overlap with but reinforce a great many international environmental instruments adopted at global, regional, sub-regional and bilateral levels.

To date, implementation of the Convention has concentrated on the meetings of the Conference of Parties, where discussions have taken place on how best to give effect to the obligations to protect and conserve biological diversity. At the first meeting of the Conference of Parties in 1995, agreement was reached that parties should take action in five focus areas relating to marine and coastal biodiversity: (1) integrated marine and coastal area management; (2) marine and coastal protected areas; (3) sustainable use of coastal and marine living resources; (4) mariculture; and (5) alien species.[57]

Assessment

One notable aspect of environmental protection in the polar regions has been the various regimes established to protect and manage certain species. However, this approach has been selective, with little attempt made at adopting a comprehensive regime. The Convention on Biological Diversity is significant globally because of the emphasis given in its Article 8 to the establishment and management of protected areas, which in the polar regions is important for the enhancement of marine protected area systems. While protected areas are relatively well developed under the Antarctic Treaty System and are provided for under the 1991 Environmental Protocol,[58] comparatively little attention has been given to *marine* protected areas in the Antarctic. Likewise, in the Arctic there is also scope for a closer assessment of the development of a more extensive system of marine protected areas. This is a matter which the Conservation of Arctic Flora and Fauna Working Group under the AEPS has been assessing, and it has been reviewed as part of the regional response by Arctic states to the Convention on Biological Diversity.[59]

OTHER GLOBAL ENVIRONMENTAL CONVENTIONS AND THEIR IMPACT ON THE POLAR MARINE ENVIRONMENT

In addition to the international environmental instruments discussed above, there remain a great many more with the potential to protect the polar marine environment indirectly. Instruments dealing with nuclear weapons testing and nuclear materials are clearly of significance.[60] The concern expressed over the depletion of the ozone layer, particularly over the polar regions, also makes the principal legal instrument dealing with this problem significant. All polar states have given strong support to the 1985 Vienna Convention for the Protection of the Ozone Layer,[61] and its subsequent Protocol.[62] Of special importance to the polar states is the issue of climate change; nuclear emergency assistance and notification has also been a matter of concern.

[57] L. Glowka and F. Burhenne-Guilmin, 'Convention on Biological Diversity', *Yearbook of International Environmental Law*, Vol. 6, 1995, p. 320.

[58] See Annex V of the Protocol, on 'Area Protection and Management'; however, as at 8 June 1999, this annex had not entered into force, still lacking acceptance or approval by four more Consultative Parties. On Annex V, see Joyner, Chapter 5 in this book.

[59] VanderZwaag, 'International Law and Arctic Marine Conservation', p. 318.

[60] See the 1963 Treaty Banning Nuclear Weapon Tests in the Atmosphere, in Outer Space and Under Water, UNTS, Vol. 480, pp. 43ff; and the 1979 Convention on the Physical Protection of Nuclear Material, ILM, Vol. 18, 1979, pp. 1,422ff.

[61] UNTS, Vol. 1,513, pp. 293ff. As at 8 June 1999, all Arctic states are parties to it.

[62] The 1987 Montreal Protocol on Substances that Deplete the Ozone Layer, as amended; published in UNTS, Vol. 1,522, pp. 3ff. As at 8 June 1999, all Arctic states are parties.

The 1992 Framework Convention on Climate Change

The United Nations Framework Convention on Climate Change,[63] also adopted at UNCED, was intended to achieve stabilisation of greenhouse gas concentrations at a level that would prevent 'dangerous anthropogenic interference with the climate system' (Article 2). To that end the Convention emphasises the need to adopt a precautionary approach to 'anticipate, prevent or minimize' the causes of climate change and to mitigate its adverse effects (Article 3). However, the initial targets set by the Convention were relatively weak and concerns were expressed over the need to strengthen the emission targets for all states in order to reduce the impact of greenhouse gas emissions. This was addressed in 1997 with the adoption of the Kyoto Protocol[64] which provides for firm reductions in greenhouse gas emissions into the first decade of the twenty-first century.

Assessment

The Framework Convention on Climate Change has clear implications for the polar regions, as the effects of global warming could have a devastating impact upon various types of ice found there. Not only would the release of fresh water from the ice cap cause a rise in sea level, but it would also impact upon the polar marine ecosystem. The warming of the polar oceans is also likely to have unforeseen impacts. To date, there is no evidence that the polar states are working towards a common position on the issue of climate change. Many are large industrial states which have significant temperate lands in addition to their polar claims and interests. The problem of climate change is a truly global issue, one in which states collectively need to work together to achieve results. As such, the success or otherwise of the Framework Convention on Climate Change is not dependent upon the polar states alone, though given their political importance they may have a clear impact upon how the international community responds.

The 1986 Nuclear Accident Conventions

The issue of nuclear emergency assistance and notification became a matter of global importance following the 1986 Chernobyl nuclear power station accident in the former Soviet Union, which had environmental consequences for the rest of Europe, including some Arctic states (Finland, Norway and Sweden). The response by the international community was to adopt the same year two international instruments: the Convention on Early Notification of a Nuclear Accident,[65] and the Convention on Assistance in the Case of a Nuclear Accident or Radiological Emergency.[66] In the case of the first Convention, parties are obliged to notify all states which may be affected as a result of a nuclear accident and promptly to provide such states with relevant information so as to minimise the effects (Article

[63] ILM, Vol. 31, 1992, pp. 849ff. [64] ILM, Vol. 37, 1998, pp. 22ff.
[65] ILM, Vol. 25, 1986, pp. 1,370ff. [66] *Ibid.*, pp. 1,377ff.

2). The second Convention establishes procedures for parties which have suffered a nuclear accident to request assistance from neighbouring states in order to respond to the emergency.

Both Conventions have clear implications for the polar regions – especially in the Arctic, where nuclear power has been widely used and where there is the potential for severe environmental impact following a nuclear accident within the region or to the south. These Conventions may also have application in the Southern Ocean, notwithstanding the effect of Article V of the Antarctic Treaty, as nuclear accidents may arise from a maritime disaster involving a nuclear-powered vessel or during the transport of nuclear fuels or radioactive waste.[67]

INTERACTION BETWEEN GLOBAL CONVENTIONS AND REGIONAL REGIMES FOR MARINE ENVIRONMENTAL PROTECTION

Chapter 1 of this book discussed the forms of regionalism that exist in the protection of the marine environment. It noted that the LOS Convention provides different models for the relationship between global conventions and those at the regional and sub-regional level, depending on the source of pollution. The restrictive model of regionalism is best evidenced in the polar regions by the way the polar states have responded, either individually or collectively, to *vessel-sourced* marine pollution. Article 211(2) of the LOS Convention envisages that, in dealing with pollution from vessels, both coastal and flag states shall enact laws and regulations that 'shall at least have the same effect as that of generally accepted international rules and standards established through the competent international organization or general diplomatic conference'.

In the Antarctic this expectation has been met by the provisions of the Environmental Protocol, Annex IV of which provides that: 'With respect to those Parties which are also Parties to MARPOL 73/78, nothing in this Annex shall derogate from the specific rights and obligations thereunder.'[68] The Annex also contains several cross-references to MARPOL 73/78 provisions dealing with the discharge at sea of oil, the disposal of garbage and the discharge of sewage.[69] The close connection between MARPOL 73/78 and Annex IV of the Environmental Protocol is further confirmed in Article 13 of the Protocol, which provides that the Antarctic Treaty parties are to keep under 'continuous review' the provisions of the Annex, including any amendments and new regulations adopted under MARPOL 73/78.[70]

A similar deference to MARPOL 73/78 can also be found in the Arctic, where the AEPS has made reference to MARPOL 73/78 standards and the need to ensure that states maintain those standards when operating in polar waters. The

[67] It should also be recalled that the United States operated a nuclear power plant at McMurdo Station from 1962 to 1972; however, any renewed attempt to establish such a facility would raise for consideration the application of the 1991 Protocol on Environmental Protection to the Antarctic Treaty.

[68] Environmental Protocol, Annex IV, Art. 14.

[69] *Ibid.*, Arts. 3, 5 and 6.

[70] For further discussion, see Joyner, Chapter 5 in this book.

'respect' for the MARPOL 73/78 provisions is also reflected in the early drafts of the Polar Code, preambles of which expressly stated that the Code is not intended to replace other requirements for ships operating in polar waters.

The existing global standards for *land-based* marine pollution, however, anticipate that states will seek to deal with the problem at the regional and sub-regional levels. To that end, Article 207(1) of the LOS Convention provides that, when responding to land-based marine pollution, states are to take 'into account internationally agreed rules, standards and recommended practices and procedures'. This certainly reflects what has occurred in the Arctic, where European states have long attempted to regulate land-based marine pollution – initially through the 1974 Paris Convention for the Prevention of Marine Pollution from Land-Based Sources,[71] and more recently through the 1992 Convention for the Protection of the Marine Environment of the North-East Atlantic.[72] This independent approach by the Arctic states to developing responses to land-based marine pollution is also reflected in the initiatives undertaken through the AEPS, under which the PAME working group has been engaged in developing the Regional Programme of Action for the Protection of the Arctic Marine Environment from Land-Based Activities.[73]

Yet another form of interaction between the global and regional marine environmental conventions can be identified: one in which the global conventions have recognised the *special status* of polar waters as a result of initiatives adopted at the regional level. This type of recognition is most prominent regarding the Antarctic, where MARPOL 73/78 and the Basel Convention have given special recognition to the waters south of 60° S in deference to the Antarctic Treaty System. A further example is the special status accorded to ice-covered waters in Article 234 of the LOS Convention; in that case, recognition was granted following Canada's adoption of the Arctic Waters Prevention Pollution Act in 1970 and subsequent support for special recognition of the Arctic marine environment by Canada and the former Soviet Union during the negotiation of the LOS Convention.[74]

CONCLUSIONS

There exist many international law sources that provide a legal basis for protection of the polar marine environment. They are 'additional sources' in the sense that they operate in addition to the framework provisions of the LOS Convention, and the more specific regional instruments that have been adopted for the polar oceans. The conventions reviewed here are, however, conventions with *global* application, designed to apply to all of the world's oceans and seas. It is

[71] ILM, Vol. 13, 1974, pp. 546ff; see Art. 3(a) for the application of the Convention to parts of the Arctic Ocean.

[72] ILM, Vol. 32, 1993, pp. 1,068ff. See the discussion by VanderZwaag, Chapter 8 in this book.

[73] For further details, see VanderZwaag, Chapter 8 in this book.

[74] For further discussion on Art. 234 of the LOS Convention, see Vukas, Chapter 2; Rothwell and Joyner, Chapter 7; and Brubaker, Chapter 10 in this book.

therefore not surprising that in only a few instances are specific provisions found dealing with the particular marine environmental conditions that exist in the polar regions. With the LOS Convention providing a framework within which marine environmental law operates at the global level, a complex web of international instruments is created for both environmental law and maritime law in the Arctic and Antarctic.

However, notwithstanding the global application of these marine environmental instruments – thus including application to the polar regions as well – there are practical differences, for a variety of reasons. The most significant difference is that the lack of general recognition of Antarctic 'coastal states', in conjunction with the limitations on the assertion of sovereignty and jurisdiction imposed by the Antarctic Treaty, constrains the application of many of the global instruments in Antarctica. In contradistinction, the Arctic is characterised by settled coastal state sovereignty, which means that Arctic states are on a sounder legal foundation when they seek to enforce the international legal regime in their polar waters. Arctic states have also had longer exposure to the threats posed to their marine environments by commercial shipping, and this is reflected by the more stringent domestic legal regimes adopted in the past to regulate pollution.[75] In addressing marine pollution issues, Arctic states have had to take into consideration the significant human settlements in the Arctic, including indigenous peoples, and the industrial activities which attract commercial attention.[76] This is to be contrasted with Antarctica, where it was only in the late 1990s that some states began to patrol their polar waters on a regular basis, and even then only in sub-Antarctic waters where sovereignty is uncontested.

Clearly, the legal regimes reviewed above in many respects fail to address adequately the particular challenges posed by protection of the polar marine environment. To that end, the proposed Polar Code should be significant, especially as it is being developed by a global body within the IMO.[77]

Various gaps notwithstanding, the legal regimes reviewed here do provide a basis for increasingly comprehensive protection of the marine environment, although still with land-based pollution as the significant remaining neglected area. Through combined enforcement by polar coastal states and also by flag and port states, these legal regimes have wide application. Ultimately, however, their effectiveness will depend upon action taken by the polar states through legal and policy responses at the regional and, indeed, national levels. In policy terms, the continuing emphasis since UNCED on the need for sustainable development at the national, regional and global levels is also influential. Marine environmental protection of the polar regions is no longer the responsibility of the polar states alone: it is increasingly becoming a truly global responsibility.

[75] See especially Brubaker, Chapter 10 in this book, for the regulation of navigation in the Northern Sea Route. [76] See the Introductory overview to this book.

[77] On recent challenges as to the bi-polar application of the future Polar Code, in particular regarding the Antarctic following decisions adopted by the IMO in May 1999 and by the Antarctic Treaty Consultative Parties in June 1999, see Brigham, Chapter 11 in this book.

4 The polar marine environment in regional cooperation

DAVOR VIDAS

In the course of the 1990s, international cooperative efforts concerning both the Arctic and the Antarctic have come to share one significant similarity: an emphasis on international protection of the polar environment in general, and the polar marine environment in particular. It may be tempting to see this as a natural consequence of the special features of polar regions, with their environment characterised by difficult ice conditions, including large areas of ice-infested waters – and increased environmental risks which human activities involve in this setting. This is what, broadly speaking, the opposite poles have in common, and is also what sets them apart from all other parts of the globe.[1] However, this 'first glance' impression of a shared environmental focus due to shared polar features may not apply when it comes to political realities and legal measures. A closer look at current international instruments and institutional arrangements for environmental protection of the two polar oceans, adopted through the respective regional cooperation arrangements, reveals a somewhat paradoxical situation.

As regards the Southern Ocean, recent assessments confirm that the overall threat of pollution of its marine environment from sources within the region appears generally low.[2] Nevertheless, the states parties to the 1959 Antarctic Treaty[3] supplemented that treaty more recently with a comprehensive environmental protection instrument: the Protocol on Environmental Protection to the Antarctic Treaty.[4] In addition to providing a comprehensive regional environmental protection regime for the Antarctic, the Protocol addressed protection of Antarctic waters from vessel-source pollution at the regional level in a special annex on 'Prevention of Marine Pollution'.

As regards the opposite pole, the Arctic countries, since they started to

[1] See the Introductory overview to this book.

[2] See, for instance, COMNAP, 'An Assessment of Environmental Emergencies Arising from Activities in Antarctica', doc. XXIII ATCM/WP 16, April 1999.

[3] UNTS, Vol. 402, pp. 71ff.

[4] The Protocol was adopted in 1991 and entered into force on 14 January 1998; text reprinted in ILM, Vol. 30, 1991, pp. 1,461ff.

cooperate within the framework of the Arctic Environmental Protection Strategy[5] (AEPS), have identified the Arctic as being in need of a thorough examination of its current status, including an inquiry into international instruments for environmental protection. Among the principal results of this intergovernmental cooperative process have been several comprehensive assessments published between 1996 and 1998 – all confirming that both actual and potential sources of regional pollution of the Arctic marine environment are far graver than those that may threaten the Antarctic.[6] Despite this, no Arctic-specific multilateral instrument for the protection of the marine environment has so far been adopted by the Arctic countries at the regional level. Indeed, Arctic policy-makers have not even concluded that there is a need for such an instrument. Their opinion has been expressed in various documents that will be analysed in this chapter. Briefly stated, they hold that, for the time being, the existing global and other instruments suffice, even though these are admittedly not always tailored for polar Arctic conditions.

For the marine environment of the Arctic – which appears more endangered in environmental terms – policy-makers have failed to see the need for new international legal regulations to strengthen protection at the regional level. By contrast, those dealing with the Antarctic – where sources of pollution would seem to pose far less threat to the environment – found it urgently necessary one decade ago to adopt new legally binding rules at the regional level. Is then the need for regional marine environmental protection in the Arctic underestimated? Or has such a need in respect of the Antarctic marine environment been overestimated? There is also a third possibility: perhaps the issue is neither solely, nor even primarily, a matter of environmental needs, whether regional or not, but rather of various other concerns. This chapter aims to look into these questions.

INTERNATIONAL COOPERATIVE PROCESSES FOR POLAR ENVIRONMENTAL PROTECTION: THE DECADE OF CHANGE

It is the differences between the Arctic and Antarctic regions, rather than their similarities, that are traditionally pinpointed by most Arctic experts.[7] These differences, it is often argued, make each polar region a separate case, and little would be gained by a comparative focus. Concerning protection of the marine environment, one author has argued that comparison of the polar regions in this respect can be misleading or inappropriate due to sharp contrasts between the

[5] Adopted at the First Ministerial Conference on the Protection of the Arctic Environment, at Rovaniemi, Finland, 14 June 1991. Text reprinted in ILM, Vol. 30, 1991, pp. 1,624ff.

[6] The two reports issued by the Arctic Monitoring and Assessment Programme (AMAP) should especially be noted: *AMAP Assessment Report: Arctic Pollution Issues* (Oslo: Arctic Monitoring and Assessment Programme, 1998); and *Arctic Pollution Issues: A State of the Arctic Environment Report* (Oslo: Arctic Monitoring and Assessment Programme, 1997).

[7] See especially G. Osherenko and O. Young, *The Age of the Arctic* (Cambridge University Press, 1989), pp. 242–4.

Arctic and Antarctic, especially since 'no regional structure exists to facilitate or promote cooperation' among the Arctic states, while 'evidence of agreement among Arctic states on a legal structure for protecting the marine environment is equally scanty'.[8] And, indeed, in the late 1980s, when this observation was made, it was fully accurate.

Any international cooperation in Arctic environmental protection during the 1980s was characterised by the conclusion of several *bilateral* agreements between the Arctic countries – not by any regional instruments.[9] However, marine environmental protection in the Arctic has predominantly been governed not by bilateral agreements, but by *unilateral* acts promulgated by the Arctic rim states, which acts are also partly based on Article 234 of the United Nations Convention on the Law of the Sea[10] (LOS Convention). In contrast, during that same period, the Antarctic Treaty Consultative Parties, working within the framework of the Antarctic Treaty System (ATS), adopted at the regional level a series of *multilateral* instruments related to the protection of the Antarctic environment, comprising several international conventions,[11] as well as various other measures. The protection of the Antarctic environment has been approached through those instruments in an issue-specific manner.

The years since the late 1980s have seen extraordinarily dynamic developments in international cooperative efforts for the protection of the environment in both polar regions. A regional cooperative structure among the Arctic countries has emerged. Initiated in the late 1980s and adopted in the form of the AEPS in 1991, Arctic regional cooperation has since 1996 been developing within the framework of the Arctic Council.[12] In the Antarctic, changes of no less importance have been underway since the late 1980s in regulating environmental protection within the Antarctic Treaty System. Thus, the polar regions today seem to share an important characteristic: the tendency to *structure* environmental protection through multilateral cooperation at the regional level.

Let us briefly review these simultaneous developments in the Antarctic and Arctic regions, and then focus more closely on the current approaches taken by these two regional cooperative processes for protecting the marine environment of their respective polar oceans from pollution.

[8] A. E. Boyle, 'Legal Regimes of the Arctic – Remarks', *American Society of International Law Proceedings*, Vol. 82, 1988, pp. 324 and 326.

[9] For a concise overview see P. Kunig, 'Arctic', in R. Bernhardt (ed.), *Encyclopedia of Public International Law*, Vol. 1 (Amsterdam: Elsevier Science, 1992), pp. 246–7.

[10] Text reprinted in ILM, Vol. 30, 1982, pp. 1,261ff. On Art. 234 of the LOS Convention ('Ice-covered areas') see the discussion by Vukas, Chapter 2; Rothwell and Joyner, Chapter 7; and Brubaker, Chapter 10 in this book.

[11] In addition to the earlier adopted 1972 Convention for the Conservation of Antarctic Seals (CCAS; reprinted in ILM, Vol. 11, 1972, pp. 251ff), the 1980s saw adoption of the 1980 Convention on the Conservation of Antarctic Marine Living Resources (CCAMLR; reprinted in ILM, Vol. 19, 1980, pp. 837ff) and the 1988 Convention on the Regulation of Antarctic Mineral Resource Activities (CRAMRA; reprinted in ILM, Vol. 27, 1988, pp. 868ff).

[12] See the Declaration on the Establishment of the Arctic Council, signed by the eight Arctic states in Ottawa, Canada, 19 September 1996; text reprinted in ILM, Vol. 35, 1996, pp. 1,387ff.

The Antarctic

The Antarctic Treaty System is what provides the mechanism for regional cooperation in respect of the Antarctic.[13] Although the area of application of the Antarctic Treaty lies within the confines of the Antarctic as a region, it is difficult to regard the ATS as 'regional' cooperation *stricto sensu*, as this term is usually understood in international law[14] – not least since the countries comprising the group of twenty-seven Antarctic Treaty Consultative Parties belong to all the other six (inhabited) continents, and various different regions. On the other hand, the ATS is *regionally applicable*. Here the determining criterion for seeing that cooperation as regional is the object of cooperation, not the geographical placement of the subjects involved in cooperation.

This peculiar situation is the consequence of the fact that the entire Antarctic Treaty System, under continuous development for almost four decades now, has been based on unresolved sovereignty issues concerning the Antarctic. The essential requirement in the development of the ATS was to build it, through cooperation, so as not to prejudice the position of any countries claiming sovereignty in the Antarctic – or those not recognising these claims.[15] This international-level governance system operates through annual Antarctic Treaty Consultative Meetings as the main policy-making body regulating all human activities in the Antarctic. The Antarctic Treaty Consultative Parties have decision-making capacity in this forum, while other, non-Consultative Parties have the right to attend meetings.[16] There are various other mechanisms for policy- and decision-making within the ATS, as well as for scientific and technical advice;[17] and a definition of the ATS regarding its normative components has been formulated in the Environmental Protocol.[18] These normative components today include several international conventions noted above and a large number of other measures that have been adopted.

The long-standing record of the issue-specific approach to Antarctic environmental protection, introduced to the ATS through recommendations and then through several international conventions, culminated in the adoption in June 1988 of the Antarctic Minerals Convention, CRAMRA.[19] Shortly thereafter, the

[13] For a recent comprehensive study of the Antarctic Treaty System, see O. S. Stokke and D. Vidas (eds.), *Governing the Antarctic: The Effectiveness and Legitimacy of the Antarctic Treaty System* (Cambridge University Press, 1996).

[14] On the understanding of 'region' and 'regional cooperation' in international law, see Boyle, Chapter 1; and Vukas, Chapter 2 in this book.

[15] See the interplay between various provisions of the Antarctic Treaty, especially Arts. IV and IX.

[16] As of 8 June 1999, there were forty-four states parties to the Antarctic Treaty; twenty-seven are Consultative Parties while the remaining seventeen are non-Consultative Parties.

[17] For an overview of various components of the ATS, and of the various balances made among these, see D. Vidas, 'The Antarctic Treaty System in the International Community: An Overview', in Stokke and Vidas (eds.), *Governing the Antarctic*, pp. 35–60.

[18] See Art. 1(e) of the Protocol.

[19] For an overview, see C. C. Joyner, 'The Effectiveness of CRAMRA', in Stokke and Vidas (eds.), *Governing the Antarctic*, pp. 152–62. See also F. Orrego Vicuña, *Antarctic Mineral Exploitation: The Emerging Legal Framework* (Cambridge University Press, 1988); and R. Wolfrum, *The Convention on the Regulation of Antarctic Mineral Resource Activities: An Attempt to Break New Ground* (Berlin: Springer-Verlag, 1991).

'CRAMRA crisis' shook the ATS: in the course of the spring of 1989, Australia and France announced that they would neither sign nor ratify CRAMRA, an attitude that was soon adopted by several other Antarctic Treaty Consultative Parties.[20] Instead, these countries proposed a new instrument that would ban any mineral activity (with the exception of scientific research) in the Antarctic, and would introduce a comprehensive environmental protection system. Following a decision of the 1989 Antarctic Treaty Consultative Meeting, a Special Meeting was convened later the same year, in order to negotiate a new legal instrument. In a record time of less than two years, this new legal instrument – the Protocol on Environmental Protection to the Antarctic Treaty – was adopted in October 1991; though it then took more than six years before eventually entering into force, in January 1998. One of the Annexes to the Protocol is specifically devoted to protecting the Antarctic marine environment from pollution from ships.[21]

The great majority of the provisions of the Protocol have been taken over from recommendations adopted earlier. In fact, part of the Protocol's basic environmental principles come from CRAMRA – the very instrument which it superseded.[22] While perhaps not bringing too much fresh regulation into the ATS, the Protocol did approach the protection of the Antarctic environment in a *comprehensive* manner. It also 'codified' the existing recommendations into a legally binding instrument. Moreover, it provided for the establishment of a new institution within the ATS, the Committee for Environmental Protection (CEP), operative as of 1998. Thus, some have seen the Protocol as one of the most advanced international legal instruments adopted to date in the field of environmental protection.[23]

Since the Protocol entered into force in 1998, the main preoccupation of the Consultative Parties (which are all parties to the Protocol) became a complex set of issues connected with *implementation* of this legal instrument.[24] In this connection, the basic problem of the specific regional situation in the Antarctic is the unresolved question of sovereignty and jurisdiction, and thus – in the context of this book – also of control and enforcement when it comes to implementation of legal instruments for marine environmental protection in the Southern Ocean. Other major issue-areas involved in the implementation of the Protocol include: institutionalisation within the ATS; further normative development of the Protocol, in particular the pending liability regime;[25] the relationship of the Protocol to other

[20] See also Rothwell and Joyner, Chapter 7 in this book.

[21] Annex IV of the Protocol; other annexes are also highly relevant for marine pollution, especially Annex I ('Environmental Impact Assessment') and Annex III ('Waste Disposal and Waste Management'). See the discussion by Joyner, Chapter 5 in this book, which contains a detailed examination of the Protocol and its annexes, as relating to protection against marine pollution in the Antarctic.

[22] See C. C. Joyner, 'The Legitimacy of CRAMRA', in Stokke and Vidas (eds.), *Governing the Antarctic*, pp. 255–67.

[23] See, for example, Joyner, Chapter 5 in this book.

[24] For an examination of the various issues involved in the implementation of the Protocol, see D. Vidas (ed.), *Implementing the Environmental Protection Regime for the Antarctic* (Dordrecht: Kluwer Academic Publishers, 2000).

[25] In Art. 16, the Protocol requires its parties to 'elaborate rules and procedures relating to liability for damage arising from activities taking place in the Antarctic Treaty area and covered by this Protocol'.

applicable environmental agreements; and issues in domestic legislation adopted in the implementation of the Protocol.

Thus, environmental protection of the Antarctic within the ATS is currently characterised by reliance on a regional multilateral instrument that is already in force. Approaches to protection of the Antarctic environment tend to be comprehensive and are adopted at the regional level, with the focus on a regional cooperative structure which facilitates implementation. The protection of the Antarctic marine environment lies within this general framework introduced by the Protocol.

The Arctic

In the Arctic, the picture of cooperation is more complex and segmented, in so far as environmental protection is addressed at various levels of cooperation and regulation. Apart from the regional level, sub-regional cooperation is an important and relatively recently added layer in environmental protection – for example, within the Barents Euro–Arctic Region (BEAR).[26] Moreover, several other forms of international environmental cooperation make up today's Arctic cooperative web, which also involves transnational, sub-national, non-governmental and extra-regional cooperation.[27] Less than one decade ago, however, any international cooperation between the Arctic countries was dominated by bilateral agreements,[28] which of course still figure prominently in this area. Finally, domestic legislation of Arctic coastal states has remained important in the overall regional picture of marine environmental protection.

The focus on Arctic affairs has grown considerably since the late 1980s. The dissolution of the Soviet Union and the easing of tensions between the two Cold War rivals have presented new opportunities for the development of Arctic cooperation – now on a multilateral, pan-Arctic regional level. The international 'green wave' offered an opportunity to divert and direct attention, including that of the Arctic countries, to the environment. These major political changes, as well as the growing awareness of the actual and potential threats to the Arctic environment, prompted a swift response from countries such as Finland: an initiative of January 1989, aimed at the convening of a conference of the eight Arctic countries on protection of the Arctic environment. For the purpose of this cooperative process, the eight Arctic countries are defined as being Canada, Denmark/ Greenland, Finland, Iceland, Norway, Russia, Sweden and the United States.

[26] On sub-regional cooperation in the protection of the Arctic marine environment, see Stokke, Chapter 6 in this book. For a comprehensive treatment of BEAR, see O. S. Stokke and O. Tunander (eds.), *The Barents Region: Cooperation in Arctic Europe* (London: SAGE, 1994).

[27] For a detailed overview, see *Co-operation in the Arctic Region – Report Submitted to the Nordic Council of Ministers* (Copenhagen: Nordic Council of Ministers, 1995); see also D. Scrivener, 'Environmental Cooperation in the Arctic: From Strategy to Council', *Security Policy Library*, No. 1 (Oslo: Norwegian Atlantic Committee, 1996), pp. 3–33.

[28] With the notable exception of the International Agreement on the Conservation of Polar Bears and Their Habitats, done at Oslo, 15 November 1973, entered into force 26 May 1976; text reprinted in ILM, Vol. 13, 1974, pp. 13ff.

Following several preparatory meetings, the first ministerial conference of the Arctic countries addressing environmental protection was held in Rovaniemi, Finland, in June 1991. It resulted in the adoption of the foundations for subsequent Arctic environmental cooperation through the mid-1990s: the Declaration on the Protection of the Arctic Environment (the Rovaniemi Declaration),[29] and the Arctic Environmental Protection Strategy (AEPS), with its Action Plan.

Since 1991, this process has evolved through several programmes and working groups which undertook data-gathering, compilation of information and assessment tasks (sometimes also called 'programmatic' activities).[30] A major task has been undertaken by the Arctic Monitoring and Assessment Programme, where two main initial activities have been monitoring the Arctic environment, particularly for contamination, and an assessment of the state of the Arctic environment. A working group on the Protection of the Arctic Marine Environment (PAME) has also been established, and its activities and results will be presented in further detail below. Other components have included the Emergency Prevention, Preparedness and Response Programme; the Conservation of Arctic Flora and Fauna Programme; and the Task Force (later Working Group) on Sustainable Development and Utilisation.

These AEPS components have reported to the ministers of the environment of the respective Arctic countries, who have reviewed their progress at intervals of a few years. Following the 1991 Rovaniemi ministerial meeting, several more ministerial conferences were held: in 1993 at Nuuk, Greenland; in 1996 at Inuvik, Canada; and in 1997 at Alta, Norway. The latter was the final meeting of the AEPS itself, whose existing programmes were thereupon integrated into the work of the Arctic Council, which was established in 1996 as a 'high level forum'.[31] During its first two years the Arctic Council was chaired by Canada, and thereafter, until October 2000, by the USA. The chairing functions and the hosting of ministerial meetings of the Council rotate sequentially among the Arctic states on a biennial basis;[32] and one Arctic Council ministerial meeting has been held so far, at Iqaluit, Canada, in September 1998.

Building on the AEPS, the Arctic Council Declaration has taken a wider view of the entire northern polar cooperation, with its two interconnected focal points of sustainable development and environmental protection in the Arctic.[33]

[29] Adopted on 14 June 1991; text reprinted in ILM, Vol. 30, 1991, pp. 1,624ff.

[30] See O. Young, 'The Arctic Environmental Protection Strategy: Looking Backward, Looking Forward' (unpublished paper based on presentations at the Danish/American Greenland Science Conference and the North Calotte Academy, April–May 1995; on file with author). See the elaboration by Stokke, Chapter 6 in this book.

[31] Para. 1 of the Arctic Council Declaration. On the Council's takeover of the AEPS programmes, see paras. 1(b) and 1(c) of the Arctic Council Declaration. See also para. 10 of the Alta Declaration on the Arctic Environmental Protection Strategy, of 13 June 1997; text available at the Arctic Council website at http://arctic-council.usgs.gov.

[32] Paras. 4 and 5 of the Arctic Council Declaration.

[33] See D. Scrivener, 'Arctic Environmental Cooperation in Transition', *Polar Record*, Vol. 35, 1999, pp. 51–8; and O. Young, 'Sustainable Development in the Arctic: Operationalizing the Arctic Council', in L. Heininen and R. Langlais (eds.), *Europe's Northern Dimension: The BEAR Meets the South* (Rovaniemi: University of Lapland, 1997), pp. 259–77. See also D. Vidas (ed.), *Arctic Development and Environmental Challenges* (Copenhagen: Scandinavian Seminar College, 1997).

This broader approach has been possible because of a strict separation from issues of military security.[34]

As to the protection of the Arctic *marine* environment, the 1991 Rovaniemi Declaration singled this out as a principal concern in the implementation of the AEPS. In the AEPS document itself, Chapter 4(2) addresses oil pollution; this is, however, restricted to a listing of the principal international instruments relevant to the Arctic, together with remarks on the need for further assessment of their adequacy to Arctic conditions, as well as for stricter standards for oil transport in the Arctic. Chapter 7 of the AEPS deals specifically with protection of the Arctic marine environment. It emphasises the need for *preventive* measures, consistent in particular with the LOS Convention, regarding marine pollution in the Arctic, irrespective of origin.

At the 1993 Nuuk Conference, the ministers of the Arctic countries agreed, in implementing Chapter 7 of the AEPS, to establish a working group to assess:

> the need for further action or instruments to prevent pollution of the Arctic marine environment and to evaluate the need for action in appropriate international fora to obtain international recognition of the particularly sensitive character of the ice-covered sea areas of the Arctic.[35]

This, then, provided the terms of reference for the PAME working group. The scope of its work was defined as the examination of the activities/sources of pollution and their impact on the Arctic marine environment. On this basis the working group was to evaluate possible options as to necessary legal instruments.[36] Accordingly, in the period between the 1993 Nuuk and the 1996 Inuvik Ministerial Conferences, PAME was particularly concerned with the choice of *approach* for protection of the northern polar marine environment from pollution. Should the protection of the Arctic marine environment be pursued by interpreting how relevant global marine protection instruments may be applied to the Arctic case (and perhaps adapted in some instances)? Or by creating a new regional regime specifically tailored to suit the requirements of the protection of the Arctic marine environment? On the basis of PAME's main report on the matter,[37] the 1996 Inuvik Ministerial Conference concluded:

> taking into consideration the nature of the threats and an assessment of existing instruments, the existing instruments provide an adequate basis for the

[34] An explanatory note to para. 1(a) of the Arctic Council Declaration makes it clear that the Council 'should not deal with matters related to military security'.

[35] See para. 2 of the 1993 Nuuk Declaration on Environment and Development in the Arctic; the text of the Declaration is available at the Arctic Council website at http://arctic-council.usgs.gov.

[36] See para. 3(1)(a) in PAME, 'Report from the First Meeting, Oslo, Norway, May 3–5, 1994' (unpublished document; on file with author). Following this, PAME held four more meetings before finalising its main report and presenting it to the 1996 Inuvik Ministerial Conference.

[37] Working Group on the Protection of the Arctic Marine Environment, *Report to the Third Ministerial Conference on the Protection of the Arctic Environment, 20–21 March 1996, Inuvik, Canada* (Oslo: Norwegian Ministry of the Environment, 1996) (PAME 1996 Report).

> protection of the Arctic marine environment and there is no urgent need to develop new legal instruments for this purpose.[38]

In other words, the protection of the Arctic marine environment was to rely on existing global and other instruments rather than any newly developed Arctic-regional ones.

Nor has this overall situation changed under the Arctic Council. PAME continued to work on several programmatic activities, in particular developing an Arctic regional programme of action on land-based pollution,[39] guidelines for offshore petroleum activities, as well as collecting information on current and potential shipping activities in the Arctic – a course charted already at the Inuvik Ministerial Conference.

Possibilities of mutual insight and communication

In considering recent inter-state processes in Arctic and Antarctic environmental protection, we should note the significant degree of overlap of the participating states: out of the eight countries directly involved in the AEPS and Arctic Council processes, Iceland is the single country outside the Antarctic Treaty System. The majority of Arctic countries (five out of eight) are also Consultative Parties to the Antarctic Treaty: Norway, Russia/Soviet Union, and the United States are original Consultative Parties, while Finland and Sweden acceded to the Antarctic Treaty and subsequently acquired Consultative Party status. Canada and Denmark are non-Consultative Parties to the Antarctic Treaty, though both have relatively long-standing involvement in Antarctic cooperation.[40] Thus, all the Arctic countries except Iceland were involved in the gradual creation of the regime for the protection of the Antarctic environment, including its maritime space; and all of these countries which are also Antarctic Treaty Consultative Parties are currently bound by the Environmental Protocol applying there.

There seem to exist two main channels for communication between the Arctic and Antarctic regional cooperative processes. One of these is at the practical, domestic decision-making level in those countries that are simultaneously a party to the Antarctic Treaty and a member of the Arctic Council. In the foreign ministries of Finland, Norway and Sweden, the same senior officials ('polar ambassadors'), or the same departmental units, are responsible for both Arctic and Antarctic affairs.[41] Canada took a similar approach when in 1994 it appointed its

[38] Para. 2(3) of the *Report of the Third Ministerial Conference on the Protection of the Arctic Environment, 20–21 March 1996, Inuvik, Canada* (Ottawa: Canadian Ministry of Indian and Northern Affairs, 1996).

[39] See below in this chapter; and VanderZwaag, Chapter 8 in this book.

[40] Denmark became party to the Antarctic Treaty in 1965, Canada in 1988.

[41] However, that simultaneous coverage of affairs of both polar regions by the same senior official, or the same departmental unit, seems more accentuated in the ministries of foreign affairs, and less so in the ministries of the environment of these countries.

first polar ambassador, who was to focus especially on environmental issues and to represent Canada at international meetings on circumpolar issues concerning both the Arctic and the Antarctic.[42] To a certain extent, this is also true for the US State Department, with the Polar Affairs Chief in the Bureau of Oceans and International Environmental and Scientific Affairs.[43]

The other channel for communication is at the international level of the regional cooperative processes. As mentioned above, most Arctic countries are also party to the Antarctic Treaty. In addition, several non-Arctic states, all Antarctic Treaty Consultative Parties, take part in the Arctic Council as observers:[44] Germany, the Netherlands, Poland and the United Kingdom. This overlap, and its inherent potential for effective exchange between the two cooperative processes, has been noted at the Antarctic Treaty Consultative Meetings, where some other countries also cultivate a vigorous interest in Arctic developments. It was thus on a Chilean initiative at the 1994 meeting that the Consultative Parties commenced considering the Arctic–Antarctic relevance 'in matters of environmental protection'.[45] At the 1995 meeting, the Consultative Parties agreed that an exchange of main documents, adopted in the respective regional processes and containing information on environmental issues, should go both ways, via the host country of the then AEPS Ministerial Meeting.[46] This has since become a regular practice at the Antarctic Treaty Consultative Meetings,[47] and has been adopted by the Arctic Council as well.

In addition to a pure exchange of main documents and briefing on major activities, this process of inter-polar communication has embodied two types of policy emphasis. On the one hand, there is an emphasis on differences: a key country in both polar contexts intervened at the 1996 Consultative Meeting, stressing the differences rather than similarities between the Arctic and Antarctic. The formulation which entered the final report from that meeting underlined:

> the need to bear in mind that, as far as co-ordination was concerned, the political and legal context governing activities in the Arctic and the Antarctic differ considerably.[48]

[42] See 'Canada Names First Ambassador to Focus on Arctic Environmental Issues', *International Environmental Reporter*, Vol. 18, 1995, p. 52.

[43] On the role of the US Polar Affairs Chief, see F. Griffiths, 'Environment in the US Discourse on Security: The Case of the Missing Arctic Waters', in W. Østreng (ed.), *National Security and International Environmental Cooperation in the Arctic – The Case of the Northern Sea Route* (Dordrecht: Kluwer Academic Publishers, 1999), pp. 194–6.

[44] This status is regulated under para. 3(a) of the Arctic Council Declaration.

[45] See paras. 53 and 54 of the *Final Report of the Eighteenth Antarctic Treaty Consultative Meeting, Kyoto, Japan, 11–22 April 1994* (Tokyo: Ministry of Foreign Affairs of Japan, 1994).

[46] See para. 35 of the *Final Report of the Nineteenth Antarctic Treaty Consultative Meeting, Seoul, 8–19 May 1995* (Seoul: Ministry of Foreign Affairs of the Republic of Korea, 1995).

[47] As of 1995, an agenda item titled 'Relevance of developments in the Arctic to the Antarctic' (renamed in 1996 to read 'Relevance of developments in the Arctic and the Antarctic') has figured regularly on the agendas of all the Antarctic Treaty Consultative Meetings.

[48] See para. 36 of the *Final Report of the Twentieth Antarctic Treaty Consultative Meeting, Utrecht, The Netherlands, 29 April–10 May 1996* (The Hague: Netherlands Ministry of Foreign Affairs, 1997).

On the other hand, there has also been a recurrent emphasis on the similarities between the two. Thus the 1998 Consultative Meeting stated:

> the view that there were several important points of convergence between the two polar areas, not the least with regard to the question of environmental protection.[49]

This sentiment has been echoed by several Arctic countries; and also the 1999 Antarctic Treaty Consultative Meeting noted that 'bipolar approaches could provide an understanding of common environmental aspects'.[50] It now remains to be seen to what extent rhetoric of this kind has influenced the development of approaches to environmental protection for the two polar regions.

APPROACHES TO POLAR MARINE ENVIRONMENTAL PROTECTION IN REGIONAL COOPERATION

We have seen that the marine environments of the two polar regions do share some special features, distinguishing the polar oceans from the remaining, warmer parts of the world's oceans. In view of these unique features, one would expect the degree of *prevention* embodied in legal measures for the protection of the polar marine environment to be commensurate with the severity of actual or potential risks posed by sources of pollution. Moreover, such preventive measures could be expected to embody the *adaptation* of more general, global standards to specific polar circumstances and the creation of special regional standards – again in line with the risks involved in human activity there. And, finally, one would assume that such preventive measures, tailored to the unique conditions of the polar areas, would accordingly apply to an area defined on the basis of the *ecosystemic* approach.

Let us take a closer look at these three aspects, all illustrative for exploring the dilemma put forward in the introductory section of this chapter:

1. the preventive measures for the protection of the (marine) environment, as applied in the polar regions;
2. the degree and modes of adaptation of general environmental protection standards to special polar conditions; and
3. the criteria employed for determining the area of application of measures for protecting the polar marine environment.

Enhanced prevention

The essence of the environmental protection measures adopted on the basis of anticipatory or precautionary approaches lies in preventing rather than

[49] See para. 99 of the *Final Report of the Twenty-Second Antarctic Treaty Consultative Meeting, Tromsø, Norway, 25 May–5 June 1998* (Oslo: Norwegian Ministry of Foreign Affairs, 1998).

[50] See para. 113 of the *Final Report of the Twenty-third Antarctic Treaty Consultative Meeting, Lima, Peru, 24 May–4 June 1999*, available at www.rree.gob.pe/conaan/meeting1.htm.

trying to 'repair' environmental degradation, which may well be irreversible.[51] Illustrative of this is the general approach of the Antarctic Treaty Consultative Parties, who have often managed to ensure that legally binding regulations are in place *before* serious environmental problems become a reality in the Antarctic – even before activities such as mining of minerals become a matter of any serious consideration. For all other activities not prohibited in the Antarctic, except for those undertaken pursuant to CCAMLR or CCAS,[52] the Protocol requires an environmental impact assessment at the planning stage.[53]

As to the Arctic, there still seems to be an opening for such an approach based on preventive measures. Viewed in global proportions, the level of pollution from sources within the Arctic still remains relatively low[54] – although its marine environment is far more exposed to the various human activities within the region than is the case with the Antarctic. Nevertheless, the direction taken by the AEPS process, and later the Arctic Council, does not seem to indicate the choice of an anticipatory, preventive approach to marine environmental protection. On the contrary, the Inuvik Ministerial Conference of March 1996 concluded:

> should implementation of various proposed actions not occur, or should they prove inadequate to address emerging problems, *then* reconsideration of further legally binding instruments should be pursued.[55]

Why have policy-makers approached marine environmental protection in the Antarctic with far more 'anticipation' and 'precaution' than in the Arctic – while the threats posed by actual and potential sources of pollution would suggest the converse? Here we should not forget that, in the case of the Antarctic environmental protection instruments, the Antarctic Treaty Consultative Parties – appearing at first glance to be wise anticipators of the threats to the environment – were in fact reacting to two acute *political* problems: first, the challenge to their legitimacy in governing the Antarctic, coming from subjects external to the ATS; secondly, and equally important, the struggle to maintain internal cohesion and balance within the ATS, especially as to the sovereignty issue.[56]

The Consultative Parties have thus had substantial incentives – in themselves often not directly or exclusively related to environmental protection – which prompted them to agree expeditiously on issues related to human activities in the Antarctic and environmental protection there. When CRAMRA was abandoned in 1989, soon after its adoption, that marked the start of negotiations on the Environmental Protocol. The Consultative Parties made a new start *not* because

[51] For a discussion on the precautionary approach, see VanderZwaag, Chapter 8 in this book, and the literature referred to therein.

[52] See para. 8 of the Final Act of the Eleventh Antarctic Treaty Special Consultative Meeting; reprinted in J. A. Heap (ed.), *Handbook of the Antarctic Treaty System*, 8th edn (Washington, DC: United States Department of State, 1994), pp. 2,016–18.

[53] Art. 8 and Annex I of the Protocol; see the discussion by Joyner, Chapter 5 in this book.

[54] *A State of the Arctic Environment Report*, p. vii.

[55] Para. 2(3) of the 1996 *Report of the Third Ministerial Conference on the Protection of the Arctic Environment*; emphasis added.

[56] For a comprehensive discussion, see Stokke and Vidas (eds.), *Governing the Antarctic*.

CRAMRA had contained insufficient environmental safeguards (these were in fact stringent),[57] but due to a complex combination of economic and political factors. In addition to the awareness that, for the foreseeable future, any mineral activities in the Antarctic would be devoid of commercial significance, the major factors were: (1) fears that CRAMRA would disturb the sensitive balance of sovereignty in the Antarctic; (2) a political-ideological critique of the Consultative Parties, from a group of developing countries in the UN; (3) pressures from environmental NGOs; and (4) domestic policy considerations which related to the above factors.

The very fact that rules for the environmental protection were adopted so rapidly in the most comprehensive ATS instrument in this field may be seen as largely the outcome of efforts made by the Consultative Parties to find an urgent solution to the 'CRAMRA crisis' within the ATS. It took them only two years to negotiate and adopt the 1991 Environmental Protocol. The Protocol addressed environmental protection in two essentially different ways: a blanket *prohibition* of mining (the one activity regulated under CRAMRA) and detailed *regulation* of all other activities in the Antarctic (save for activities already regulated under CCAMLR, CCAS and the International Convention for the Regulation of Whaling).[58] In its Article 7, the Protocol states unambiguously: 'Any activity relating to mineral resources, other than scientific research, shall be prohibited'. This single provision is essentially a response to the many criticisms voiced against CRAMRA. First, it made the sovereignty issue redundant, insofar as a 'delimitation' connected with mineral rights was no longer required. Secondly, it neutralised the criticism from developing countries which, since 1989, had been demanding in the UN General Assembly that a ban on minerals activities be introduced in the Antarctic. Thirdly, the Consultative Parties could present themselves as environmentally highly conscious, more so than anywhere else on the globe, thereby satisfying many of the demands for which environmental NGOs had campaigned. This latter point was instrumental in the domestic policy concerns of several Consultative Parties.

An additional related aspect deserves mention here: the extent to which environmental policy- and law-making is based on the available syntheses of the current state of *scientific knowledge* on the polar environment. In the bi-polar comparison, that aspect may appear somewhat contradictory. As to the Arctic, several comprehensive studies on the state of the Arctic environment have been compiled through international cooperation in recent years. Most important among these are the two state-of-the-art reports issued by AMAP in 1997 and 1998,

[57] See W. M. Bush, 'The 1988 Wellington Convention: How Much Environmental Protection?', in J. Verhoeven, P. Sands and M. Bruce (eds.), *The Antarctic Environment and International Law* (London: Graham & Trotman, 1992), pp. 69–83; F. Orrego Vicuña, 'The Effectiveness of the Protocol on Environmental Protection to the Antarctic Treaty', in Stokke and Vidas (eds.), *Governing the Antarctic*, pp. 197–8; Sir Arthur Watts, *International Law and the Antarctic Treaty System* (Cambridge: Grotius Publications, 1992), p. 276; and Wolfrum, *The Convention on the Regulation of Antarctic Mineral Resource*.

[58] See para. 7 of the Final Act of the Eleventh Special Antarctic Treaty Consultative Meeting.

respectively: *State of the Arctic Environment* and *AMAP Assessment Report*.[59] For the Antarctic, by contrast, no comprehensive synthesis of the state of its environment is as yet available. It was only recently, at the 1996 Consultative Meeting, that it was indicated in discussions between the Consultative Parties that there may be a need for a 'State of the Antarctic Environment' report.[60] Pursuant to the establishment of the CEP in 1998, the matter has figured on its agenda, since the Protocol lists as one of the CEP's functions to provide advice on the state of the Antarctic environment.[61] In the course of the past few years, both the Scientific Committee on Antarctic Research and an 'intersessional open-ended contact group' established by the CEP and chaired by Sweden have been involved in discussing how best to structure such an assessment.[62] However, actual work on the preparation of a 'State of the Antarctic Environment' has not yet commenced; and a recent estimate on the possible completion of such a potential report is 'around year 2003–2004'.[63] The Consultative Parties have spent considerable time discussing the approach to be used in compiling such an assessment, and airing concerns about the costs and time involved in such work.[64] Incidentally, the length of time spent discussing how to structure a future 'State of the Antarctic Environment' report was by mid-1999 already longer than the two-year period required to negotiate and adopt the Protocol itself. Be that as it may, the current stage of preparations for assessing the state of the Antarctic environment is rather comparable to the stage at which the Arctic countries were during the initial phase of the AEPS, even prior to the actual beginning of the operation of AMAP.

The approach employed in the Arctic would appear reasonable, albeit somewhat (bureaucratically?) cautious and slow: first assess the state of the environment in question, and then start considering how to formulate measures to protect it, on the basis of a synthesis of available scientific knowledge. The approach used in the Antarctic, however, looks somewhat strange: first adopt one of the most stringent international treaties for environmental protection, and then, on the eve of its entry into force, inquire as to the actual state of the environment which that regulation is intended to protect.

59 Other major syntheses, published shortly before AMAP reports, include J. R. Hansen, R. Hansson and S. Norris (eds.), *The State of the European Arctic Environment* (Copenhagen: European Environment Agency, 1996); and C. Bernes (ed.), *The Nordic Arctic Environment – Unspoilt, Exploited, Polluted?* (Copenhagen: Nordic Council of Ministers, 1996).

60 See para. 163 of the *Final Report of the Twentieth Antarctic Treaty Consultative Meeting*.

61 Art. 12(1)(j) of the Protocol.

62 See especially Sweden, 'Report on the Work of the Intersessional Contact Group on SAER', doc. XXIII ATCM/WP 5, March 1999; and Scientific Committee on Antarctic Research, 'Reporting on the State of the Antarctic Environment: The SCAR View', doc. XXIII ATCM/WP 6, March 1999. For initial views see New Zealand, 'On the Need for a State of the Antarctic Environment Report', doc. XXI ATCM/WP 32, May 1997; and Scientific Committee on Antarctic Research, 'State of the Antarctic Environment Report', doc. XXI ATCM/WP 19, April 1997.

63 'Report on the Work of the Intersessional Contact Group on SAER', para. 47.

64 See para. 141 of the *Final Report of the Twenty-First Antarctic Treaty Consultative Meeting, Christchurch, New Zealand, 19–30 May 1997* (Wellington: New Zealand Ministry of Foreign Affairs and Trade, 1997); and para. 58 of the Report of the Committee for Environmental Protection, in *Final Report of the Twenty-Second Antarctic Treaty Consultative Meeting*, Annex E.

Clearly, the application of preventive measures – whether 'anticipatory' or 'precautionary' approaches – does not depend solely on the state of the environment and the availability of scientific knowledge about it. Relevant factors include a variety of considerations, including, but not limited to, environmental policy.

In the Arctic Ocean case, it might be justifiable to say that the state of the environment still creates a precondition for using an anticipatory, preventive approach, and that the recently created political circumstances of regional cooperation could be regarded as an incentive.[65] However, when it comes to adopting new *legally binding* commitments among the eight Arctic countries, the situation may be the direct opposite. Here the importance of the Arctic socio-economic and strategic setting emerges, linked in particular with considerations of military security and economic issues. For the former reason, the USA has remained very reserved and restrictive towards any new Arctic international commitments; vital interests of the US Navy pose a major obstacle to a legally binding Arctic regional instrument in the sphere of the marine environment.[66] This might, as Griffiths observes, 'actually bolster land-based pollution prevention activity', since that is an aspect of Arctic marine pollution 'whose regulation may be least threatening to the interests of the US Navy'.[67]

In this connection, while concluding that there is no need for a regional instrument to protect the Arctic marine environment, the Inuvik Ministerial Conference set as the main priority for PAME:

> to address the development of an Arctic Regional Programme of Action on marine pollution resulting from land-based activities.[68]

In September 1998, PAME presented its draft final Regional Programme of Action for the Protection of the Arctic Marine Environment from Land-Based Activities (RPA) to the first ministerial meeting of the Arctic Council for adoption.[69] As issues of land-based pollution in the Arctic are discussed in Chapter 8 of this book, it suffices here to ask how far such a regional programme can go. This is where the other important aspect – the economic one – is likely to surface. For instance, Russia, clearly unable to comply with the requirements set by the 1974 Convention for the Prevention of Marine Pollution from Land-Based Sources[70] (Paris

[65] See the discussion by O. S. Stokke, 'Arctic Environmental Cooperation After Rovaniemi – What Now?', in L. Lyck (ed.), *Nordic Arctic Research on Contemporary Arctic Problems* (Aalborg University Press, 1992), pp. 228–30.

[66] See the discussion by Griffiths, 'Missing Arctic Waters', pp. 192–201.

[67] *Ibid.*, p. 198.

[68] See para. 6 of the 1996 Inuvik Declaration, in conjunction with paras. 2.3 and 2.3.5 of the 1996 Inuvik Report. It was the initiative of Canada, backed by the United States, to conduct a government-designated expert meeting to initiate the development of an Arctic Regional Programme of Action on land-based sources of pollution; see *ibid.*, para. 2.3.1.

[69] See Part II(E) and Annex 6 of the Report of Senior Arctic Officials to the Arctic Council, Iqaluit, Canada, 17–18 September 1998; and para. 24 of the Iqaluit Declaration of the First Ministerial Meeting of the Arctic Council, 18 September 1998. Both the Report and the Iqaluit Declaration are available at the Arctic Council website at http://arctic-council.usgs.gov.

[70] Text reprinted in ILM, Vol. 15, 1976, pp. 1,113ff.

Convention), was never a party to it; neither is it a party to the successor 1992 Convention for the Protection of the Marine Environment of the North-East Atlantic[71] (OSPAR). The cost-side of environmental protection is crucial in that context; and any ambition of the Arctic RPA will have to take this into account. This is recognised in the Iqaluit Declaration of the Arctic Council, stating that the Russian-hosted Partnership Conference:

> would seek funds to remediate regional priority pollution sources and activities identified in the RPA and Russian NPA-Arctic.[72]

All funding for both AEPS and Arctic Council activities thus far has been provided on a *voluntary* basis, largely by the participating countries hosting various secretariats.

Environmental protection standards and special polar conditions

Adapting environmental protection standards to the conditions existing in polar oceans means obtaining the recognition, from global arrangements and institutions, of the polar marine environment as subject to special risks; and then adapting global environmental protection standards or creating new region-specific standards. How does the current state of environmental protection measures for the polar oceans fare in this respect?

In their 1991 Environmental Protection Strategy, the Arctic countries agreed to:

> Undertake joint actions in relevant international fora to further strengthen recognition of the particularly sensitive character of ice-covered parts of the Arctic Ocean.[73]

They also noted the need for strict standards in the transportation of oil in the Arctic, and that these should be developed within the framework of the IMO.[74] When the 1993 Nuuk Conference established the PAME Working Group, one of its major stated tasks was:

> to evaluate the need for action in appropriate international fora to obtain international recognition of the particularly sensitive character of the ice-covered sea areas of the Arctic.[75]

In respect of shipping activities, PAME mapped several options to be considered, including the possible establishment of the Arctic Ocean as a 'Special Area' under

[71] Status as at 8 June 1999. Text of OSPAR Convention reprinted in ILM, Vol. 32, 1993, pp. 1,069ff.

[72] See para. 25 of the 1998 Iqaluit Declaration of the Arctic Council. 'Russian RPA-Arctic' refers to a Russian Programme of Action for the Protection of the Arctic Marine Environment from Land-based Activities.

[73] AEPS, Chapter 7(iii). 'Ice-covered parts' of the Arctic Ocean is of course a relative term, as there are seasonal variations of up to 6 million km^2. For the differences between the average winter maximum and average summer minimum extent of Arctic sea ice, see Brigham, Chapter 11 in this book, Figure 11.1 and accompanying text.

[74] AEPS, Chapter 4(2).

[75] See para. 2 of the Nuuk Declaration.

the relevant Annexes to MARPOL 73/78,[76] and as a 'Particularly Sensitive Sea Area' under the IMO.[77] However, no action has been taken within the AEPS process that would result in international recognition of the particularly sensitive character of the ice-covered sea areas of the Arctic, following options mapped by PAME. Quite the contrary, the Inuvik Ministerial Conference declined to include those in its assessment of the need for future action.

Adaptation of global environmental standards to the special conditions of the Arctic Ocean thus far has proceeded independently of, or in parallel with, the AEPS and the Arctic Council as regional processes. One example is Article 234 of the LOS Convention, which opens the way for higher standards in prevention and control of vessel-source pollution in ice-covered areas to be adopted through domestic legislation.[78] Another is the current discussion within the IMO of a draft International Code of Safety for Ships in Polar Waters (Polar Code), aimed at harmonising standards and rules for polar shipping.[79] Both Article 234 and the Polar Code process have originated in initiatives of several key Arctic states, and, although they may seem to be aimed at 'polar' conditions, their actual applicability is limited to the Arctic.[80]

At the opposite pole, there are several precedents of global instruments according *special* treatment to the Antarctic maritime area – defined by Article VI of the Antarctic Treaty as the area south of 60° South latitude. In 1990, for example, the IMO designated the major part of the Antarctic waters – 'the sea area south of latitude 60° S' – as a special area deserving greater environmental protection under MARPOL Annexes I and V.[81] In addition, in 1992 the Special Area status was extended to the same area under MARPOL Annex II.[82] Also the 1989 Basel Convention on the Control of Transboundary Movements of Hazardous Wastes and Their Disposal[83] contains an explicit provision for the area south of 60° S; according to its Article 4(6):

> The Parties agree not to allow the export of hazardous wastes or other wastes for disposal within the area south of 60 South latitude, whether or not such wastes are subject to transboundary movement.

[76] International Convention for the Prevention of Pollution from Ships, done in London, 2 November 1973, as modified by the Protocol of 17 February 1978; texts reprinted in ILM, Vol. 12, 1973, pp. 1,319ff (Convention) and ILM, Vol. 17, 1978, pp. 546ff (Protocol).

[77] See PAME 1996 Report, p. 127. The latter option is based on the 1991 IMO Assembly Resolution A.720(17), 'Guidelines for the Designation of Special Areas and the Identification of Particularly Sensitive Sea Areas'.

[78] For a discussion of Art. 234 and domestic legislation, see Brubaker, Chapter 10 in this book.

[79] For a detailed examination, see Brigham, Chapter 11 in this book.

[80] See Vukas, Chapter 2; Rothwell, Chapter 3; Rothwell and Joyner, Chapter 7; Brubaker, Chapter 10; and Brigham, Chapter 11 in this book.

[81] Resolution MEPC.42(30); see the amended texts of Annexes I (Regulation 10) and V (Regulation 5) in *MARPOL 73/78. Consolidated Edition, 1997* (London: International Maritime Organisation, 1997), pp. 52–6 and 363–5.

[82] Resolution MEPC.57(33); see the amended text of Annex II (Regulation 1) in *ibid.*, pp. 223–5.

[83] Text reprinted in ILM, Vol. 28, 1989, pp. 657ff.

In contrast to the Arctic experience, the origin of this adaptation of global standards to Antarctic circumstances lies in Antarctic *regional* cooperation which requires *coordinated* action of its members, not in individual or joint initiatives of several parties only. For quite some time already, coordination of positions when Antarctic matters have been discussed in fora other than within the ATS has been a notable trend in the 'external behaviour' of the Consultative Parties. This has been the case in the sphere of environmental protection as well. Thus in 1989 – one year prior to IMO's designating the waters within the Antarctic Treaty area as a special area under MARPOL 73/78 – the Consultative Parties adopted at their fifteenth Consultative Meeting a recommendation that:

> those [Consultative Parties] that are parties to MARPOL 73/78 consider taking actions within the International Maritime Organization (IMO) to secure designation of the waters south of 60° South Latitude as a special area under Annexes I and V of that Convention, provided that the establishment of reception facilities otherwise called for in these Annexes not be considered either necessary or desirable in the Antarctic Treaty area.[84]

Along the same lines the operative paragraph of Resolution 2 (1995) on 'Nuclear Waste Disposal', adopted at the Seoul Consultative Meeting, urges the Consultative Parties to coordinate their positions in any negotiations on the disposal of nuclear waste, 'with the objective of the inclusion of provisions prohibiting the transfer of nuclear waste to, and the disposal of nuclear waste in, the Antarctic Treaty Area'.[85] This resolution was adopted in connection with the IAEA initiating, in autumn 1994, preparations for a convention on the safety of radioactive waste management.[86]

This coordination requirement, relating to a broader category of 'global environmental agreements' and not to any specific one, is confirmed in the final report of the 1994 Consultative Meeting, where it is stated:

> The meeting agreed that the requirements for coordination were specific to each of the agreements and that the primary responsibility for ensuring such coordination lay with the Parties to the Antarctic Treaty that were Parties to the other agreements.[87]

As noted above, the Arctic countries decided not to opt for regulating the protection of the Arctic marine environment at the regional level, while for the Antarctic the Consultative Parties have concluded that the converse was needed, and therefore adopted the Environmental Protocol. What, then, has been achieved by regulating the protection of the Antarctic marine environment at the regional level? Does the Protocol actually add to or modify the provisions of existing global

[84] Para. 5 of Recommendation XV-4; text reprinted in Heap (ed.), *Handbook of the Antarctic Treaty System*, p. 2,074.

[85] See text in *Final Report of the Nineteenth Antarctic Treaty Consultative Meeting*, Annex C, p. 97.

[86] *Ibid.* [87] Para. 55 of the *Final Report of the Eighteenth Antarctic Treaty Consultative Meeting*.

instruments, thereby adapting these for waters characterised by sea ice conditions? Such an analysis would necessarily include MARPOL 73/78, to which Annex IV of the Protocol expressly refers – not least because shipping activities (and possible fuel spills) are primary potential sources of marine pollution.[88]

And what of the *specificity* of the Protocol rules? Did the perceived need for regionally specific rules result in a modification of existing global provisions, or in the creation of entirely new regulation by the Protocol? This is also related to another aspect – the *stringency* of the Protocol rules: have the Consultative Parties established marine protection regulation with standards stricter than those otherwise applicable globally? The analysis of the Protocol's role in the protection of the Antarctic environment from marine pollution is the subject of Chapter 5 in this book; and several studies have been undertaken on the particular relationship between Annex IV of the Protocol and global regulation, primarily MARPOL 73/78.[89] Here we should note that a general sovereign immunity clause contained in Article 11 of Annex IV of the Protocol applies also to ships owned or operated by states – and a significant number of the ships operating in the Southern Ocean fall into just that category. Moreover, the provisions of Annex IV on discharge of sewage (Article 6) do not apply if that would 'unduly impair Antarctic operations'.

Annex IV of the Protocol is limited to a minimum common standard agreeable to the Consultative Parties; and the same area to which it applies has already been designated as a Special Area under Annexes I, II and V of MARPOL 73/78. As observed by one commentator, in this context any practical utility of Annex IV of the Protocol is:

> dependent upon the number of States active in Antarctica which are not parties to MARPOL, or which, although parties to MARPOL, have not accepted optional Annex V to it.[90]

Of the twenty-eight states currently bound by the Protocol including its Annex IV,[91] *all* are simultaneously parties to MARPOL 73/78. As to Annex V of MARPOL, which relates to the prevention of pollution by garbage from ships, only two of those states (Chile and India) have not yet accepted it.

There is one other striking difference between the Arctic and Antarctic regional processes: prior to approaching any possible regional-level regulation of marine environmental protection, the Arctic countries formed a working group – PAME – whose tasks included assessing the existing international regulation

[88] See COMNAP, 'An Assessment of Environmental Emergencies Arising from Activities in Antarctica'.

[89] See, in particular, Orrego, 'The Effectiveness of the Protocol', pp. 194–6; T. Scovazzi, 'The Application of the Antarctic Treaty System to the Protection of the Antarctic Marine Environment', in F. Francioni (ed.), *The Environmental Law for Antarctica* (Milan: Guiffrè Publishing, 1992), pp. 125–9 and 133; and D. R. Rothwell, 'A Maritime Analysis of Conflicting International Law Regimes in Antarctica and the Southern Ocean', *Australian Year Book of International Law*, Vol. 15, 1994, pp. 175–8. For a comparison between the two instruments, see a document by the United Kingdom, 'A Comparison of Annex IV of the Protocol with MARPOL 73/78', doc. XVIII ATCM/INFO 75, 14 April 1994.

[90] Scovazzi, 'The Application of the Antarctic Treaty System', p. 128.

[91] These are all the twenty-seven Consultative Parties plus one non-Consultative Party (Greece).

applicable to the Arctic marine environment.[92] The Antarctic Treaty Consultative Parties never made a joint, coordinated assessment to that effect. Some individual parties did initiate discussion on this point, but well after the Protocol itself had been adopted.[93]

Area of application

Just what is 'the Arctic'? When the question of determining the precise delimitation of the area understood as 'the Arctic' for the purposes of the PAME Working Group came up, the Group decided that for the time being it was not necessary to agree on any specific geographical definition of the Arctic.[94] Eventually, PAME opined that:

> it was not necessary to have a single uniform geographical definition and the delineation of the Arctic was therefore left to the Arctic countries themselves.[95]

What then of the opposite pole, and the long-standing controversy over how to define 'the Antarctic'?[96] We should begin by noting that an unambiguous determination of the area of application of the Environmental Protocol is made difficult by the fact that the Protocol lacks a specific provision as to its territorial scope. On the one hand, this apparent oversight appears to stem from the fact that the Protocol is meant to supplement the Antarctic Treaty; thus, in the absence of any provision to the contrary, its area of application will be identical to that of the Antarctic Treaty, i.e. south of 60° South.[97] Moreover, the essence of the Protocol lies

92 The 1996 PAME Report is the result of a two-year coordinated effort by the Arctic countries.

93 See Chile, 'Relation Between the Protocol on Environmental Protection to the Antarctic Treaty and Other International Agreements of a Global Scope', doc. XVIII ATCM/WP 31, 13 April 1994; United Kingdom, 'A Comparison of Annex IV of the Protocol with MARPOL 73/78'; Chile, 'Relation Between the Protocol on Environmental Protection to the Antarctic Treaty and Other International Agreements of a Global and Regional Scope', doc. XIX ATCM/WP 20, 10 May 1995; United Kingdom, 'The Relationship Between the Protocol on Environmental Protection to the Antarctic Treaty and Other International Agreements of a Global or Regional Scope', doc. XX ATCM/WP 10, Rev. 1, April 1996; and Chile, 'Relationship Between the Protocol on Environmental Protection to the Antarctic Treaty and Other International Environmental Protection Treaties', doc. XX ATCM/WP 30, May 1996.

94 See, for instance, PAME, 'Report from the Second Meeting, London, September 1994', p. 2.

95 PAME 1996 Report, p. 21.

96 The Antarctic Treaty contains no definition of either 'the Antarctic' or 'Antarctica'; moreover, the single substantive Article of the Treaty which lacks any mention of the word 'Antarctica' is precisely Article VI, on the area of application of the Treaty. For the background to this, see P. Beck, 'Preparatory Meetings for the Antarctic Treaty 1958–59', *Polar Record*, Vol. 22, 1985, p. 658. Compare also the differences in the areas of application of the Antarctic Treaty (Art. VI), CCAS (Art. 1(1)), CCAMLR (Art. I(1) and (4)) and CRAMRA (Art. 5).

97 Art. 4 of the Protocol, in conjunction with Art. VI of the Antarctic Treaty. See the comment by W. M. Bush, *Antarctica and International Law: A Collection of Inter-State and National Documents*, Vol. I, Booklet AT91C (Dobbs Ferry, NY: Oceana Publications, 1992), p. 2; at another place Bush comments: 'the area south of 60 degrees south latitude . . . is the same as the area of operation of the protocol': *ibid.*, Booklet AT91D, p. 11. Orrego's view also seems to be in line with such comment, though with an additional measure of precaution: see Orrego, 'The Effectiveness of the Protocol', p. 182.

in its Article 3, which comprises 'all activities in the *Antarctic Treaty area*'.[98] The Protocol uses the formulation 'Antarctic Treaty area' throughout the text of its provisions. The Consultative Parties have declared at several of their recent gatherings (both formal and informal) that they agree that the area of application of the Protocol is that of the Antarctic Treaty itself.[99]

On the other hand, it appears contrary to the main (proclaimed) purpose of the Protocol to confine it to such a geographic limit that seems inadequate in the context of the Protocol's environmental protection provisions.[100] Article 3 of the Protocol demonstrates the contradiction of being limited to 'activities in the Antarctic Treaty area', while at the same time introducing the concept of the 'protection of the Antarctic environment and dependent and associated ecosystems'. The latter are linked to the *natural*, not the political, boundary of the Antarctic. In implementing the environmental principles of the Protocol:

> monitoring shall take place to facilitate early detection of the possible unforeseen effects of activities carried on *both within and outside the Antarctic Treaty area* on the Antarctic environment and dependent and associated ecosystems.[101]

This would indicate that the Antarctic Convergence should be considered the appropriate boundary for the seaward extent of the area of application of the Environmental Protocol. We find a precedent within the ATS itself, since this natural boundary of the Antarctic region was taken as the relevant one in determining the area of application of CCAMLR.[102]

An inquiry of the IMO to the Antarctic Treaty Consultative Parties several years ago is directly connected with this problem. In 1995, the IMO asked for the Consultative Parties' comments on the proposal to amend the boundary of the Antarctic Special Area under MARPOL 73/78 as it *now* stands (i.e., the area south of 60° South), by replacing it with the boundary of the Antarctic area as defined in CCAMLR, which is largely correspondent with the Antarctic Convergence as the biological boundary of the region.[103] The IMO's inquiry even alluded to the possibility of amending the Environmental Protocol itself, upon its entry into force.[104] The first official comment of the Antarctic Treaty Consultative Parties was confined to the statement that 'the matter should be decided by IMO itself', as well as suggesting to the IMO that it consider taking up the matter with the CCAMLR Commission.[105] In a later reply to the IMO's inquiry, it was clearly stated that the Protocol:

[98] Art. 3(1) of the Protocol; emphasis added. See also the comment by Bush, *Antarctica and International Law*, Vol. I, Booklet AT91C, p. 2.

[99] See, for instance, the wording included in the letter from the Chairman of the XIX Antarctic Treaty Consultative Meeting to the IMO, of 3 June 1995, cited below in this chapter.

[100] Similarly, Bush, *Antarctica and International Law*, Vol. I, Booklet AT91C, pp. 2–3.

[101] Protocol, Art. 3(2)(e); emphasis added.

[102] CCAMLR, Art. I(1) and (4).

[103] See *Boundary of MARPOL Antarctic Special Area*, doc. XIX ATCM/INFO 83, 12 May 1995, submitted by the Secretariat at the request of the IMO, p. 1.

[104] *Ibid.*, p. 2.

[105] See para. 54 of the *Final Report of the Nineteenth Antarctic Treaty Consultative Meeting*.

> applies to the area of the Antarctic Treaty, that is, 'the area south of 60° South Latitude, including all ice shelves'. Any change in this area would require amendment of the Antarctic Treaty. This is not contemplated.[106]

Therefore, while there may be a trend towards granting the Southern Ocean the status of a special *environmentally* protected area under several global regimes, to date they have all chosen the *political* (60° South) rather than the ecological boundary (the Antarctic Convergence) for determining that area. This may indicate that the actual proclamation of a special environmental protection area in the Southern Ocean is dictated at least as much by the political strength and impact of the Antarctic Treaty Consultative Parties also in the *global* international fora, and their desire to demarcate the ATS boundary, as it is by the felt need to protect the environment.

Assessment

The year 1989 marked a watershed for environmental protection in the polar regions, although the formal outcomes were not to emerge until 1991. The coincidence in *timing* of the trends of development relative to the Antarctic and the Arctic is remarkable. As to the *legal nature* of the existing regulation, the Antarctic cooperation appears far more advanced on the regional level, with a comprehensive international environmental agreement in force. (Of course, the legally binding nature of an international instrument should not be understood as a quality *per se*; however, it regularly indicates the high level of consensus reached among the parties as to the subject matter under regulation.) Arctic environmental protection regulation has so far fallen short of this, but today's level of cooperation contrasts sharply to the pre-1980s situation, when there was no regional cooperation in the Arctic whatsoever.

When we draw together the three strands of approaches to protection of the polar marine environment, as presented above in this chapter, we can see more clearly the possible separation of the legal form and the proclaimed purposes of international agreements, from their actual contents and the intentions of the parties.

The Antarctic Treaty Consultative Parties have adopted a legally binding instrument for the protection of the Antarctic environment, but the main incentives for doing this so quickly were political rather than environmental in nature. This we can see from the adoption of the more narrow political-legal boundary, instead of the wider natural boundary, for the area of application of the Environmental Protocol – an instrument with the proclaimed objective of 'comprehensive protection of the Antarctic environment and dependent and associated

[106] Letter from the Chairman of the Nineteenth Antarctic Treaty Consultative Meeting to the IMO, 3 June 1995; Annex I to the *Final Report of the Nineteenth Antarctic Treaty Consultative Meeting*. The letter was sent by the Chairman with the mandate of the Consultative Meeting; see *ibid.*, para. 54.

ecosystems'.[107] Moreover, the sense of adopting a special regime for the protection of the Antarctic marine environment would lie in adapting global standards to specific polar circumstances – which has obviously not been achieved by Annex IV to the Protocol. And, finally, while the Protocol to protect the Antarctic environment was adopted in 1991, the need for an overall assessment of the actual state of that environment was not felt by the Consultative Parties until five years later. Even with the Protocol already in force for quite some time, such an assessment cannot be expected to become available before the early twenty-first century.

As to the Arctic, the process of international cooperation has taken a different approach: first to prepare a synthesis of the state of the environment, and thereafter to proceed with closer examination of applicable regulation. That approach would seem logical, since one first has to know what to regulate, then examine the existing regulation and finally adopt additional regulation, if needed. The contrast is, however, given by a comparison: in the environmental protection of the fairly distant Antarctic, the Arctic states – of which all except Iceland are simultaneously parties to the Antarctic Treaty – have in the context of Antarctic cooperation supported the need for urgent adoption of anticipatory, legally binding measures. Conceivably, they have also deemed global international instruments for the protection of the marine environment to be insufficient, since a special annex on this matter had to be added to the Environmental Protocol. However, in their 'own backyard', the Arctic – an area far more exposed to industrial activity and human pressure – those same countries do not seem to feel an equal need for urgency and anticipation in protecting the marine environment on the regional level. When it comes to adopting international legal instruments on marine environmental protection in the Arctic, these countries have opined that the existing instruments suffice, and that there is no particular need for a regional instrument adapted to Arctic sea ice conditions. Here we should bear in mind the importance of military-strategic and economic considerations.

The final result in terms of substantive protection of the marine environment actually remains quite similar: whereas in the Arctic context the countries concerned cannot agree on a new regional-specific international instrument, those in the Antarctic have adopted Annex IV to the Protocol which, albeit formally regional, in substance neither adds to nor greatly modifies the existing global instrument, namely MARPOL 73/78. Possible differences, it might be argued, should now become apparent in the implementation of the Protocol. However, all the twenty-eight parties to the Environmental Protocol are currently parties to MARPOL 73/78 (as are all eight Arctic countries). Perhaps implementation mechanisms might function better under the Protocol, i.e. within the ATS, than under MARPOL 73/78. On the other hand, as we have noted, several provisions of Annex IV to the Protocol do contain significant 'escape clauses' for future implementation on the ATS level.[108]

[107] Art. 2 of the Protocol. [108] For further discussion, see Joyner, Chapter 5 in this book.

CONCLUSIONS: A PARADOX OF SIMILARITIES OR A CONSEQUENCE OF DIFFERENCES?

Why is there such contrast between the regional-level regulation of marine environmental protection in the Arctic and the Antarctic? Is it simply a paradox, given their comparable polar conditions? Or should it be explained by accounting for their differences?

Features of the respective regional cooperative processes seem to offer part of the explanation. The Antarctic Treaty System is a form of international cooperation which has, over almost four decades of operation, acquired some priorities and values of its own, not necessarily identical to the priorities and values of each individual state party to the Antarctic Treaty. The ATS has succeeded in forming its own *identity*, and its members share the feeling of belonging to a club. As long as the individual priorities of these 'club members' – like a substantial interest in acquiring economic benefit – do not prevail over the common 'club interests', the latter are seen as worth protecting and facilitating, since that is the method of both securing a decision-making position within the ATS and preventing 'non-members' from disturbing the decision-making process. These are the circumstances that have allowed another value to surface – the value of the Antarctic as a *symbol* of a pristine environment. It has been pointed out that:

> Because of this symbolic role, human activities in the Antarctic are evaluated not only by the actual pressure exerted on the environment but also by the attitude demonstrated.[109]

In the Arctic, no such common identity has yet been formed. Potential conflict rather than actual cooperation has dominated the Arctic political scene for almost half a century, from the end of World War II to the end of the Cold War. From the very start, the fledgling regional cooperative processes in the Arctic have not only been hampered by such political and military–strategic heritage, but have had to face the realities of the ongoing economic exploitation of the region's rich natural resources, as well as respond to the needs and demands of the Arctic population.[110] The emergence of the Arctic as a symbol is thus hindered by the actual interests involved in its socio-economic and political rather than natural setting.

In coping primarily with the problem of preserving the values and priorities of the 'club', the Antarctic Treaty System has been supplemented by an international treaty – the Environmental Protocol – based largely on the gradual evolution of environmental protection instruments within the ATS. That evolution has been sensitive to the solutions adopted by the global instruments. Neither in

[109] O. S. Stokke and D. Vidas, 'Introduction', in Stokke and Vidas (eds.), *Governing the Antarctic*, pp. 5–6.

[110] On aspects of post-Cold War Arctic cooperation, see W. Østreng, 'The Post-Cold War Arctic: Challenges and Transition During the 1990s', in Vidas (ed.), *Arctic Development and Environmental Challenges*, pp. 33–49.

this respect is Arctic cooperation fully formed as yet: it is still searching for adequate solutions.

The other part of the explanation for this quite peculiar situation in the protection of the marine environment in the two polar regions seems to come not from the respective regional cooperation processes as such, but from their placement in relation to other levels of regulation. The *composition* of regulatory pictures in the Arctic and Antarctic differs sharply. This is largely the consequence of different sovereignty situations in the two polar regions. In this respect, the Arctic resembles all other regions in the world, with recognised sovereign states possessing their parts of coastline of the Arctic Ocean.[111] By contrast, the Antarctic is unique when it comes to sovereignty: a continent where only a few countries have claimed sovereignty but none of those claims has ever received general recognition.

Therefore, the regulatory picture for the protection of the Arctic (marine) environment is a *diffuse* one: it consists of all the various levels of regulation – domestic, sub-regional, regional and global – in a composition today comparable to various other regions in the world. Part of marine environmental protection is dealt with at the domestic level of Arctic countries as coastal states, where national prerogatives are jealously protected.[112] Part is dealt with through sub-regional processes (including bilateral), as with the Barents Sea cooperation.[113] Here, transnational and sub-national contacts also surface – unknown in the Antarctic. And, finally, global treaties largely apply in the Arctic as in other 'usual' regions; however, due to the special polar conditions and the nature of the potential sources of pollution, Arctic-specific provisions or instruments have also been drafted at the global level (Article 234 of the LOS Convention, and the Polar Code).

The Antarctic regulatory picture is unique in its emphasis on *regionally centralised* regulation. There of course exists domestic legislation of Antarctic Treaty parties applying to activities of their nationals in the Antarctic, but this is based on personal, not territorial jurisdiction. Moreover, the pattern of this legislation is to implement the instruments agreed by the Antarctic Treaty Consultative Parties through their regional ATS cooperation.[114] Nor is there any real sub-regional level in the protection of marine environment. Everything centres around a single, comprehensive instrument: the Environmental Protocol.[115] Finally, as to the global instruments, these do apply to the Antarctic and its environment. However, the centralised role of the regional ATS level modifies this application as well. Through the decision-making impact of its members in global fora and in negotiations, the ATS has always managed to serve as a 'filter' for the application of global instru-

[111] We should note the unique status of Svalbard under the 1920 Paris Treaty concerning Spitsbergen; published in *League of Nations Treaty Series*, Vol. 2, pp. 7ff.

[112] See especially Brubaker, Chapter 10; also Rothwell and Joyner, Chapter 7 in this book.

[113] See the analysis by Stokke, Chapter 6 in this book.

[114] See Rothwell and Joyner, Chapter 7 in this book.

[115] See a detailed examination by Joyner, Chapter 5 in this book.

ments, especially when these become polar-specific. Thus it came about that the Antarctic was included as a Special Area under several annexes of MARPOL 73/78. More recently, this has determined the destiny of the Polar Code in the IMO, reduced to apply to shipping in Arctic waters only.

Beyond the *combined* effects of those two possible explanations – the level of developed identity of their respective regional cooperation and the different composition of respective regulatory pictures – today's situation for the marine environmental protection in the Arctic and the Antarctic is in each case a result of quite different causes. In the Antarctic, regulation of environmental protection has been *stimulated* by various other policy considerations. In the Arctic, various policy considerations have *prevented* regional environmental regulation; these considerations include the unwillingness of key regional states to change their current behaviour or their inability to pay for such change.

There is, however, one thing the two regional processes do have in common: in their decision-making on marine environmental protection, neither has in fact been driven primarily by considerations of environmental protection.

5 Protection of the Antarctic environment against marine pollution under the 1991 Protocol

CHRISTOPHER C. JOYNER

On 4 October 1991 the Protocol on Environmental Protection to the Antarctic Treaty was adopted and opened for signature by the Antarctic Treaty Parties in Madrid.[1] Coming after two years of negotiations, this instrument with its attendant five annexes represents one of the most comprehensive multilateral environmental agreements yet promulgated. It embodies a legal blueprint for protection and preservation of the Antarctic. No less important, the Protocol also signals a profound shift – indeed a reversal in course – in Antarctic Treaty Consultative Parties' aspirations for the Antarctic. In the late 1980s the policy direction of the Consultative Party group still appeared headed towards possible exploration and potential exploitation of Antarctic minerals. By 1991, however, that course had been diverted toward a general commitment of legal obligation to protecting and conserving the continent and its circumpolar seas.

A critical aim of the Protocol is to prevent marine pollution in the Antarctic Treaty area. This chapter examines how and to what extent the Protocol contributes to the general international law against marine pollution *as applied in the Southern Ocean.* To that end, the first section of the chapter briefly considers how marine pollution occurs in the Antarctic, and what legal framework is already in place for dealing with it internationally. The second section assesses the Protocol as a legal instrument for preventing marine pollution. Particular attention here is given to the innovative anti-pollution qualities of the Protocol, its obligatory characteristics and its enforcement capabilities. Deficiencies of the Protocol are also pointed out, with a view to highlighting loophole provisions that might adversely impinge upon the ability to prevent marine pollution in Antarctic waters. The third section evaluates the five annexes to the Protocol to ascertain their respective roles in strengthening the international law against marine pollution, as applied in the Antarctic. Particular focus here is on Annex IV, which specifically addresses marine pollution in the Antarctic Treaty area. Finally, conclusions are

[1] The Protocol entered into force on 14 January 1998; text reprinted in ILM, Vol. 30, 1991, pp. 1,461ff.

suggested about the ways in which the Protocol contributes to preventing marine pollution in Antarctic seas.

MARINE POLLUTION IN THE ANTARCTIC

The wreck of the *Bahia Paraiso* in January 1989 clearly signalled the dangerous consequences of increased shipping along continental ice shelves. The ship, carrying several hundred tourists and supplies for an Argentine station, ran aground and tore open its hull offshore Antarctica. Some 250,000 gallons of diesel oil spilled into the frigid waters, killing seals, penguins, krill and other marine life near the US Palmer Station on the Antarctic Peninsula. In the process several US marine scientific projects were ruined as well.[2]

There is little question that the most unpredictable and potentially most damaging anthropogenic pollution contaminants in Antarctic seas will come from fossil-fuel spills from vessels sailing in the area. The treacherous waters, severe weather conditions, remoteness of the area and increased shipping traffic in the region suggest that such spills might almost be inevitable. As more vessels carrying greater numbers of tourists visit Antarctic waters, the prospects for accidents and resultant marine pollution are likely to grow.

Environmental impacts of oil spills in Antarctic waters will be grave, since biological decomposition of petroleum is slowed in frigid temperatures. Antarctic wildlife, primarily that which lives in or near the sea, will be affected by such oil spills. Oil contamination in Antarctic waters will severely impact upon phytoplankton and krill stocks, upon which most higher species prey, thus adversely affecting the Antarctic food chain. Moreover, oil spills could seriously disrupt marine scientific research in the Antarctic, which relies upon a relatively pristine environment for establishing baselines for monitoring global ecological change.

Legal framework for marine pollution in the Antarctic

The law regulating marine pollution draws from the same sources and follows the same law-making processes as international law generally. Thus, the legal framework governing marine pollution in the Southern Ocean is undergirded by the sources of international law, including primarily international treaties, international customary law and general principles of law. Notably, however, the international legal framework for regulating marine pollution also draws considerably from so-called 'soft law'– instruments that are formally non-binding and are often adopted through the decisions of international organisations. Not surprisingly,

[2] P. A. Penhale, 'Research Team Focuses on Environmental Impact of Oil Spill', *Antarctic Journal of the United States*, Vol. 24, 1989, p. 9. See also J. Wilford, 'Sunken Ship's Oil Spill Held a Peril to Antarctic Wildlife', *New York Times*, 9 February 1989, p. 7A.

though, international law governing marine pollution has developed since the early 1950s mainly through a process of multilateral actions, complemented by national actions.[3]

The pillars of international marine pollution law today stand mainly as certain normative principles accepted as customary law and as various international conventions intended to regulate pollution activities in the oceans. First, to undergird contemporary international environmental law, a customary norm has arisen embodying the general obligation to protect the marine environment from pollution. This general norm has evolved from the entire body of legal instruments adopted relating to marine pollution, and is clearly and explicitly codified in Article 192 of the 1982 UN Law of the Sea Convention (LOS Convention): 'States have the obligation to protect and preserve the marine environment.'[4]

As a second principle, states are obliged not to transfer pollution from one area to another or to transform one type of pollution to another. As stipulated by Article 195 of the LOS Convention:

> In taking measures to prevent, reduce and control pollution of the marine environment, States shall act so as not to transfer, directly or indirectly, damage or hazards from one area to another or transform one type of pollution into another.

Finally, environmental monitoring and assessment also emerge as autonomous and effective obligations for protection of the marine environment. International norms on marine pollution thus tend to have certain basic contents. They include mainline fundamental rules (rules that prohibit activities or establish standards) and enforcement rules, normally coupled with safeguards. Marine pollution norms usually provide for rules on global and regional cooperation, including environmental monitoring and assessment, technical assistance and varying kinds of responsibility for pollution, liability and settlement of environmental disputes. This general framework is set out in the 1991 Environmental Protocol for specific application to the circumpolar Antarctic seas.

With respect to international agreements, in addition to the LOS Convention, the principal conventions forming the foundation for marine pollution law in the Antarctic ocean are the 1973 International Convention for the Prevention of Pollution from Ships, as amended by its 1978 Protocol (MARPOL 73/78), and the 1972 Convention on the Prevention of Marine Pollution by Dumping of Wastes and Other Matter (London Convention).[5] As a framework agreement prohibiting pollution in Antarctic waters, the 1991 Environmental

[3] See also Rothwell and Joyner, Chapter 7 in this book. For general treatments of the international law of marine pollution, see generally D. Brubaker, *Marine Pollution and International Law: Principles and Practice* (London: Belhaven Press, 1993); D. B. Macgraw (ed.), *International Law and Pollution* (Philadelphia, PA: University of Pennsylvania Press, 1991); and G. J. Timagenis, *International Control of Marine Pollution* (Dobbs Ferry, NY: Oceana, 1980).

[4] For a detailed discussion see Vukas, Chapter 2 in this book.

[5] For an overview see Rothwell, Chapter 3 in this book.

Protocol was designed such that its provisions, although not always expressly referred to, are cross-linked with norms established by these international regulatory instruments.

MARINE POLLUTION LAW IN THE 1991 ENVIRONMENTAL PROTOCOL

The Protocol is intended to supplement the Antarctic Treaty[6] and be consistent with other components of the Antarctic Treaty System.[7] The Protocol obliges parties to consider the Antarctic (defined as the area south of 60° South latitude, inclusive of ocean space) as a 'natural reserve devoted to peace and science' and commits them to comprehensive protection of the region's environment.[8] This duty of comprehensive protection explicitly embraces the obligation to prevent marine pollution from occurring in the area.

The Protocol contains certain fundamental rules that articulate basic obligations, set specific prohibitions, and fix acceptable standards for activities in Antarctic marine areas. In this regard, the Preamble to the Protocol reaffirms the special responsibility of the Antarctic Treaty Consultative Parties 'to ensure that Antarctica shall continue forever to be used exclusively for peaceful purposes and shall not become the scene or object of international discord'; and recalls 'the designation of Antarctica as a Special Conservation Area . . . to protect the Antarctic environment and dependent and associated ecosystems'. To this end, the key principle supporting the Protocol comes in Article 3, which in its first paragraph provides that:

> The protection of the Antarctic environment and dependent and associated ecosystems and the intrinsic value of Antarctica, including its wilderness and aesthetic values and its value as an area for the conduct of scientific research . . . shall be fundamental considerations in the planning and conduct of all activities in the Antarctic Treaty area.

The Protocol mandates that marine pollution must be *prevented* from befouling the Antarctic marine ecosystem. The principle thus becomes that proper planning and prudent conduct of activities are necessary and essential to prevent such pollution from occurring. To give effect to this principle, Article 3(2)(a) of the Protocol asserts that:

> activities in the Antarctic Treaty area shall be planned and conducted so as to limit adverse impacts on the Antarctic environment and dependent and associated ecosystems.

This provision makes plain the Consultative Parties' concern over the threat of marine pollution in southern circumpolar waters. The Southern Ocean, at least that portion falling within 60° South latitude, may properly be considered part of

[6] Protocol, Art. 4(1). [7] *Ibid.*, Art. 5. [8] *Ibid.*, Art. 2.

the 'Antarctic environment'. It may also reasonably be inferred that, within the context of the Protocol, 'dependent and associated ecosystems' could extend as far northwards as the Antarctic Convergence, the biological boundary generally accepted by bio-oceanographers for designating 'the Antarctic', as well as sometimes by diplomats, i.e. for setting the northernmost jurisdictional reach of the 1980 Convention on the Conservation of Antarctic Marine Living Resources.[9] Put another way, dependent and associated ecosystems in the Antarctic Treaty area include much ocean space south of 40° South latitude. Hence, activities that might generate pollution within that area, particularly those that adversely impact upon Antarctic ecosystem, must be prevented.

The obligation of *planning* the conduct of activities to prevent marine pollution is elaborated in Article 3(2)(b) of the Protocol:

> [A]ctivities in the Antarctic Treaty area shall be planned and conducted so as to avoid:
> 1. adverse effects on climate or weather patterns;
> 2. significant adverse effects on air or water quality;
> 3. significant changes in the atmospheric, terrestrial (including aquatic), glacial or marine environments;
> 4. detrimental changes in the distribution, abundance or productivity of species or populations of species of fauna and flora;
> 5. further jeopardy to endangered or threatened species or populations of such species; or
> 6. degradation of, or substantial risk to, areas of biological, scientific, historic, aesthetic or wilderness significance.

The intent and relevance of this provision seem clear. Certain activities are considered potentially deleterious to the health and well-being of the circumpolar marine ecosystem. To prevent such activities from producing harmful effects like marine pollution, prudent planning becomes necessary. Planning of activities thus should be undertaken to avoid 'adverse' impacts on water quality – a clear and direct reference to marine pollution. Planning is intended to preclude significant changes from being introduced into the marine environment, especially those that could be caused by pollution activities; planning is viewed as a means to deter 'detrimental' alterations from perturbing 'the distribution, abundance or productivity' of life in the region; planning aims to prevent activities that might jeopardise the well-being of threatened or endangered species in the region (which include whales, seals and several species of fin fish in the Southern Ocean); and, planning must be done to dissuade or correct activities that might degrade or pose substantial risk to areas of biological, scientific or wilderness significance. These qualities apply equally to circumpolar Antarctic waters and to the continent.

To facilitate planning along these lines, Article 3(2)(c) introduces another requirement as a principle: sufficient *information* must be furnished to permit

[9] See the discussion by Boyle, Chapter 1; and Vidas, Chapter 4 in this book.

prior assessment of potential impacts upon the Antarctic environment, including the marine ecosystem:

> [A]ctivities in the Antarctic Treaty area shall be planned and conducted on the basis of information sufficient to allow prior assessments of, and informed judgements about, their possible impacts on the Antarctic environment and dependent and associated ecosystems and on the value of Antarctica for the conduct of scientific research.

The Protocol mandates that planning – presumably done prudently, properly and based on the requirement that 'sufficient' scientific information is made available to allow a scientific determination of risk assessment – remains essential for preventing marine pollution in Southern circumpolar waters.

Finally, subparagraph (d) of Article 3(2) provides for still another legal principle: 'regular and effective monitoring . . . to allow assessment of the impacts of ongoing activities, including the verification of predicted impacts'. Thus, should the fundamental obligation of prudent planning fail for reasons either of commission or omission, *monitoring* operations by states party are expected to detect breaches and identify violators.

Article 3 thus furnishes a set of fundamental, legally binding principles for deterring marine pollution. These principles include the following:

1. obligations to meet specific environmental standards and to limit adverse impacts on the marine environment;
2. obligations to give priority to scientific research in the Antarctic, including the marine ecosystem, and to preserve the Antarctic for global research;
3. obligations to ensure that human activities are planned and carried out on the basis of information sufficient to permit prior assessments of their possible impacts on the marine ecosystem; and
4. obligations to conduct environmental monitoring of the marine ecosystem in order to detect possible violations of these obligations.

Consequently, the provisions in Article 3 of the Protocol furnish significant contributions for broadening and strengthening the international law of marine pollution, as applied in the Southern Ocean.

Pollution from mineral or hydrocarbon exploitation

The Protocol places a prohibition on all mining activity in Antarctica: 'Any activity relating to mineral resources, other than scientific research, shall be prohibited.'[10] For marine pollution law, this ban on Antarctic mining is highly salient, since it applies also to drilling activities offshore Antarctica. By banning mining for

[10] Protocol, Art. 7.

minerals on the continent and drilling for offshore hydrocarbons in the Antarctic continental shelf, several potentially grave threats to the Antarctic marine environment are precluded. No atmospheric pollution or mining sludge will be produced from onshore mining operations. No oil blowouts from offshore wells can occur in Antarctic waters. Vessel-source pollution from tanker accidents or intentional discharges is obviously curtailed, since no tankers will be transporting petroleum from Antarctica. And pollution effluents that might have been produced by increased numbers of resident personnel supporting operations facilities needed on shore will be eliminated, since no exploitation operations can go forward. Article 7 thus serves as a preclusive anti-pollution law that directly prevents marine pollution by prohibiting those very activities that would undoubtedly produce such pollution.

This ban, however, is not necessarily permanent. While no period for a moratorium is specified, Article 25(1) allows for modification or amendment of the Protocol at any time by unanimous agreement of *all* Consultative Parties. Granted, this does not perforce mean that modifications will be called for, or, even if so, enter into force. Nor does it mean that the ban will be lifted. What it does mean is that modifications some day might be possible. Moreover, fifty years after the Protocol enters into force (i.e., after 14 January 2048), it could be possible for the mining prohibition to be lifted if such a proposal were adopted at a review conference by a majority of the parties to the Protocol, including three-quarters of current Consultative Parties, and then ratified by three-quarters of Consultative Parties, 'including ratification, acceptance, approval or accession by all States which are Antarctic Treaty Consultative Parties at the time of adoption of this Protocol'.[11] Such a provision ostensibly ensures that no minerals development on Antarctica or in its circumpolar waters can lawfully take place within the foreseeable future. Put simply, this prohibition means that no degradation of Antarctic seas is likely to occur from minerals or hydrocarbon development or related activities on or around the continent, nor are natural habitats of Antarctic living marine resources likely to be disrupted or destroyed by these activities for at least fifty years.[12]

Committee for Environmental Protection

Two institutions are provided for in the Protocol to give effect to its legal principles. According to Article 10, Antarctic Treaty Consultative Meetings will

[11] *Ibid.*, Art. 25(4). At the time of the adoption of the Protocol, twenty-six states were Consultative Parties: Argentina, Australia, Belgium, Brazil, Chile, China, Ecuador, Finland, France, Germany, India, Italy, Japan, Korea (Republic of), the Netherlands, New Zealand, Norway, Peru, Poland, the Russian Federation, South Africa, Spain, Sweden, the United Kingdom, the United States and Uruguay. In 1998, Bulgaria became a Consultative Party, thus increasing the current (as of 8 June 1999) number of Antarctic Treaty Consultative Parties to twenty-seven states in total.

[12] However, pursuant to a US proposal, any state has the right to withdraw from the Protocol (presumably giving it the right to mine without regulation) if an amendment lifting the ban has not entered into force within three years of the date of its adoption: see Art. 25(5)(b) of the Protocol.

make *decisions* for implementing the Protocol regime. Article 11 of the Protocol goes on to stipulate the establishment of a Committee for Environmental Protection (CEP), while Article 12 spells out its main functions as giving *advice* and formulating recommendations to the parties regarding implementation of the Protocol and its annexes. However, the CEP is given no decision-making authority.

The CEP will undoubtedly perform valuable functions by supplying advice to ensure that environmental rules, including those intended to prevent marine pollution, are interpreted uniformly and consistently by all parties.[13] In addition, the Committee might serve as a forum for investigating controversial environmental matters, for assisting in the proper preparation of environmental impact statements, and for proposing common interpretations of key terms and threshold levels in the Protocol.[14] All the same, the CEP lacks real authority to enforce compliance with the Protocol's anti-pollution provisions or to define mandatory environmental conservation zones, or to send out inspection or monitoring agents to conduct oversight of human activities that might produce marine pollution around Antarctica.

Enforcement

Enforcement is critical in marine pollution law. It is the process by which a regulation is made effective or the process designed to compel obedience to a legal rule. While viewed as a continuous process, enforcement as a practical matter occurs in phases: (1) a violation is reported or discovered; (2) an investigation occurs and evidence is gathered of the violation; (3) the evidence is evaluated and sanctions are determined for the violation; and (4) the process of giving effect to the sanction is determined.

The Protocol contains certain enforcement rules that assign responsibility for compliance and enforcement to the states party, who are expected to impose penalties and methods for punishing contravention of fundamental rules, or for effective application of those rules. Compliance rests with governments that are party to the Protocol. Parties are obliged under Article 13 to take 'appropriate measures' to ensure compliance with the anti-pollution provisions of the Protocol.

[13] Specifically, in accordance with Art. 12(1) of the Protocol, the Committee is to furnish advice on: 'a) the effectiveness of measures taken pursuant to this Protocol; b) the need to update, strengthen or otherwise improve such measures; c) the need for additional measures, including the need for additional Annexes, where appropriate; d) the application and implementation of the environmental impact assessment procedures set out in Article 8 and Annex I; e) means of minimising or mitigating environmental impacts of activities in the Antarctic Treaty area; f) procedures for situations requiring urgent action, including response action in environmental emergencies; g) the operation and further elaboration of the Antarctic Protected Area system; h) inspection procedures, including formats for inspection reports and checklists for the conduct of inspections; i) the collection, archiving, exchange and evaluation of information related to environmental protection; j) the state of the Antarctic environment; and k) the need for scientific research, including environmental monitoring, related to the implementation of this Protocol.'

[14] Antarctic and Southern Ocean Coalition, 'A Critique of the Protocol to the Antarctic Treaty on Environmental Protection', *ASOC Information Paper No. 1*, XVI ATCM (8 October 1991), p. 4.

Further, the Protocol provides for inspections of stations, installations, equipment, ships and aircraft within the Antarctic Treaty area to be carried out 'to promote the protection of the Antarctic environment and dependent and associated ecosystems, and to ensure compliance with this Protocol' (Article 14), as well as procedures for mandatory dispute settlement (Articles 18, 19 and 20) and advance environmental impact assessment for proposed activities in the Antarctic (Article 8). Inspections will be used to detect any violations of the standards set by the Protocol or its annexes to deter and prevent marine pollution.

THE ANNEXES

The annexes are integral parts of the Protocol.[15] There are currently five annexes, dealing with, respectively: environmental impact assessment; conservation of fauna and flora; waste disposal and waste management; marine pollution; and protected areas. These annexes are to be implemented in furtherance of the environmental protection of Antarctica. In this regard, they 'form an integral part' of the Protocol and are to be adopted in line with Article IX of the Antarctic Treaty.[16] Moreover, each annex supplies significant contributions for enhancing the international law of marine pollution in the Southern Ocean.

Annex I: environmental impact assessment

Annex I represents a significant accomplishment of the Protocol. It sets out procedures for environmental impact assessment, which is considered the acid test for the environmental protection capability of the Protocol.[17] Such assessments are essential for deciding whether certain activities might actually pollute the circumpolar Antarctic environment. The assessment procedures elaborated in Annex I (and previously iterated in Article 8 of the Protocol) provide for assessing human activities on a graduated impact scheme. Human activities are divided into those having 'less than a minor or transitory impact'; those having 'a minor or transitory impact'; or those having 'more than a minor or transitory impact'.[18] Regrettably, however, these terms are neither precisely defined nor quantifiably explained in either Article 8 of the Protocol or in Annex I. Hence interpretation and implementation of environmental impact assessment procedures primarily devolves to the discretion and responsibility of each state party.

Annex I will have important implications for regulating the discharge of effluents from land-based Antarctic stations into the circumpolar marine environment. The environmental impacts of sewage, chemical wastes, and port maintenance facilities will have to be assessed under Annex I in order to determine their

[15] Protocol, Art. 9(1).

[16] Art. IX of the Antarctic Treaty concerns the Consultative Party process and avers adoption of measures through approval by all Antarctic Treaty Consultative Parties (since interpreted as meaning *consensus*).

[17] See Protocol, Art. 8.

[18] Annex I, Arts. 1, 2 and 3.

harmful nature for the offshore environs. Similarly, environmental impact assessments will have to be made for any new facilities that might produce effluent discharges into marine areas.

Annex I would have had greater legal reach in regulating pollution in the Antarctic marine environment if mining or drilling activities for prospecting, exploring or exploiting hydrocarbons had been permitted in or around the Antarctic continental shelf. There is little question that if such activities were permitted, the possibility for marine pollution would be considerable. These activities, however, are prohibited by the Protocol. Even so, scientific research activities involving drilling into the continental shelf are permissible, and these will require environmental impact assessments in accordance with Annex I. Similarly, if explosives or chemical tests are used in marine scientific experiments, their impact will have to be gauged under Annex I.

No question exists that Annex I properly pertains to activities that might produce pollution in circumpolar Antarctic seas. The mandate for this annex flows from Article 3 of the Protocol, which affirms the principle that activities shall be 'planned and conducted on the basis of information sufficient to allow prior assessments of, and informed judgements about, their possible impacts on the Antarctic environment' (Article 3(2)(c)). Environmental impact assessment becomes critical for the planning process, and will be essential for detecting activities potentially harmful to the marine environment.

Annex I also sets out a three-stage evaluation procedure for performing environmental impact assessment of activities, including those that might cause marine pollution. First, Article 1 specifies that preliminary assessments for proposed activities will be conducted by parties 'in accordance with appropriate national procedures'. Article 2 goes on to mandate an Initial Environmental Evaluation (IEE) for activities 'likely to have no more than a minor or transitory impact' by the party proposing the activity; then, for all activities deemed 'likely to have more than a minor or transitory impact', Article 3 requires the preparation of a Comprehensive Environmental Evaluation (CEE), which is then to be circulated to all parties for comment, as well as made publicly available. In any event, according to its Article 7, application of Annex I is exempted in cases of emergency relating to the safety of human life or of ships, aircraft or equipment and facilities of high value, or protection of the environment which require some activity be taken absent completion of the procedures set out in the Protocol, i.e. Annex I.

The incorporation of these procedures marks a significant step towards broader environmental protection of the circumpolar marine environment. Even so, greater responsibility for environmental impact assessment should have been designated to the Committee for Environmental Protection. One would think that more extensive CEP involvement would improve the quality and consistency of the assessment process in each stage, particularly as regards the body of scientific information required to arrive at a near-accurate assessment. Instead, the burden of assessment remains largely with *national* governmental agencies which are planning the activities. Still, according to Article 3(5) of Annex I, no final decision to proceed can be taken for a proposed activity that might impact upon the marine environment until the draft CEE has been considered by an Antarctic Treaty Consultative Meeting on the advice of the CEP.

Annex II: conservation of Antarctic fauna and flora

Annex II reasserts the need for the conservation of Antarctic fauna and flora and essentially updates the 1964 Agreed Measures for the Conservation of Antarctic Fauna and Flora, which apply most directly to land and ice shelves of the continent.[19] While the nine articles in Annex II contain little that is new, it is nonetheless important for integrating the conservation of Antarctic fauna and flora into a more comprehensive, comprehensible environmental protection framework and for re-emphasising the critical need for wildlife conservation in the Antarctic.

With respect to marine pollution law, Annex II supplies two notable contributions. First, provision is made in Article 3(4) that the 'Specially Protected Species' list in Appendix A to the Annex 'shall be accorded special protection by the Parties'. The two species so listed for protection are the Fur Seal and the Ross Seal. It follows then that parties are obliged to prevent any activity, including forms of marine pollution, which might threaten or jeopardise the special protection accorded these seal species.

Secondly, provision is made in Article 4 to prevent species pollution of the Antarctic marine environment: 'no species of animal or plant life not native to the Antarctic Treaty area shall be introduced onto land or ice shelves, *or into water* in the Antarctic Treaty area except in accordance with a permit' (emphasis added). The introduction of alien species that might disrupt the Antarctic marine ecosystem is thus prohibited, in effect banning from Antarctic waters all forms of living marine pollution, such as the use of certain non-indigenous bacteria to ingest petroleum slicks from oil spills.

Annex III: waste disposal and waste management

Annex III pertains to waste disposal and waste management, and contributes much of substance to the international law of marine pollution. This annex grew out of the 1975 Code of Conduct for Antarctic Expeditions and Station

[19] For the text of the 1964 Agreed Measures, see J. A. Heap (ed.), *Handbook of the Antarctic Treaty System*, 8th edn (Washington, DC: US Department of State, 1994), pp. 2,048ff. Of note, three improvements are made by Annex II in the conservation regime for Antarctica. First, protection is extended to terrestrial and freshwater invertebrates (Art. 6(1)(b)); secondly, a prohibition is placed on the presence of dogs in Antarctica after 1 April 1994 (Art. 4(2)); and, thirdly, significant damage to native terrestrial plants is included within the definition of 'harmful interference' to the Antarctic environment (Art. 1(h)(v)).

Activities[20] and from Recommendation XV-3, which attempted to upgrade the 1975 Code.[21] In general, Annex III contains notable improvements over the 1975 Code, as it places greater emphasis on retrograding waste and other materials from the continent, and standardises collection and circulation of information on waste management. Particularly important in the context of this chapter, Annex III bears directly on preventing activities that might overtly pollute the Antarctic marine ecosystem.

Under Annex III, Article 8, waste is classified into five main groups: Group 1, sewage and domestic liquid wastes; Group 2, other liquids and chemicals, including fuels and lubricants; Group 3, solids to be combusted; Group 4, other solid wastes; and Group 5, radioactive material. The annex mandates that parties remove all Group 2, 4, and 5 wastes if generated after its entry into force. Moreover, parties are obliged to remove Group 1 wastes 'to the maximum extent possible' from the Treaty area.

With specific regard to marine pollution, Article 5(1) of Annex III pertains to the disposal of wastes at sea. Sewage and domestic liquid wastes may be discharged directly into the sea, provided that three conditions are met: (1) the capacity of the marine environment to assimilate such wastes is taken into account; (2) the discharge is situated 'wherever practicable' under conditions for 'initial dilution and rapid dispersal'; and (3) 'large' quantities of these wastes – defined as generated from a station where the weekly occupancy in the austral summer is thirty persons or more – are treated at least by maceration. Certain products that could threaten the health of the marine ecosystem are prohibited from being introduced into any part of the Antarctic Treaty area. As specified in Article 7 of Annex III, these include polychlorinated biphenyls, non-sterile soil, polystyrene beads, chips or like forms of packaging, or pesticides (other than those required for scientific, medical or hygiene purposes).

Finally, in order to coordinate efforts and reduce the impact of waste on the Antarctic environment, inclusive of circumpolar waters, a waste management system is to be set up. Each party shall annually review and update its waste reduction, storage and disposal efforts, specifying each site for field camps and for each ship.[22] Each party is also obliged to prepare an inventory of locations of past activities, before that information is lost, so that such locations can be taken into account in planning future scientific programmes (such as snow chemistry, pollutants in lichens or ice core drilling).[23]

Still, Annex III is not without problems and flaws. For one, the provisions contain numerous qualifiers, such as 'to the maximum extent practicable', 'as far as

[20] Recommendation XII-4, 'Man's Impact on the Antarctic Environment: Code of Conduct for Antarctic Expeditions and Station Activities', reprinted in Heap (ed.), *Handbook of the Antarctic Treaty System*, p. 2,062.

[21] Recommendation XV-3, 'Human Impact on the Antarctic Environment: Waste Disposal', reprinted in Heap (ed.), *Handbook of the Antarctic Treaty System*, pp. 2,063ff.

[22] Annex III, Art. 8(2).

[23] *Ibid.*, Art. 8(3).

practicable', and 'as soon as practicable'.[24] Such indefinite parameters make it difficult to hold operators accountable for actions that produce marine pollution. Also objectionable is the acceptance (in Article 3(1)) of incineration as a permissible form of waste disposal. The fact is that incineration pollutes the air and also produces contaminated, toxic ash that must be disposed. Moreover, incinerated particulate matter inevitably falls out into circumpolar waters, thus causing marine pollution along coastal areas.

Sewage and liquid waste disposal present still more problems. Annex III relies on maceration (softening by soaking in a liquid over time) as a principal means for dealing with such waste products.[25] This method, however, fails to account for heavy metals, bacteria, viruses and other chemical contaminants remaining in the waste matter, which if emitted into the seas would obviously pollute the local marine environment. Moreover, Article 5(1) explicitly permits discharge of liquid wastes directly into the sea. Accepting the obligation not to pollute the Antarctic marine environment, a preferable strategy would have sludge from these waste processes retrograded from the continent, rather than be dumped at sea.[26]

Annex IV: prevention of marine pollution

Annex IV directly pertains to the 'Prevention of marine pollution', and is specifically linked by its Articles 3, 5, 6, 13 and 14 to MARPOL 73/78. This fourth Protocol annex deals with discharges from ships, in particular oil, noxious liquids, garbage and sewage. In a real sense, this annex strengthens the jurisdictional reach and legal breadth of MARPOL 73/78 into the Antarctic region. Annex IV applies 'with respect to each Party, to ships entitled to fly its flag and to any other ship engaged in or supporting its Antarctic operations, while operating in the Antarctic Treaty area'.[27]

Certain provisions highlight the necessity for vessel retention capacity, emergency response and operator preparedness. Annex IV prohibits in Article 3 'any discharge into the sea of oil or oily mixture', save in circumstances permitted under Annex I of MARPOL 73/78. In Article 4, the marine pollution annex forbids 'the discharge of any noxious liquid substance, and any other chemical or other substances, in quantities or concentrations that are harmful to the marine environment'. In addition, Article 5 goes on to prohibit by name the disposal into the sea of two other categories of substances: (1) plastics, 'including but not limited to synthetic ropes, synthetic fishing nets, and plastic garbage bags'; and (2) all forms of garbage, 'including paper products, rags, glass, metal, bottles, crockery, incineration ash, dunnage, lining, and packing materials'. These provisions are specifically intended to link Annex V of MARPOL 73/78 to prohibitions against

[24] See *ibid.*, Arts. 1(2), 1(4), 2(2), 3(1), 3(2), 4(2), 4(3) and 8(3).

[25] See *ibid.*, Art. 5(1)(b) (Disposal of waste in the sea).

[26] Antarctic and Southern Ocean Coalition, 'A Critique of the Protocol', p. 8.

[27] Annex IV, Art. 2.

dumping garbage in high seas areas south of 60° South latitude. In addition, parties are also obliged in Article 6 to 'eliminate all discharge into the sea of untreated sewage . . . within 12 nautical miles of land or ice shelves', which directly links the Protocol to Annex IV of MARPOL 73/78. Beyond that distance, any sewage discharge should be made 'at a moderate rate and, where practicable, while the ship is *en route* at a speed of no less than 4 knots'.

An obvious loophole, however, could undermine this fiat. As the *chapeau* to paragraph 1 of Article 6 provides, this prohibition applies 'except where it would unduly impair Antarctic operations'. But the terms 'unduly', 'impair' and 'Antarctic operations' are left undefined. Furthermore, determination of where and when those conditions exist is apparently left to the discretion of vessel operators. Such an open-ended provision is an open invitation to private abuse and national non-enforcement.

Annex IV also contains certain pollution abatement rules that pertain to the prevention of and the cleaning up of spills, be they intentional discharges or accidents. Contingency planning, international or regional coordination, and, if necessary, intervention on the high seas are all considered aspects of marine pollution abatement. To these ends, Article 9(1) binds parties to ensure that all ships flying their flags have 'sufficient capacity' on board for the retention of garbage while within the Antarctic Treaty area. Likewise, Article 9(2) specifies that parties must ensure that their ships entering Antarctic waters are fitted with tanks ('adequate facilities') of sufficient capacity for the retention of all sludge, dirty ballast, tank washing water, oily residues, and garbage from ships.

Parties must also have concluded arrangements to discharge oily residues and garbage at a reception facility after leaving the area. In addition, Article 9 stipulates that parties at whose ports ships depart for or arrive from Antarctica are committed to ensuring that adequate facilities are provided to receive the above-mentioned garbage without causing undue delay or placing inequitable burdens on countries near Antarctica. According to its Article 10, the objectives of Annex IV are to be taken into account in the design, construction, manning and equipping of ships engaged in or supporting Antarctic operations.

An important consideration in marine pollution law concerns the sovereign immunity accorded to warships. The standard approach in the law of marine pollution is to adopt a formula that provides for complete exemption from the application of a convention for warships and other ships owned or operated by a state and used only on governmental non-commercial service, so far as the main rules and enforcement provisions of that convention are concerned. Exemption may be linked to an obligation of each state to ensure that its vessels act in a manner consistent, as far as is reasonable and practicable, with the provisions or the objectives of the convention concerned.

The Antarctic marine pollution annex attempts to close gaps for sovereign immunity left in previous instruments. It achieves only partial success, however. Article 11 provides for sovereign immunity as it asserts that the annex 'shall not apply to any warship, naval auxiliary or other ship owned or operated by a State'. Still, parties are obliged to 'ensure by the adoption of appropriate measures not impairing the operations or operational capabilities of such ships owned or operated by it, that such ships act in a manner consistent, so far as is reasonable and practicable' with the annex (Article 11(1)).

The problem here seems obvious: most vessels operating in circumpolar Antarctic waters *are* state-owned or operated – so most vessels will be affected by this qualified exception of sovereign immunity. To the extent that determination of 'appropriate measures' and 'reasonable and practicable' conditions for compliance is left principally to the discretion of vessel operators, the possibility of violations becomes widened and the prospects for enforcing compliance are narrowed. That situation remains unfortunate, as it fails to uphold the principal purposes of Annex IV and may in fact permit pollution to occur in the Antarctic without proper accountability.

Any future Protocol parties which are non-parties to MARPOL 73/78 are symbiotically obligated to MARPOL 73/78 provisions under the Protocol's Annex IV.[28] Under Article 3 of that annex, discharges of oil or oily mixtures into the sea are prohibited, 'except in cases permitted under Annex I of MARPOL 73/78'. Similarly, Article 5 of Annex IV allows the disposal into the sea of food wastes if passed through a comminuter or grinder, made 'as far as practicable from land and ice shelves', with exceptions only for those cases 'permitted under Annex V of MARPOL 73/78'. The definition of 'sewage' used in Article 6 of the Protocol's Annex IV is that from Annex IV of MARPOL 73/78. Article 13 of Annex IV binds Protocol parties 'to keep under continuous review' provisions of the annex to ensure the prevention and reduction of and a response to pollution in the marine environment, 'including any amendments and new regulations adopted under MARPOL 73/78'. Finally, for those Protocol parties also party to MARPOL 73/78, Article 14 ensures that nothing in Annex IV 'shall derogate from the specific rights and obligations thereunder' in the MARPOL instrument.

The conclusion here is evident: Annex IV for the Antarctic seas south of 60° South latitude and MARPOL 73/78 for all ocean space are intended to complement and strengthen each other. Those Protocol parties that are not party to MARPOL 73/78 are thus contracted into obligations not to pollute Antarctic ocean space that mirror the principles and rules of the MARPOL 73/78 regime.[29] State parties to both instruments become linked into mutually reinforcing obligations aimed at preventing marine pollution in Antarctic waters. Thus, in these interacting ways, parties are brought into a stronger, more comprehensive legal regime for suppressing pollution of the high seas in Antarctic waters.

So long as vessel operators encounter such great difficulty in managing their wastes, they will continue to dump garbage at sea. Regrettably, MARPOL

[28] Currently, however, there is little practical difference: after New Zealand's accession to MARPOL 73/78 in September 1998, all the current parties to the Protocol are also parties to MARPOL 73/78 (status as of 8 June 1999). [29] *Ibid.*

73/78 does little to discourage these violations from occurring. In Antarctic waters, however, enforcement for compliance with Annex IV of the Protocol is left to each contracting party to exercise over ships flying its own flag or supporting that government's Antarctic operations.[30] In addition, Protocol parties also party to MARPOL 73/78 are explicitly bound to the rights and obligations of that latter instrument, notwithstanding any provisions in Annex IV.[31] For the Antarctic, then, the Protocol strengthens the obligation and duty of its parties to enforce anti-dumping restrictions upon vessels flying their flags. Even so, the extent to which this obligation is exercised and fulfilled remains dependent on the political will of the governments of those parties to do so.

Annex V: area protection and management

At the Sixteenth Antarctic Treaty Consultative Meeting in Bonn on 7–18 October 1991, the United States and the United Kingdom proposed a fifth annex to the Protocol that simplified and significantly expanded the future scope of the Antarctic protected area system.[32] As adopted, this annex supplies an integrated approach to the creation and management of protected areas in the Antarctic.

In Annex V, the five existing categories of protected areas under the Antarctic Treaty are consolidated into two. According to Article 3, any area may be designated as an Antarctic Specially Protected Area (ASPA) to protect outstanding environmental, scientific, historic, aesthetic or wilderness values, any combination of those values, or ongoing or planned scientific research. The second category, set out in Article 4, consists of Antarctic Specially Managed Areas (ASMAs) to coordinate multiple-use activities occurring in the same area, ostensibly to avoid possible conflicts, improve cooperation between parties and minimize environmental impacts.[33]

With particular importance for marine pollution law, Annex V provides that 'any area, including any marine area', can be a candidate for designation as either an ASPA or an ASMA.[34] In fact, such areas may specifically include areas threatened by marine pollution activities. Article 3(1) asserts that any (marine) area may be designated an ASPA 'to protect outstanding environmental, scientific, historic, aesthetic or wilderness values'. Protection from pollution activities would clearly qualify in this regard.

[30] Annex IV, Art. 2. [31] *Ibid.*, Art. 14.

[32] Annex to Recommendation XVI-10: Annex V to the Protocol on Environmental Protection to the Antarctic Treaty: Area Protection and Management; text reprinted in Heap (ed.), *Handbook of the Antarctic Treaty System*, pp. 2,117–18 and 2,125–9.

[33] While permits will not be required, a detailed management plan will be necessary for each area (Annex V, Art. 5). Importantly, the concept of 'sufficient size to protect the values for which the special protection or management is required' is included (Art. 5(2)) which improves upon the previous 'minimum size requirement'. Management for tourist visits is also included. Perusal of Annex V implies that the CEP will play a pivotal advisory role in the management plan process, as also will the Scientific Committee on Antarctic Research.

[34] Annex V, Art. 3(1). See also *ibid.*, Art. 4(1).

Similarly, as stipulated in Article 4(2)(a), ASMAs may include (marine) 'areas where activities pose risks of mutual interference or cumulative environmental impacts'. Here again, threats from marine pollution activities might be a cause for the designation of such a management area. Finally, in drawing up management plans for both ASPAs and ASMAs, parties are obliged to provide 'clear descriptions of the conditions under which permits may be granted' for several activities, including 'the disposal of wastes'.[35] This consideration holds particular import for ensuring that pollution is prevented in any marine areas so designated.

Liability for marine pollution

Effective, enforceable anti-pollution law can be enhanced by rules regarding responsibility and liability for damage done by pollution, including appropriate forms of compensation for that environmental damage. Even so, convention law on marine pollution has moved only hesitantly in the area of liability for pollution damage. Not surprisingly, the Protocol has been earmarked by similar hesitation. While the Protocol contains in its Article 16 a general undertaking of its parties 'to elaborate rules and procedures relating to liability for damage arising from activities taking place in the Antarctic Treaty area and covered by this Protocol' to be 'included in one or more Annexes', specifics for that liability regime have yet to be resolved. However, a liability regime remains essential for completing the Protocol.

The principal objective of a liability regime is deterrence. The sensitive nature of the Antarctic region makes it difficult, if not impossible, to restore a damaged area to its previous condition. Thus, a liability regime must be designed that provides strong incentives to take measures aimed at preventing damage. At a minimum, such incentives should include requirements for immediate and ongoing action to mitigate impacts of environmental damage, as well as the fundamental obligation to pay compensation if the environment cannot be restored to its pre-damaged state.

Liability connotes assigning responsibility and accountability to some party for an unwanted event. A principal purpose of imposing liability is to penalise and thus deter commission of a criminal act. In civil cases, liability can help to allocate the costs of compensation and restitution. In any case, the effectiveness of a liability regime remains dependent on having a mechanism to enforce legal obligations for compensating injured parties. The Antarctic Environmental Protection Protocol, as of mid-2000, still lacked both the diplomatic framework and the legal substance of that enforcement mechanism.

The extent of liability and how precisely such liability will be applied to operators or governments has not yet been agreed upon by the Consultative Parties. Should all operators be held fully accountable under absolute liability for

[35] See *ibid.*, Art. 5(3)(i)(viii) and (j)(vii).

all acts of marine pollution that cause damage to the environment? Or should a system of strict liability prevail, under which certain conditions or circumstances might excuse such pollution damage?[36] While environmentalists would obviously prefer absolute liability, strict liability appears more practicable and likely to be adopted by the Consultative Parties. This is simply because the notion of strict liability more closely parallels their national self-interests and provides greater legal flexibility given the harsh environmental and weather conditions in the frigid southern seas.

The Protocol thus remains unfinished business. Rules and procedures 'relating to liability for damage arising from activities taking place in the Antarctic Treaty area' still need to be elaborated and adopted, not only for marine pollution in particular but for all environmental damage there in general.[37]

While negotiating a liability regime for the frozen south remains law-making in progress, formulating such an annex has proved a protracted endeavour. In the period between 1993 and 1998, eight iterations of a draft annex on liability were presented to the Consultative Parties by a legal expert group chaired by Professor Rüdiger Wolfrum.[38] At the Twenty-Second Consultative Meeting, held in Tromsø, Norway, on 25 May–5 June 1998, the Consultative Parties decided that the legal expert group, by submitting its report, had fulfilled its task and completed its work; and that the further negotiation of an annex or annexes on liability be undertaken in Working Group I of the Antarctic Treaty Consultative Meeting.[39] Deliberations over liability, now in the policy rather than the legal forum, continued at the Twenty-Third Consultative Meeting, held in Lima, Peru on 24 May–4 June 1999; a new listing of key issues has been made, but a consensus of parties on how to solve these remains equally distant.[40]

What has emerged is the realisation that genuine consensus is required for making such a liability annex politically and legally viable, and that such consensus is only willing to accept a strict liability regime. Also apparent is the critical need to convert international commitment among the Consultative Parties into national law and enforced compliance – a process that has hardly begun. Hence, given the complex law and politics and high national interests at stake, completion

[36] For a recent thorough study of these and related issues, see R. Lefeber, 'The Prospects for an Antarctic Environmental Liability Regime', in D. Vidas (ed.), *Implementing the Environmental Protection Regime for the Antarctic* (Dordrecht: Kluwer Academic Publishers, 2000).

[37] See R. Lefeber, 'The Legal Need for an Antarctic Environmental Liability Regime', in Vidas (ed.), *Implementing the Environmental Protection Regime for the Antarctic.*

[38] For a summary of the work of the group of legal experts on liability and for an overview of the seven main pending issues, see *Report of the Group of Legal Experts*, doc. XXII ATCM/WP 1, 14 April 1998. For an overview and analysis, see M. Skåre, 'A Liability Annex or Annexes to the Environmental Protocol: A Review of the Process within the Antarctic Treaty System', in Vidas (ed.), *Implementing the Environmental Protection Regime for the Antarctic.*

[39] See paras. 61–84 of *Final Report of the Twenty-Second Antarctic Treaty Consultative Meeting, Tromsø, Norway, 25 May–5 June 1998* (Oslo: Norwegian Ministry of Foreign Affairs, 1998); and Decision 3 (1998), 'Liability', in *ibid.*, p. 69.

[40] See paras. 75–98 of *Final Report of the Twenty-Third Antarctic Treaty Consultative Meeting, Lima, Peru, 24 May–4 June 1999*, available at www.rree.gob.pe/conaan/meeting1.htm.

of the negotiations for a liability annex acceptable to all seems to remain a distant diplomatic ambition.

THE BALANCE SHEET

Over the past three decades, the Antarctic Treaty Consultative Parties have increasingly applied international law to protect the Antarctic from activities leading to environmental degradation. Among those protective measures has been the creation of a special legal regime designed to prevent pollution of the circumpolar marine environment. Not only have efforts aimed to prohibit degradation of the continent under reinforced international environmental law, Antarctic Treaty Parties have also sought to negotiate specific instruments for regulating pollution activities that could adversely affect the Antarctic marine ecosystem. The Environmental Protocol brings these regulations into a neater, tighter, more manageable legal package.

The Protocol contributes to the general law against marine pollution by applying to the Antarctic Treaty area core legal norms, principles and conventions intended to prevent and prohibit marine pollution. Certain new legal principles are also made imperative for activities in the Antarctic marine ecosystem. Requirements have been introduced to ensure that the conduct of activities must be carefully planned, that environmental impacts of proposed activities must be adequately assessed, and that sufficient information must be obtained for deciding which activities may be undertaken where. All these principles are notable contributions that enhance anti-pollution law for the Southern circumpolar marine environment. These are fundamental procedures that, if properly implemented, should go far towards preventing marine pollution in the Antarctic Treaty area, as well as in the world ocean.

No less significant is the fact that the Protocol, coupled with its five annexes, directly addresses all the prominent forms of marine pollution that threaten the ocean space: sewage, petroleum, metallic effluents, chlorinated hydrocarbons and radionuclides. Similarly, the Protocol and its annexes are made to apply to all sources of pollution affecting the Southern Ocean ecosystem – whether from land, dumping, vessels, the atmosphere or the seabed. Such an all-encompassing approach serves well to undergird the legal framework of anti-pollution law in the region.

Important, too, is that while the Protocol embodies the hallmarks of a regional regime for regulating marine pollution, it also incorporates global conventional rules. In this way, a symbiotic relationship is effected. On the one hand, the specific anti-marine pollution regime for the Southern Ocean region is bolstered by international norms prohibiting pollution of the marine environment. On the other hand, the body of global international law forbidding pollution of ocean space is also augmented and strengthened by innovations contained in the Protocol, in particular those set out in its Annex IV.

The Consultative Parties intentionally tied the Protocol to binding anti-pollution obligations and principles set out in international conventions of a global character, particularly MARPOL 73/78 and – although without an explicit reference – the 1972 London [Dumping] Convention and the 1982 LOS Convention. This integration of salient legal norms from global international agreements into a specific Antarctic regional context furnishes a more cogent and coherent legal framework for regulating marine pollution in the circumpolar South.

Ultimately, however, the effectiveness of international anti-pollution law rests with the degree of genuine commitment by national governments. Governments have made laws protecting the Antarctic marine environment from pollution, and governments must enforce those laws against nationals who violate them. In the final analysis, degradation of the Antarctic marine environment will not occur on account of weak law. The law prohibiting marine pollution in Antarctic waters is robust, present and plain. Moreover, if the past is prologue, additional environmental law will be devised by the Consultative Parties as new needs for it are recognised. This clearly is the case for promulgating an additional annex, or annexes, on liability that sets penalties for and compensation from nationals of parties for pollution damage done to the marine environment.

The blame for violations must fall to the lack of political will among the Consultative Parties to monitor their nationals' activities, enforce compliance and compel compensation for liability. The bottom line is simple: the Protocol can be only as effective as the governments of its parties are willing to make it. If the Antarctic is to be preserved in a condition free of marine pollution, then marshalling and sustaining that necessary political will must be a critical consideration.

6 Sub-regional cooperation and protection of the Arctic marine environment: the Barents Sea

OLAV SCHRAM STOKKE*

Over the past decade the states governing the Arctic territories have taken on a variety of commitments regarding marine environmental management. As the first three chapters of this book have shown, several global regimes have emerged thus far. At the regional level, the Arctic Environmental Protection Strategy (AEPS) has generated a range of programmatic activities, vastly improving the level of knowledge about the nature and gravity of environmental hazards in the high North.[1] The focus of this chapter is on sub-regional marine environmental protection, more specifically the bilateral Russian–Norwegian Environmental Commission and the multilateral Barents Euro–Arctic Region. The aim is to bring out *whether* and *how* these sub-regional cooperative processes can complement efforts at the regional and global levels.

There are several reasons for including the Barents Euro–Arctic Region in a study of protection of the marine environment, although the 1993 Kirkenes Declaration,[2] on which the latter structure is based, made no mention of marine areas when delineating the *spatial* scope of the cooperation. The unsettled maritime delimitation of the Barents Sea between Russia and Norway is the main reason for not mentioning marine cooperation.[3]

For one thing, much of the marine pollution in the Barents Sea area originates from land-based activities which fall clearly within the cooperative domain of the Declaration. This goes for matters such as leakages from land-based storages of radioactive waste and riverborne or atmospheric pollution from, e.g., the metallurgical industry on the Kola Peninsula and elsewhere. But, more importantly, when the *functional* range of the cooperation was being spelt out, the marine

* I would like to thank Bernt Bull, Steven Sawhill, Davor Vidas, Budislav Vukas and Oran Young for their very helpful comments.

1 For an analysis of the AEPS as regards marine pollution, see Vidas, Chapter 4 in this book.

2 Declaration on Cooperation in the Barents Euro–Arctic Region, adopted at the Conference of Foreign Ministers, Kirkenes, Norway, 11 January 1993; text reproduced in *UD Informasjon*, No. 1 (Oslo: Norwegian Ministry of Foreign Affairs, 1993).

3 See R. Castberg, O. S. Stokke and W. Østreng, 'The Dynamics of the Barents Region', in O. S. Stokke and O. Tunander (eds.), *The Barents Region: Cooperation in Arctic Europe* (London: SAGE Publications, 1994), pp. 71–84.

environment figured prominently from the outset: the prevention of dumping of radioactive waste was among the very first issues mentioned in the Declaration. In practice, ensuring the health of the Barents Sea has been a key target for the Environmental Committee under the Barents Council; moreover, among the first decisions of this Council was the establishment of a Committee on the Northern Sea Route. For its part, the Russian–Norwegian Environmental Commission (hereinafter referred to as the bilateral Environmental Commission) gave a central place to protection of the marine environment from the very start.[4]

After sketching the conceptual terrain demarcated by 'sub-regional' and 'effectiveness', this chapter offers a brief account of the main marine environmental problems faced in the Barents Sea area. There then follows a discussion of whether and how this sub-regional cooperation links up effectively with other efforts to solve these problems.

SUB-REGIONAL COLLABORATION IN THE BARENTS REGION: EFFECTIVENESS CONCERNS

The term 'Barents Region' came into use in the early 1990s in connection with a range of bilateral and multilateral cooperative networks under development in the Barents Sea area, across the East–West divide. The first wave of sub-regional institutions was bilateral in nature. A bilateral Environmental Commission was set up under a 1988 Soviet–Norwegian agreement; this commission has since served as the major instrument of coordination between the Norwegian Ministry of Environment and the Soviet, later Russian, lead environmental agency – now named the State Committee for Environmental Protection.[5]

The Barents Euro–Arctic Region (BEAR), established at a ministerial conference held in Kirkenes, Norway, in 1993, is notable for its two-tiered structure. At the first level, there is the Regional Council, composed of municipal representatives from the three North Norwegian *fylker* of Nordland, Troms and Finnmark, the northernmost Swedish and Finnish *län* Norrbotten, Västerbotten, Lappland and Oulu – and Murmansk and Arkhangelsk *oblasti* as well as the Karelian Republic and Nenets Autonomous area, all in Russia.[6] The Regional Council also includes one representative from the indigenous peoples – a Saami delegate. The second layer

[4] O. S. Stokke, 'A Green Partnership? – Norway, Russia and the Northern Environment', *International Challenges*, Vol. 14, 1994, pp. 11–23.

[5] See 'Overenskomst mellom Kongeriket Norges regjering og Unionen av Sovjetiske Sosialistiske Republikkers regjering om samarbeid på miljøvernområde' (Agreement Between the Governments of the Kingdom of Norway and the Union of the Soviet Socialistic Republics on Cooperation in Environmental Matters), Oslo, 15 January 1988, in force the same day, published in *Overenskomster med fremmede makter* (Oslo: Norwegian Ministry of Foreign Affairs, 1988). In September 1992, that agreement was replaced by a new one, under the same name but now concluded with the Russian Government; it covered additional areas and focused on common *measures* in the environmental area and not solely on generating a common fund of information; published in *Overenskomster med fremmede makter* (Oslo: Norwegian Ministry of Foreign Affairs, 1992), pp. 1,532–5.

[6] Participation in the Regional Council has expanded over time: Västerbotten, Oulu and Nenets were not among the original members.

consists of the Barents Council, made up of government representatives from Russia and from the Nordic countries (Denmark, Finland, Iceland, Norway and Sweden) as well as one representative from the European Commission. Membership in the Council is in fact open to any state wishing to take an active part, but the chairmanship will rotate between the four states governing the counties involved in the cooperation.[7]

While the bilateral Environmental Commission and the BEAR have been especially important in efforts to address protection of the Barents Sea, there are also other sub-regional mechanisms worth mentioning, partly in interaction with the two former. For instance, the trilateral Arctic Military Environmental Cooperation (AMEC), involving the defence ministries of Russia, Norway and the United States,[8] has been of some significance in efforts to cope with the dumping of radioactive material in the Barents Sea.

Layers of regionality

The cooperative processes discussed in this chapter link two different layers of regionality. At the *international* level, regional initiatives seek to involve clusters of states in closer interaction and joint framing of problems and opportunities.[9] Within the Barents Region, the governmental Barents Council reflects this layer of regionality, as does the bilateral Russian–Norwegian Environmental Commission. At the *transnational* level, sub-state region-builders strive to coordinate behaviour and establish common terms of reference in adjacent territories separated by national borders. Thus, the BEAR Regional Council comprises representatives of county authorities and indigenous peoples; and similarly, a Permanent Working Group on Local Cooperation under the bilateral Environmental Commission brings together on a regular basis the regional environmental bureaucracies of the border counties Finnmark (in Norway) and Murmansk (in Russia).

But regionality is also a *domestic* phenomenon – and the evolution of the BEAR initiative in particular cannot be understood without reference to how the territories involved in this process all seek recognition as units distinct from other parts of their respective nations – in ways which have triggered special administrative measures designed to ensure comparable standards of living, including tax relief and other efforts to stimulate economic activity. In Norway sparse settlement, a gradual population decline, harsh climatic conditions, and an economy based on rich but volatile fish stocks are important reasons for those special measures. For its part, the Soviet plan economy used to employ various means to draw workers to Arctic regions; the steep demographic growth during the first half of this

[7] Terms of Reference for the Council of the Barents Euro–Arctic Region, Arts. 2 and 6; see text in *UD Informasjon*, No. 1 (Oslo: Norwegian Ministry of Foreign Affairs, 1993), p. 13.

[8] Declaration on Arctic Military Environmental Cooperation, signed in Bergen, 26 September 1996; text available at www.denix.osd.mil/denix/Public/Intl/AMEC/declar.html.

[9] For legal discussions of this layer of regionality in the protection of the marine environment, see Boyle, Chapter 1; and Vukas, Chapter 2 in this book.

century in Murmansk county and the Soviet North in general shows the effectiveness of this strategy – which is now history.

Another feature which makes the Barents Sea area stand out domestically, especially for the coastal states, is its strategically sensitive location. The role of strategic submarines in the nuclear balance made this area a key front in the East–West military rivalry. As a result, conflict avoidance has been a key priority in this area, as reflected in certain self-imposed restraints in Norwegian and NATO military activities close to the common border; and by the establishment of bilateral political institutions in areas such as fisheries where rational management requires coordinated behaviour.

This distinctiveness of the northern areas formed part of the motivation for launching initiatives to expand sub-regional cooperation following the thaw in East–West relations. The bilateral Environmental Commission was spurred on by growing worries in Norway about transboundary fluxes of airborne pollution emanating from the metallurgical industry on the Kola Peninsula, but also about possible problems associated with growing offshore petroleum activity in the Russian part of the Barents Sea; later on, nuclear waste came into focus. The multilateral BEAR initiative fed on the same worries, but also linked up to broader concerns related to the economic and geopolitical situation of the Barents Sea area. Despite special administrative incentives, the economies in the northern parts of Norway, Sweden and Finland had been slow for several years. Stimulating contacts with northwest Russia, crisis-ridden, but rich in natural resources and gradually opening up to Western economies, seemed a promising avenue.

A sub-region of what? Linkages to other levels of cooperation

While we have seen that the Barents Sea area is marked by a certain measure of regional distinctiveness, it can be useful to look into the various *linkages* between this set of cooperative processes and adjacent ones.[10] If the Barents Sea area is a sub-region, what is it subordinate to? There are at least three broad answers to this question.

An *Arctic* answer is that Barents Sea cooperation is primarily linked to the wider flow of cooperative initiatives concerning the High North in the aftermath of the Cold War. The Gorbachev initiative, launched in the 'Murmansk speech' of 1987, sparked off a truly hectic period for Arctic policy-makers and bureaucrats.[11] The bilateral Environmental Commission belongs to the first wave of political responses to this challenge. Similarly, the scientific community was quick to reintroduce an earlier plan for a circumpolar body to foster greater coordination in

[10] On the notion of regime linkages, see O. R. Young, 'Institutional Linkages in International Society: Polar Perspectives', *Global Governance*, Vol. 2, 1996, pp. 1–24.

[11] For an overview of the Gorbachev initiative and early responses to it, see D. Scrivener, 'Gorbachev's Murmansk Speech: Soviet Initiative and Western Responses', *Security Policy Library*, No. 1 (Oslo: Norwegian Atlantic Committee, 1988).

this sector, notably improved physical access to the entire circumpolar area – and in 1990 they succeeded in establishing the non-governmental International Arctic Science Committee. The subsequent Finnish initiative to set up a cooperative apparatus for protection of the Arctic environment produced the Arctic Environmental Protection Strategy of 1991, including a set of working groups that generated a considerable amount of programmatic activities, especially environmental monitoring and mapping of international cooperative mechanisms relevant to the northern environment.[12] Next in line were the Canadians with their Arctic Council initiative, which after years of pushing and shoving saw the light of day in 1996.[13] In this context, the Norwegian initiative for a Barents Region, launched in 1992 and institutionally completed with the Kirkenes Declaration the year after, is one of a series of ambitious diplomatic undertakings seeking to civilise political interaction in the High North.

Indeed, this rapid development has now given rise to cost-efficiency concerns, as the sheer multitude of cooperative arenas may easily imply duplication of work. The national administrative layers responsible for Arctic affairs are generally thin in the governments involved, and the host of international-level initiatives in the past decade, each with a separate set of meetings and programmes, is beginning to create a measure of bureaucratic fatigue. Such cooperative overload highlights the need to avoid duplication of responsibilities and tasks. The BEAR initiative would seem particularly well placed to promote a sensible division of labour between the numerous cooperative processes relevant to the protection of the marine environment: it brings together actors at several political and administrative levels and in different issue-areas, and it is closely involved in the proliferation of societal and transnational contacts as well.

Important as this northern link is, however, it should not lead us to ignore how the Barents Region is placed in the wider *European* region. The clearest expression of this southern linkage is the fact that the European Commission is represented in the Barents Council; more recently, there has been the Finnish initiative to strengthen the Northern Dimension of the EU. Early on, domestic critics of the Norwegian initiative argued that BEAR was partly a project designed to improve the EU's image, especially in the northernmost parts of the country. The European BEAR argument in Norway has emphasised that the EU would be helpful or even necessary for regional problem-solving, given the awesome dimensions of some of the transboundary environmental problems in northwest Russia.

But there is also an *Atlantic*, or western, linkage defining the focus and the resources of the Barents Region. With the demise of the Cold War and the strategic rivalry with the former Soviet Union less pronounced, there is a possibility that the United States will gradually reduce its political and military presence in the European Arctic; and the western Nordic states in particular have based their

[12] For more details on this cooperation, see Vidas, Chapter 4 in this book.

[13] See Declaration on the Establishment of the Arctic Council, signed by the eight Arctic states in Ottawa, Canada, 19 September 1996; text reprinted in ILM, Vol. 35, 1996, pp. 1,387ff.

security policy precisely upon this presence. While the sheer size of the Russian Northern Fleet and the possibility of a strengthened Russia in the not-too-distant future suggest that US strategic interests in this region are fairly stable, Norway's insistence on a high degree of openness regarding participation in the BEAR institutions reflects in part the desire to provide the region with a solid westward linkage. The United States is an observer in the Barents Council; and, as we shall see, Norway has alerted NATO to the hazards associated with Soviet and Russian handling of nuclear waste – and drawn the United States into several projects aimed at enhancing nuclear safety in the Russian northwest.[14]

There is a political side to these institutional linkages northwards, southwards and westwards: they reflect the generally cautious approach to eastward cooperation taken by Norway, the state that initiated the BEAR process. Fearing bilateralism in a region marked by legal disputes and asymmetric power relations, Norwegian governments have traditionally favoured broad Western participation in cooperative arrangements with its huge eastern neighbour. The various linkages discussed here also reflect competing images of what this region is – or should be. It is no secret that the BEAR initiative stirred up considerable controversy in the Norwegian foreign policy establishment, with those emphasising the Atlantic ties highly sceptical to what they perceived as an institutional creation overly oriented towards Europe.[15] For their part, regional actors such as county authorities or representatives of the Saami population, and also those primarily oriented towards the environmental strand of the Barents cooperation, have tended to focus instead on the way BEAR links up to the circumpolar processes, especially the AEPS and the Arctic Council.

Sub-regionality, therefore, is partly a matter of cooperative direction, or orientation, and partly a matter of adapting to the reality of a great many ongoing cooperative processes. This forms the framework for any discussion of the effectiveness of sub-regional arrangements: functional overlaps imply vulnerability to charges that the process in question is wasteful and redundant. A political initiative designed to survive must carve out a niche for itself and avoid duplication of activities already dealt with elsewhere. The rest of this chapter will trace the effectiveness of sub-regional cooperation in the Barents Sea area by addressing three questions. First, what are the main marine environmental problems faced in this area? Secondly, to what extent and how has the sub-regional cooperation addressed those particular problems? And, thirdly, how do these efforts complement those flowing from other levels of cooperation relevant to the Barents Sea environment, such as global or circumpolar processes?

[14] For a broader discussion of international and national approaches to nuclear security in northwest Russia, see Stokke, Chapter 9 in this book.

[15] For details, see J. M. Kvistad, *The Barents Spirit: The Process of Regionalization and Norwegian Foreign Policy in the Barents Euro–Arctic Region. A Bridge-Building Project in the Wake of the Cold War* (Cand. polit. thesis, 1994, available from the Department of Political Science, University of Oslo).

THREATS TO THE BARENTS SEA MARINE ENVIRONMENT

The Barents Region area, stretching from the cold, damp Arctic rim to the more fertile inland of southern Arkhangelsk and Karelia, has extremely varied environmental conditions. The coastal zone, where most of the population and most industrial and military activities are located, is also the most vulnerable to human pressures. While the Barents Sea is among the most productive in the world, low temperatures slow down evaporation and may serve to reduce the bacteriological breakdown of pollutants such as petroleum. Terrestrial and marine ecosystems are generally simple in the Barents area, implying that the disruption of one link of the food chain can severely affect the rest of the system. Let us look more closely at some of the gravest environmental dangers in the region and the extent to which they generate threats to the marine environment.

Land-based activities[16]

A significant cause for environmental worry in the Barents Region is the nuclear activity of the Russian Northern Fleet. In the years after World War II, the military complex appropriated vast land areas on the Kola Peninsula for seven naval bases, from the Murmansk Fjord in the east to the Zapadnaya Fjord some 40 kilometres from the Norwegian border. Neither the safety practices of those operating the numerous nuclear installations at these bases nor the quality of the storage facilities for various types of radioactive waste, including spent nuclear fuel, are very reassuring, and numerous leakages have been reported.[17] Another case in point is the civilian Kola nuclear power plant in Polyarny Zori, the only one in the European Arctic and generating as much as two-thirds of the electric power consumed in Murmansk *oblast*. The two oldest reactors are of a type which, according to Western experts, should be shut down immediately,[18] due to lack of physical containment and low redundancy of safety precautions. This notwithstanding, most of the nuclear contamination found in the Barents Sea area originates either outside the region, from reprocessing plants in Great Britain and France, or from atmospheric nuclear tests conducted in the 1950s and 1960s.[19]

The Yenisey and, even more so, the Ob rivers are the main channels for riverborne pollution into the Barents and Kara Seas, including organochlorines, heavy metals, hydrocarbons and radioactivity. Some of the largest and most heavily industrialised centres in Russia are found on the banks of rivers branching into the

[16] See also VanderZwaag, Chapter 8 in this book.

[17] See V. N. Lystsov, 'The Yablokov Commission Report on Soviet Radioactive Waste Dumping at Sea: Additional Comments', *Arctic Research of the United States*, Vol. 8, 1994, pp. 270–2.

[18] M. Rosen, Assistant Director-General for Nuclear Safety in the International Atomic Energy Agency, cited in 'Newsbriefs', *IAEA Bulletin*, Vol. 36, 1994, p. 81; see also Report to the Storting, *St.meld. 34 (1993–94), Atomvirksomhet og kjemiske våpen i våre nordlige nærområder* (Oslo: Norwegian Ministry of Foreign Affairs), p. 26.

[19] See AMAP, *Arctic Pollution Issues: A State of the Arctic Environment Report* (Oslo: Arctic Monitoring and Assessment Programme, 1997). After the 1963 Test Ban Treaty, such tests have been conducted underground, resulting in far less release of radioactive material into the environment.

Arctic seas: the mining and metallurgical centre of Norilsk, the West Siberian oil and gas complex, the Kuzbas coal basin, and the nuclear reprocessing plant in Mayak near Chelyabinsk in the Urals.

As to atmospheric pollution, the smelter and the roasting shop in Pechenga municipality on the Kola Peninsula, near Russia's border with Norway and Finland, pour out more than 200,000 tonnes of sulphur dioxide each year, as well as large amounts of nitrogen oxides and heavy metals. The smelter-works in Monchegorsk further south on Kola emits similar amounts, but without creating nearly as great problems in neighbouring countries, since pollution levels diminish with distance from source. As is the case with the much larger amounts of atmospheric pollution originating in industrial centres in western and central Europe, a significant part of this eventually falls into the Barents Sea. However, the capacity of the ocean to dilute and disperse renders the marine environmental effects far less severe than the terrestrial ones.[20]

Dumping

Most outside attention has been directed to Russia's comprehensive dumping of radioactive waste in the Barents and Kara Seas. As discussed in greater detail in Chapter 9 of this book, dumped materials range from low-level liquid waste, which originates in cooling and incineration facilities of radioactive installations, to low- and medium-level solid waste and the most intensely radioactive objects, several nuclear reactors still containing spent nuclear fuel.[21] While there has been no deliberate dumping of reactors and solid waste since the 1980s, considerable concern attends the accumulation of spent nuclear fuel and other less radioactive types of waste. This problem will only mount in the years to come, as a large number of submarines will be taken out of operation in line with the Strategic Arms Reduction Treaty regime, and temporary – and highly deficient – storage facilities for removed fuel units are currently filled to capacity. At the same time, Russia is still not ready to prohibit dumping of low- and medium-level liquid waste, due to the lack of satisfactory treatment technology.[22]

Offshore activity

Both Norway and Russia are engaged in offshore drilling for petroleum in the Barents Sea.[23] In the fishing industry there has been some concern about the

[20] See, in general, Joint Group of Experts on the Scientific Aspects of Marine Pollution, 'The State of the Marine Environment', *Regional Seas Reports and Studies* (Nairobi: United Nations Environment Programme, 1990), p. 88; and C. Bernes (ed.), *The Nordic Arctic Environment: Unspoilt, Exploited, Polluted?* (Copenhagen: Nordic Council of Ministers, 1996).

[21] A. V. Yablokov, V. K. Karasev, V. M. Ruyantsev, M. Y. Kokeyev, O. I. Petrov, V. N. Lystsov, A. F. Yemelyanenkov and P. M. Rubtsov, *Facts and Problems Related to Radioactive Waste Disposal in Seas Adjacent to the Territory of the Russian Federation* (Albuquerque: Small World Publishers, 1993).

[22] See Stokke, Chapter 9 in this book.

[23] For an overview of current petroleum activity in the Barents Region, see A. Moe, 'Oil and Gas: Future Role of the Barents Region', in Stokke and Tunander (eds.), *The Barents Region*, pp. 131–44.

impact of seismic detonations, as studies suggest that, on a local scale, eggs and larvae are killed and fish are scared off.[24] The part of the Barents Sea currently being explored is an important spawning and growth area for the Arctic cod stock, which supplies one of the most valuable commercial fisheries in the world.

Regular pollution arising from petroleum activity in the Barents Sea will probably add little to the total amount of oil pollution in the region, which is largely brought in by ocean currents from other marine areas,[25] but a major accident involving large-scale oil spills could have severe environmental effects. The probability of such an accident is unknown, but is presumably higher than in temperate zones; and regional differences in both equipment standards and industrial safety levels would indicate that the risk is particularly high in the Russian part of the Barents Sea.[26] Should an accident occur, climate and weather conditions, darkness and long distances will hamper rescue and restoration.

Vessel-source pollution

When petroleum activity in the Barents Sea area reaches the development and production stages, it may stimulate a considerable increase in regional ship transport. Natural conditions such as ice presence and shallow depths will render such navigation particularly dangerous, especially if it occurs in the eastern part of the Barents Sea or involves navigating the Northern Sea Route from Murmansk and eastwards to Dudinka or through the Northeast Passage.[27] According to Russian sources, the number of accidents involving ships navigating the Northern Sea Route from 1954 to 1990 was as high as 800, of which 40 per cent occurred in the Kara Sea, where ship density is the highest.[28]

Even current activities pose threats to the marine environment, in that the many nuclear submarines based in the North are prone to accidents. In 1985, partly because safety routines were violated, a dramatic explosion occurred on a submarine in a naval base near Vladivostok in the Russian Far East.[29] Four years later, the submarine *Komsomolets* went down near Bear Island off the coast of

[24] Bernes (ed.), *The Nordic Arctic Environment.*

[25] G. Futsæter, G. Eidnes, G. Hølmø, S. Johansen, H. P. Mannvik, L. K. Sydnes and U. Witte, 'Report on Oil Pollution', *The State of the Arctic Environment: Reports* (Rovaniemi: Arctic Centre, University of Lapland, 1991), pp. 270–334. On a world scale, less than 5 per cent of the oil pollution entering the oceans derives directly from platform activities; see R. B. Clark, *Marine Pollution* (Oxford: Clarendon Press, 1986), pp. 31–2.

[26] On the inadequate attention to environmental hazards in the northwest Russian petroleum industry, see R. Vartanov, A. Roginko and V. Kolossov, 'Russian Security Policy 1945–96: The Role of the Arctic, the Environment and the NSR', in W. Østreng (ed.), *National Security and International Environmental Cooperation in the Arctic – The Case of the Northern Sea Route* (Dordrecht: Kluwer Academic Publishers, 1999), pp. 53–102.

[27] For a recent overview of the environmental aspects of increased shipping activity in the Northern Sea Route, see Østreng (ed.), *National Security and International Environmental Cooperation.* See also Brubaker, Chapter 10; and Brigham, Chapter 11 in this book.

[28] W. Østreng, 'International Use of the Northern Sea Route: What is the Problem?', in Østreng (ed.), *National Security and International Environmental Security*, pp. 10–11.

[29] Ten people died and widespread radioactive pollution ensued; see Yablokov *et al., Facts and Problems.*

Norway with forty-two crew members; while radioactive leakage from the wreck is negligible as yet, there is some worry about the speed of the corrosion affecting the reactor section. While not as severe, several other accidents have occurred on Russian nuclear submarines in recent years, demonstrating the environmental hazards associated with the dense nuclear activity in the Barents Sea.[30]

Challenges ahead

A decade of regional cooperative investigations on the state of the Arctic marine environment has substantiated and confirmed prior perceptions. While scientists emphasise that spatial and temporal differentiation must be further clarified and note the limits set by methodological differences between studies in various areas, the Arctic Ocean – including the Barents Sea – is believed to be considerably less polluted than other major seas.[31] Worldwide attention to Soviet dumping of radioactive waste notwithstanding, this goes for nuclear contamination as well.[32]

As we have seen, however, this relatively clean bill of health should not lead regional decision-makers to underestimate the importance of the health of the Barents Sea: the situation is under constant threat from ongoing and future activity in the region, and indeed beyond it. While production levels are low today due to the economic transition, Russia's large-scale process industries on the Kola Peninsula and in Arkhangelsk *oblast* are still responsible for huge discharges of pollutants such as heavy metals, oil, radioactive material and nutrients that are subsequently river-borne into the Barents Sea. As the Russian economy recovers, those discharges are likely to grow further. Also, inadequate safety practices imply that the nuclear installations in the region, both marine and land-based, pose the constant risk of a severe accident involving widespread radioactive contamination. The accumulation of spent nuclear fuel and other types of radioactive waste will only accelerate in the years to come, whereas treatment and storage facilities are badly lacking in northwest Russia. Similarly, the growing offshore petroleum activity in the Barents Sea and the possible increase of commercial shipping in the Barents Sea and along the Northern Sea Route call for sustained attention to the environmental risks associated with those activities and to the range of remedial measures available.

[30] In January 1998, one officer reportedly died and four crew members were hospitalised after a non-nuclear gas leakage during routine operation of the submarine reactor; see *Dagsavisen* (Oslo), 30 January 1998, p. 9.

[31] For a summary of these investigations, see AMAP, *A State of the Arctic Environment Report*; and Working Group on the Protection of the Arctic Marine Environment (PAME), *Report to the Third Ministerial Conference on the Protection of the Arctic Environment, in Inuvik, Canada, 20–21 March 1996* (Oslo: Norwegian Ministry of the Environment, 1996); see also Bernes (ed.), *The Nordic Arctic Environment*.

[32] Joint Russian–Norwegian Expert Group for Investigation of Radioactive Contamination in the Northern Areas, *Dumping of Radioactive Waste and Investigation of Radioactive Contamination in the Kara Sea: Results from 3 Years of Investigations (1992–1994) in the Kara Sea* (Østerås: Norwegian Radiation Control Authority, 1996).

MANAGING THE MARINE ENVIRONMENT: SUB-REGIONAL CONTRIBUTIONS

Linking sub-regional arrangements to broader regional or global processes already underway can be vital for effectiveness purposes. The complementarity required here implies not only that the sub-regional cooperation provides means which are different from or additional to those generated elsewhere; but also that those means would have been difficult to provide through processes other than the sub-regional. Thus, we need to identify more specifically the *types* of linkages that relate the Barents Region to other cooperative processes – and here we shall pay particular attention to three categories: normative, structural and programmatic linkages.

A *normative* linkage involves the body of substantive behavioural principles and rules inherent or promulgated within some international arrangements – and implies that norms developed within one regime affect the normative contents, or even the compelling force, of another.[33] For instance, the precautionary principle, which received global recognition in the 1985 Vienna Convention for the Protection of the Ozone Layer,[34] was subsequently endorsed in the 1992 Rio Declaration[35] and applied in a range of other contexts. Much the same goes for the ecosystem principle, introduced in the 1980 Convention on the Conservation of Antarctic Marine Living Resources,[36] and increasingly reflected in regional international environmental agreements.[37] The appearance of such principles, identically or similarly phrased, in different contexts suggests that a measure of emulation or replication has occurred; and this may in turn serve to strengthen the legitimacy of that principle or rule in international society.[38] In other cases, normative linkages may imply that rules are specified or extended geographically; but they could also juxtapose competing principles or rules.

Another type of linkage involves the *structural* component of regimes – the way they differentiate among current and potential participants as to substantive rights and duties or procedural roles in decision-making.[39] There is a

[33] On the factors which affect the legitimacy, or compelling force, of rules and institutions, see T. M. Franck, *The Power of Legitimacy Among Nations* (New York: Oxford University Press, 1990).

[34] UNTS, Vol. 1,513, pp. 293ff.

[35] Declaration on Environment and Development, A/CONF.151/26 (Vol. I), Rio de Janeiro, 3–14 June 1992, reproduced in ILM, Vol. 31, 1992, pp. 874ff.

[36] P. W. Birnie and A. E. Boyle, *Basic Documents on International Law and the Environment* (Oxford: Clarendon Press, 1995), p. 628; for a comprehensive discussion of the effectiveness of the CCAMLR regime, see O.S. Stokke, 'The Effectiveness of CCAMLR', in O. S. Stokke and D. Vidas (eds.), *Governing the Antarctic: The Effectiveness and Legitimacy of the Antarctic Treaty System* (Cambridge University Press, 1996), pp. 120–51.

[37] P. W. Birnie and A. E. Boyle, *International Law and the Environment* (Oxford: Clarendon Press, 1992), p. 444.

[38] On the validating role of coherence among rules or institutions, see Franck, *The Power of Legitimacy Among Nations*, pp. 134–8.

[39] On the distinction between the normative and structural components of international regimes, see O. S. Stokke and D. Vidas, 'The Effectiveness and Legitimacy of International Regimes', pp. 13–31.

marked tendency, for instance, in multilateral regimes targeting the environment to provide for observer status to non-governmental organisations – which in some cases may influence their approach to solving the problems addressed. While no such provisions are found in the Terms of Reference for the Barents Council, in practice a whole series of environmental and other organisations have been involved at meetings on ministerial or working-group levels.

Programmatic linkages may emanate from both the structural and the normative components of a regime, referring to problem-solving activities within one regime feeding into or otherwise affecting those of another. In environmental and resource management regimes, various types of activities are often generated with a view to clarifying or facilitating the solution of problems addressed by the regime. Regional fisheries regimes, for instance, tend to require that results from marine biological investigations feed into decision-making processes regarding quotas or technical regulations;[40] other programmatic activities can be capacity-enhancing measures such as joint technology development, financial transfers or joint implementation. Examples of supportive programmatic linkages include the role of the International Council for the Exploration of the Sea in laying a basis for scientific recommendations generated within the EU Common Fisheries Policy.[41] In other cases, programmes within different cooperative arrangements may compete for the same scarce funds for research and development.

With those three types of linkages in mind, we can proceed to assess more accurately the contributions of the Barents Region to the protection of the Arctic marine environment.

Normative contributions

In the process leading up to the establishment of BEAR, the environmental ministers of the Nordic countries and Russia invoked several globally established norms when laying down the basic principles for environmental politics in the region,[42] confirmed in the 1993 Kirkenes Declaration:

1. activities within one country's jurisdiction shall not threaten the environment of neighbouring countries;[43]

[40] For an account of the contribution to problem-solving of the Barents Sea fisheries regime, stressing this component, see O. S. Stokke, L. G. Anderson and N. Mirovitskaya, 'The Barents Sea Fisheries', in O. R. Young (ed.), *The Effectiveness of International Environmental Regimes: Causal Connections and Behavioural Mechanisms* (Cambridge, MA: MIT Press, 1999), pp. 91–154.

[41] See R. R. Churchill, 'EC Fisheries and an EEZ – Easy!', *Ocean Development and International Law*, Vol. 23, 1992, pp. 145–64.

[42] Joint Declaration from the Meeting of the Ministers of Environment of the Nordic Countries and the Russian Federation, held in Kirkenes, Norway, 3–4 September 1992, available from the Norwegian Ministry of the Environment, Oslo; on file with author.

[43] This is the second half of Principle 21 of the (Stockholm) Declaration of the UN Conference on the Human Environment, UN doc. A/CONF.48/PC 9, 13 and 17, Stockholm, 5–16 June 1972, reproduced in Birnie and Boyle, *Basic Documents*, pp. 1–8.

2. environmental concerns shall be integrated into all economic activity;[44]
3. the polluter should in principle pay the costs of abatement;[45] and
4. decisions should be taken in conformity with the precautionary principle.[46]

Let us note immediately that the normative linkages involved here are replicative, in that no specification, extension or adaptation is provided. Moreover, the practical relevance of such declarations, in the Barents Region as elsewhere, appears tempered by the pattern of affectedness by pollution, and the willingness to cover abatement costs. As to the polluter pays principle, for instance, the sub-regional programmatic initiatives recorded below regarding modernisation of the Kola smelter-works or Russian nuclear waste management suggest that the opposite principle is equally compelling – i.e., that the victim of pollution should cover considerable parts of the abatement costs. Although exceptions occur, this is quite typical of international environmental arrangements; due to the weak status of environmental liability in international law, cooperation will rarely be achieved without contributions also from the victims of transboundary pollution.[47]

Also regarding more specific commitments pertaining to the marine environment, the cooperative mechanisms of the Barents Sea area tend to echo norms fashioned in other contexts – and usually rather feebly at that. As to *land-based pollution*, specific rules generated at the sub-regional level are either absent, generally phrased or simply repetitions of themes developed elsewhere. At the first meeting of the bilateral Environmental Commission in 1988, the Soviet party informed the Norwegian delegation of an already launched purification programme which would reduce emissions of sulphurous air pollution from one of the Pechenga plants by 49 per cent in 1993 from a 1980 base level.[48] While this statement was seen as encouraging by the Norwegian side, as the Soviet Union had rejected a 50 per cent emission target within the broader LRTAP process under the

[44] Principle 3 of the Rio Declaration on Environment and Development.

[45] The 'polluter pays' principle was clearly expressed in a 1972 OECD Council Recommendation on Guiding Principles Concerning the International Aspects of Environmental Policies; see OECD doc. C(72)128 (1972), reproduced in ILM, Vol. 14, 1975, pp. 236ff. Later on, the principle was endorsed in European Community law and various international agreements; see P. Sands, *Principles of International Environmental Law: Frameworks, Standards and Implementation* (Manchester University Press, 1995), pp. 213–17.

[46] The precautionary principle – that preventive measures should be taken whenever there are threats of serious or irreversible damage, even in the absence of full scientific certainty – is spelt out in Principle 15 in the 1992 Rio Declaration. On the emergence of the precautionary principle, see in general Birnie and Boyle, *International Law and the Environment*, pp. 97ff; and Sands, *Principles of International Environmental Law*, pp. 208–12.

[47] See in general T. Gehring and M. Jachtenfuchs, 'Liability for Transboundary Environmental Damage: Towards a General Liability Regime?', *European Journal of International Law*, Vol. 4, 1993, pp. 92–106.

[48] See 'Protokoll fra første møte i den blandete norsk-sovjetiske kommisjon for samarbeid på miljøvernområdet', Oslo, 23–26 August 1988, p. 5; available from the Norwegian Ministry of the Environment, Oslo; on file with author.

Economic Commission for Europe,[49] it would be an exaggeration to term it a normative commitment; as we shall see, the subsequent lack of real progress regarding modernisation of the metallurgical plants in Pechenga confirms this point.

As to *nuclear* contamination, global linkages to the London [Dumping] Convention and the International Atomic Energy Agency (IAEA) are the most relevant. The Nordic countries had hoped to include in the Kirkenes Declaration a prohibition of the dumping of all types of nuclear waste, thus adding dynamics to the London process,[50] but the Russian delegates admitted that they were not in a position to give such promises on behalf of the Navy.[51] Instead, the parties committed themselves in more general terms to drawing up an action plan to remove nuclear contamination. Similarly, regarding nuclear accidents, the 1988 Soviet–Norwegian Agreement on Early Notification of Nuclear Accidents and Exchange of Information on Nuclear Facilities simply reiterated accident notification commitments already elicited from both parties in the 1986 IAEA Convention.[52] In practice, moreover, those provisions failed to facilitate cooperative measures during the several grave nuclear accidents involving Soviet submarines in the late 1980s. Resentment on this account was particularly strong in connection with the fire onboard and sinking of the naval submarine *Komsomolets* in the Norwegian Sea in 1989.[53] On the other hand, unlike its global counterpart, the 1988 bilateral agreement on early notification also provides for annual exchange of information on the operational condition of nuclear installations in order to facilitate an adequate response in the case of accident.[54] While the Norwegian side was eager to interpret the latter provision as broadly as possible,[55] it applies solely to facilities used for

[49] Under the 1979 Geneva Convention on Long Range Transboundary Air Pollution, the 1985 Helsinki Protocol commits signatories to reduce emissions by 30 per cent. The Soviet Union, however, interpreted this to refer to transboundary fluxes only; see V. Kotov and E. Nikitina, 'Norilsk Nickel: Russia Wrestles with an Old Polluter', *Environment*, Vol. 38, 1996, pp. 6–11 and 32–7.

[50] The Consultative Meeting of the London Convention instituting the general prohibition (to which Russia filed a reservation) was held eleven months after the Kirkenes meeting.

[51] The Norwegian daily *Nordlys* reporting from the meeting, 12 January 1993. If successful, such a declaration would have extended the scope of the prohibition agreed to under the London Convention the same year, as Russia filed a reservation to the latter; see Stokke, Chapter 9 in this book. On the other hand, unlike amendments to the annexes of the London Convention, the Kirkenes Declaration is not a legally binding instrument.

[52] See 'Avtale mellom regjeringen i kongeriket Norge og regjeringen i Unionen av Sovjetiske Sosialistiske Republikker om tidlig varsling av atomulykker og om utveksling av informasjon om atomanlegg', Oslo 15 January 1988, published in *Overenskomster med fremmede makter* (Oslo: Norwegian Ministry of Foreign Affairs, 1988); and the Convention on Early Notification of a Nuclear Accident, Vienna, 26 September 1986, entered into force 27 October 1986, reproduced in Birnie and Boyle, *Basic Documents*, pp. 300–6. In the bilateral agreement, both the scope of application and the information to be provided in case of accident are defined by reference to the IAEA Convention (Arts. I and V, respectively).

[53] For details on the Komsomolets incident, see Yablokov *et al.*, *Facts and Problems*.

[54] Soviet–Norwegian Agreement on Early Notification, Art. VI.

[55] Norwegian Foreign Minister Kjell Magne Bondevik to the Norwegian Parliament, *Storting*, 17 October 1990; see *Stortingforhandlinger*, No. 3, 1990–1 (Oslo: Norwegian Storting), p. 201.

peaceful purposes. Dissatisfied with this – given the circumstance of more than 200 Russian naval reactors located in these waters, many of them quite outdated – Norway has taken a stepwise approach, explicating in a new bilateral agreement on early notification, signed in 1993, that nuclear-powered icebreakers were covered by the information exchange commitments.[56]

Regarding prevention of *oil pollution*, there is a strong potential for triggering a southward linkage to the regime based on the 1974 Paris Convention for the Prevention of Marine Pollution from Land-Based Sources.[57] Its Commission has passed several binding decisions pertaining to offshore petroleum activities, including discharges from platforms and refineries, and the introduction of best available technology requirements.[58] Russia is currently the only state bordering on the northeast Atlantic which is not a signatory to this Convention. Getting Russia to join has been a priority issue for Norway since the first meeting of the bilateral Environmental Commission; so far to no avail. A Russian–Norwegian agreement concerning cooperation on the combating of oil pollution in the Barents Sea was signed in 1994, introducing notification commitments in emergency situations and requiring the two states to elaborate a joint contingency plan.[59] Since Russia is not a party to the 1990 International Convention on Oil Pollution Preparedness, Response and Cooperation,[60] this bilateral arrangement adds substantially to the regulatory commitments pertaining to offshore petroleum activities in the Barents Sea. For its part, *vessel-source pollution* has not been subject to regulation activities within the sub-regional processes dealt with in this chapter, so the commitments developed within the global framework of MARPOL 73/78 are not enhanced, spatially or functionally, at the regional level.[61]

Hence, sub-regional contributions to the generation of principles or rules relevant to the Barents Sea marine environment have been very modest: the norms that have been promulgated typically echo binding commitments already entered

[56] See 'Avtale mellom Norge og Russland om tidlig varsling av atomulykker og om utveksling av informasjon om atomanlegg', Bodø, 10 January 1993, entered into force 6 August 1995, published in *Overenskomster med fremmede makter* (Oslo: Norwegian Ministry of Foreign Affairs, 1995), pp. 762–5, Art. I(2); on the Norwegian desire for such an explication, see Minister of the Environment Thorbjørn Berntsen's statement to the Storting, 19 March 1992, *Stortingsforhandlinger*, No. 37, 1991–2 (Oslo: Norwegian Storting), p. 2,902.

[57] This Convention, like the 1972 Oslo Convention for the Prevention of Marine Pollution by Dumping from Ships and Aircraft, has now been replaced by the 1992 Convention for the Protection of the Marine Environment of the North-East Atlantic (OSPAR), Paris, 22 September 1992, reproduced in ILM, Vol. 32, 1993, pp. 1,069ff. Annex III to OSPAR Convention deals with pollution from offshore sources.

[58] Oslo and Paris Commissions, *Annual Reports: Activities of the Oslo and Paris Commissions September 1992–June 1995* (London: The Secretary, Oslo and Paris Commissions, 1995), p. 52.

[59] Overenskomst mellom Regjeringen i Kongeriket Norge og Regjeringen i Den russiske føderasjon angående samarbeid om bekjempelse av oljeforurensning i Barentshavet, Moscow, 28 April 1994, in force 30 January 1996, *Overenskomster med fremmede makter* (Oslo: Norwegian Ministry of Foreign Affairs, 1996), pp. 94–8, Arts. IV and I respectively.

[60] ILM, Vol. 30, 1991, pp. 733ff.

[61] See Boyle, Chapter 1 in this book, who notes that, in general, the scope for environmental regionalism is smaller for vessel-source pollution than for dumping or land-based pollution – because the former case highlights the freedom of navigation. See also Brubaker, Chapter 10 in this book, in relation to the Northern Sea Route.

into elsewhere. Except in the area of oil pollution, the Barents Region has so far been a feeble creature in terms of environmental standard-setting.

Structural contributions

In structural terms, the two sub-regional processes discussed here differ. BEAR, unlike the Russian–Norwegian Environmental Commission, is multilateral in participation and comprehensive in scope. This makes it far better placed to allow integration of various issue-areas with regard to regional challenges; and it may also provide access to a broader range of Russian decision-makers in sectors contributing to environmental problems. As noted, the bilateral Environmental Commission comprises representatives of the lead environmental agencies in the two states. In many of the gravest environmental problems in the region, including pollution from the Pechenga smelters and nuclear contamination by the Northern Fleet, the State Committee for Environmental Protection is hardly the most influential player on the Russian side. Far more important are other bureaucratic entities, like the Ministry of Economy, the Ministry of Atomic Energy, the Ministry of Defence and the Navy.[62] In general, the domestic clout of the Russian environmental bureaucracy sets a limit for what can be achieved through the bilateral Environmental Commission. A prime motive for launching BEAR was to ensure that international coordination in the region would involve a broader range of governmental institutions than previously: like the EU Council of Ministers, the Barents Council comprises either foreign ministers or sectoral ministers, depending on the matters addressed. The comprehensive scope of the Barents Region cooperation seems to have enabled a subtle political linkage between the realisation of broader economic cooperation in the Barents Region – a point very attractive to Russia – and progress in the environmental area. The clearest evidence of such linkage is found in the 1992 Russian–Nordic Ministerial Declaration on the environment:

> The Ministers recognize that solving the existing major transboundary environmental problems will be central in realizing the potential for broader cooperation in the Barents Region.[63]

The expansion of cooperative participants triggered by the Barents Region initiatives occurs not only horizontally, in terms of the number of specialised agencies involved, but also vertically, through the blend of governmental and county-level representation. This holds true for the bilateral Environmental Commission, with its working group on local problems, but is even more pronounced with BEAR,

[62] For a recent assessment of the relative bureaucratic influence of the environmental agency as compared to other agencies involved in nuclear contamination issues, see Office of Technology Assessment, *Nuclear Wastes in the Arctic: An Analysis of Arctic and Other Regional Impacts from Soviet Nuclear Contamination* (Washington DC: Congress of the United States, 1995).

[63] Joint Declaration from the Meeting of the Ministers of Environment of the Nordic Countries and the Russian Federation, 1992.

which assigns to the Regional Council a salient role in defining the practical priorities of the cooperation – within the fairly broad parameters delineated in the Kirkenes Declaration and subsequent ministerial meetings. The main vehicle here is the Barents Programme, a rolling catalogue of projects of transboundary relevance which are identified, approved and implemented by the individual national constituencies – and ceremonially adopted by the Regional Council.[64] Projects are financed jointly by interested governments, either directly or through international financing instruments such as the Nordic Environmental Finance Corporation. This falls short of expectations that the Regional Council would have financial muscle of its own, to be provided by lump-sum transfers from the respective governments. So far only Norway has been prepared to set aside substantial funds not earmarked for particular projects.[65]

As regards protection of the marine environment, the county-level component of BEAR has contributed primarily by drawing attention to the land-based sources of pollution in the Barents Sea, especially the possibilities of coordinating measures to cope with sewage problems from the larger cities.[66] Given the limited funds available to the Regional Council, however, its Committee of the Environment recommended in 1997 that the focus should be not on the large and costly items on the regional environmental agenda, such as marine pollution and radioactive waste, but rather on measures less directly related to the marine environment, like projects to protect biodiversity and enhance environmental awareness;[67] currently, elaboration of a local Agenda 21 is a topical issue. A similar pattern appears on the bilateral level: while the working group on local environmental collaboration has promoted broader ecosystemic projects, such as registration of agricultural emissions in the Pasvik watercourse in order to develop common environmental standards and management, activity has concentrated primarily on terrestrial matters.

Another notable structural feature of BEAR is the fact that a representative of the indigenous Saami population has a seat on the Regional Council; previous efforts on the part of the Saami to achieve representation in the formal bodies of the Nordic cooperation had been futile. A structural linkage northwards is visible here, with the Barents Region on the receiving end. Already in 1989, the Nordic Saami Council, along with two other associations for indigenous peoples, had been invited to participate as observers in the Rovaniemi process. This was confirmed in subsequent ministerial meetings under the Arctic Environmental

[64] Barents Euro-Arctic Region, The Regional Council, 'Barentsprogrammet 1994/1995', available from Barentssekretariatet, Kirkenes, Norway, pp. 12–13; on file with author.

[65] See Barents Euro-Arctic Region, The Regional Council, 'The Barents Programme 1997–99: The Regional Council', available from Länsstyrelsen i Norrbottens län, Luleå, Sweden; on file with author.

[66] See 'Barentsprogrammet 1994/95'.

[67] 'Regional Council's Action Plan on Environment', presented to the Seventh Meeting of the Environment Task Force, Helsinki, 15–16 April 1997, available from the Department of the Environment, Office of the Finnmark County Governor, Vadsø, Norway; on file with author.

Protection Strategy; and later on also in the initiative to establish an Arctic Council. In BEAR, recognition of the legitimacy of indigenous participation has been brought one step further by the fact that no differentiation is made between the Saami representative and the county-level leadership assembled in the Regional Council; on the other hand, the Saami do not have a seat in the intergovernmental Barents Council. It should be noted here, however, that measures to protect the marine environment are unlikely to top the agenda of representatives of the indigenous peoples.[68]

We have already noted the institutionalisation of a European linkage in the BEAR structure through the participation of a representative of the European Commission in the Barents Council. Indeed, all the frontier regions in Europe straddling the old East–West divide have explicitly aimed at arousing the interest of the European Union – but only the BEAR and the Council of Baltic Sea States have succeeded in eliciting a formal EU response.[69] This southward linkage is probably seen as symbolically attractive in Moscow and in the Nordic capitals alike; but, in practice, EU representation at Barents Council meetings has been at a fairly low level and has not significantly affected regulative or programmatic undertakings.[70] Any real EU involvement in BEAR projects follows from the Finnish and Swedish membership, and thus influence the identification of priorities under Interreg programmes. Those are likely to play some role in funding of measures to abate land-based pollution to the marine environment.[71]

Yet another structural aspect of the BEAR should be noted here – its openness to any interested state.[72] In addition to the seven current members of the Barents Council, nine extra-regional states attend meetings as observers: Canada, France, Germany, Italy, Japan, the Netherlands, Poland, the United Kingdom and the United States. Openness also applies to cooperation under the bilateral Environmental Commission, which, in the case of practical measures to enhance nuclear security in northwest Russia, has deliberately been expanded to involve other participants as well, including the IAEA and NATO. This has specific implications for the programmatic activities generated in the Barents Region.

[68] None of the projects supported by the Environmental Committee under the Regional Council targets indigenous peoples in particular; see 'The Barents Programme 1997–99'; and 'Barentsprogrammet 1994/95', pp. 28–31.

[69] See N. Veggeland, 'The Barents Region as a European Frontier Region', in Stokke and Tunander (eds.), *The Barents Region*, pp. 201–12.

[70] This perception seems widely shared among Norwegian participants in the process; see A. Gjertsen, *Norge og Russland – fragmenterte pragmatikere?* (Cand. polit. thesis at the Department of Political Science, University of Oslo, 1997), pp. 95–6, based on interviews with senior officials.

[71] 'Interreg' is a Community initiative for border regions that grants assistance for cross-border cooperation on schemes such as infrastructure projects, cooperation between public utilities, joint ventures and businesses, and cooperation on environmental protection. In 1996, the EU Commission established two new Interreg programmes, Barents and North-Calotte, based, respectively, in Luleå, Sweden, and Rovaniemi, Finland. By providing fresh funds, Norway became an associate member of both.

[72] Terms of Reference for the Council of the Barents Euro–Arctic Region, Art. 2.

Programmatic contributions

Regarding programmatic efforts to *assess* marine environmental problems in the Barents Sea area and elaborate programmes to enhance the ability to *avoid* them, the bilateral Russian–Norwegian cooperation has been of particular significance; and, as we shall see, the availability of funds for those programmes has been facilitated by the BEAR initiative. Under the bilateral Environmental Commission, various expert and working groups have been appointed, and collaborative programmes which specify objectives, methods, time-frames and the division of labour have been drawn up. Two trends are notable regarding the programmes drawn up so far. First, the cooperation has continually included new areas. Particularly important in terms of marine environmental problems was the 1991 meeting of the Commission, when radioactive pollution was placed on the agenda: the Norwegian delegation tabled reports on alleged dumping of nuclear waste in the Barents Sea and called for dialogue between relevant authorities. Secondly, there is a growing integration of the various issue-areas included in the collaboration, apparent also in the organisational structure. As from 1992, activities previously delegated to separate working groups were combined in a new marine environment group, seeking to relate studies and management of ocean pollution with the activity hitherto organised in the seabird and the anadromous fishes groups; and a working group established in 1991 coordinates the collaboration between regional environmental protection authorities in a large number of subject areas.

As to *land-based* sources of marine pollution, a special Working Group on Airborne Pollution has coordinated the assessment activities under the bilateral Environmental Commission. This formed a focus of attention during the first years of cooperation, modelling the spread and pervasiveness of, e.g. radioactive pollution, and developing an environment monitoring and modelling programme for the border areas. Atmosphere and precipitation in an area of 10,000 km^2 have been examined over time with respect to concentrations of sulphur dioxide and the heavy metals;[73] methods and equipment were identical on either side of the Norwegian–Russian boundary. The purpose has been to combine models for the spread and fallout of atmospheric pollutants over a radius of 300 kilometres from the sources of emission and compare this to available material on *critical loads*, i.e. concentrations not believed to be harmful to the ecosystems in question. A summary report focusing on terrestrial effects concluded that it is especially the episodic nature of inflow which makes airborne pollution in the area so damaging.[74] Capacity-enhancement efforts have concentrated on modernisation of the Pechenga smelter-works. As early as 1978, the governor of Finnmark county had raised the problem of pollution from the nickel industry with Murmansk author-

[73] On this project, see B. Sivertsen, T. Makarova, L. O. Hagen and A. A. Baklanov, *Air Pollution in the Border Areas of Norway and Russia. Summary Report 1990–91* (Oslo: Norsk Institutt for Luftforskning, 1992), p. 4. [74] *Ibid.*, p. 13.

ities, but at that time the issue was not infused with much political energy in Norway. It was not elevated to the inter-governmental level until the establishment of the bilateral Environmental Commission. By then, the Finnish firm Outokumpu had already been negotiating a comprehensive modernisation project with the Soviets for several years, involving both the Pechenga and the Monchegorsk smelters.[75] Hence, when Norway decided to focus politically on air pollution from Kola, it was natural to coordinate these efforts with the Finns; in 1990, those two states committed themselves to subsidising the modernisation, provided the best available technology was employed. The Norwegian government set aside NOK300 million (some US$40 million) for the purpose, and offered to provide NOK2.5 million for the pilot study.[76] In addition, the Nordic financial infrastructure has been activated, including the Nordic Investment Bank and the Nordic Environmental Finance Corporation. Following a preliminary agreement in 1990, Outokumpu, supported by the Norwegian Elkem group, tabled a detailed offer in October 1991 with a price tag of NOK3.5 billion.[77] Nearly a year later, this offer was turned down by the Russians as too expensive, and a Nordic–Russian *ad hoc* working group began work on developing less costly solutions. Having received bids from around the world, the Russian authorities in 1994 settled on a solution promoted by a consortium headed by the Norwegian Kværner Group and the Swedish firm Boliden; but so far lack of willingness to set aside Russian funds for the project has impeded further progress. The Pechenga project has been a symbol of the new, practical spirit of cooperation in the Barents region, and the many setbacks in this area form the most conspicuous disappointment in that cooperation to date.

Programmes relevant to the problem of *radioactive pollution* have generally been far more successful, and both the bilateral Environmental Commission and the BEAR cooperation have made notable contributions – while at the same time linking up to ongoing activities under other institutional umbrellas. Under a bilateral expert group on radioactive contamination established in 1992, Norwegian and Russian researchers have undertaken a series of joint cruises in the Kara and Barents Seas to measure concentrations of radioactivity in water masses and sediments at sites where nuclear reactors have been dumped.[78] During the first cruise, the scientists were denied access to several of those sites, including three bays east of Novaya Zemlya believed to be of particular interest. The prime motive for requiring such access was to measure radioactivity close to the dumped materials and to inspect the condition of the structures which seal it off from the sea. Russian naval authorities were said to have intervened and blocked these investigations. The issue was given high priority in Norway, even to the extent of

[75] R. Castberg, 'Common Problem – Differing Priorities: Nordic–Russian Environmental Collaboration and the Nickel Works on the Kola Peninsula', *International Challenges*, Vol. 13, 1993, pp. 23–33.

[76] The Tromsø daily *Nordlys*, 21 June 1990.

[77] Minister of the Environment Thorbjørn Berntsen to the Storting, 15 January 1992; see *Stortingsforhandlinger*, No. 25, 1991–2 (Oslo: Norwegian Storting), p. 2,436.

[78] Joint Russian–Norwegian Expert Group for Investigation of Radioactive Contamination in the Northern Areas, *Dumping of Radioactive Waste.*

making such access a precondition for carrying out a follow-up cruise in 1993; and, gradually, the joint cruises were allowed to enter the dumping sites. The second cruise was permitted to inspect three out of four identified sites; and Abrosimov Bay, where as many as eight of sixteen reactors had been disposed of, was inspected on a third cruise in 1994. The general conclusion from these studies is that concentrations of radioactivity are as yet very low.[79] The measurement results have been fed into global as well as regional processes for elaborating measures to cope with the problems of dumped radioactive waste.

While of lesser import so far, the BEAR process too has generated funds for assessment activities. The Barents Programme, developed under the auspices of the Regional Council and endorsed by the Barents Council in 1995, included a highly ambitious project designed to establish a permanent monitoring and notification system for the Barents Sea.[80] In terms of cost, this is the largest project supported by Norway under the Barents Programme. In general, assessment of radioactive contamination is an area where the Barents Region is a tributary to broader processes like those under the 1972 London [Dumping] Convention and the 1991 Arctic Environmental Protection Strategy. First, as shown in Chapter 9 of this book, the *global* dumping regime under the London Convention lacks adequate funds to support organised investigations of this magnitude;[81] its main source regarding scientific assessment of this kind is the IAEA, which was also represented on the joint cruises in the Kara and Barents Seas.[82] The organisation of these investigations within a bilateral cooperative framework, one which Russian environmental authorities had over the years come to know and appreciate, was probably helpful here. The gradual and even reluctant granting of access to the most interesting dumping sites suggests that sustained pressure was necessary to overcome objections from agencies, especially the Russian Navy, sceptical to a Western presence in these waters.[83] Secondly, the investigations organised under bilateral Russian–Norwegian cooperation spurred the establishment of an IAEA Arctic Seas and Assessment Programme to assemble available information on sources, levels of concentrations and ecosystemic transport, as well as impact assessment and remedial action.[84] And as nuclear cooperation with Russia expanded beyond the environmental bureaucracy and increasingly required the involvement of Russian military agencies as well, Norway was eager to establish a set of *Atlantic* programmatic linkages in this area. On the assessment side, this

[79] *Ibid.* [80] 'Barentsprogrammet 1994/95', pp. 50–1.

[81] See also O. S. Stokke, 'Beyond Dumping? The Effectiveness of the London Convention', *Yearbook of International Co-operation on Environment and Development 1998/99* (London: Earthscan, 1998), pp. 39–50.

[82] See Joint Russian–Norwegian Expert Group for Investigation of Radioactive Contamination in the Northern Areas, *Dumping of Radioactive Waste.*

[83] On shifting attitudes in Russia to openness about nuclear affairs in the North, see Stokke, Chapter 9 in this book; see also Office of Technology Assessment, *Nuclear Wastes in the Arctic.*

[84] On this IAEA programme, see K.-L. Sjoeblom and G. S. Linsley, 'The International Arctic Seas Assessment Project: Progress Report', *IAEA Bulletin*, Vol. 37, 1995, pp. 25–30.

resulted in a large, cooperative NATO study of defence-related environmental problems, involving experts from twenty-three states, including Russia.[85]

The Barents Region has also triggered broader *capacity-enhancement* projects aimed at enabling Russia to manage without disposing of radioactive waste in the marine environment. The largely positive experience from joint assessment activities has encouraged efforts to cooperate also on specific measures to prevent nuclear contamination, including the development and improvement of storage facilities in the region and in Mayak. The 1995 Memorandum of Understanding on Russian–Norwegian Cooperation on Nuclear Safety provides a bilateral structure for identifying and implementing such projects.[86] Implementation of this instrument has been hampered, however, by disagreement regarding customs exemption for the nuclear safety equipment supplied, as well as the liability situation if that equipment should fail; an agreement reached in May 1998 closed those controversies for a restricted set of projects.[87] One trilateral project involving nuclear agencies from Norway, Russia and the United States is in the process of increasing the capacity of a treatment plant in the Atomflot base of the Murmansk Shipping Company; inadequate treatment facilities for low-level liquid nuclear waste are seen as the main barrier to a Russian repeal of its reservation to the prohibition of dumping of nuclear waste under the London Convention.[88] The Arctic Military Environmental Cooperation – which has, since 1996, involved the defence ministries of the same three states – has gained momentum and promises to be a major mechanism for addressing the issue of military radioactive waste in the region. AMEC was recently linked financially to a United States Cooperative Threat Reduction Program which is allocated considerable funds from the US Department of Defense budget.[89] The rationale for spending US defence money on Russian radioactive waste management is that inadequate storage and treatment facility may impede the process of scrapping Northern Fleet submarines removed from service.[90]

Regarding *offshore petroleum activity*, a working group for oil pollution

[85] North Atlantic Treaty Organisation, *Cross-Border Environmental Problems Emanating from Defence-Related Installations and Activities: Volume 1, Radioactive Contamination (Final Report)* (Brussels: North Atlantic Treaty Organisation, 1995).

[86] 'Memorandum om norsk-russisk samarbeid på atomsikkerhetsområdet', Oslo, 4 October 1995, in force the same day, in *Overenskomster med fremmede makter* (Oslo: Norwegian Ministry of Foreign Affairs, 1995), pp. 784–6.

[87] See Agreement Between the Government of the Kingdom of Norway and the Government of the Russian Federation on Environmental Cooperation in Connection with the Dismantling of Russian Nuclear Powered Submarines Withdrawn from the Navy's Service in the Northern Region, and the Enhancement of Nuclear and Radiation Safety, Moscow, 26 May 1998 especially Arts. 2, 5 and 9; in force the same day, available from the Norwegian Ministry of Foreign Affairs; on file with author.

[88] See also Stokke, Chapter 9 in this book.

[89] See 'Program Plan and Report on Proposed Obligations for the Arctic Military Environmental Cooperation Program FY 1999', available from the US Department of Defense, Washington DC; on file with author.

[90] For an analysis of AMEC, see S. G. Sawhill, 'Cleaning-Up the Arctic's Cold War Legacy: Nuclear Waste and Arctic Military Environmental Cooperation', *Cooperation and Conflict*, Vol. 35, 2000, pp. 5–36.

prevention under the bilateral Environmental Commission provided the main forum for working out the 1994 agreement on oil pollution mentioned above. That agreement aims at facilitating joint exercises in early warning and emergency preparedness, the exchange of information, the development of technology and other practical measures.[91] A set of Arctic linkages appears to have influenced development in this area. First, a working group on emergency preparedness and response under the Arctic Environmental Protection Strategy, focusing on the development of joint contingency planning, added momentum to the bilateral process. This notwithstanding, the parties had great difficulties in hammering out provisions, especially regarding responsibility for triggering a cooperative response, as this was perceived as interfering with the long-standing dispute over the delimitation line in the Barents Sea. This was a major reason why the Oil Pollution Agreement took more than five years to negotiate,[92] and a second Arctic linkage provided the means required to overcome the difficulties: a similar boundary dispute had not stopped the United States from elaborating an oil spills contingency plan with the Soviet Union concerning the Bering and Chukchi Seas in 1989. The productively unclear definition in the latter agreement of 'area of responsibility' for each party as 'the internal waters or territorial sea, and the sea areas beyond the territorial sea in which that party exercises its sovereign rights and jurisdiction in accordance with international law', reinforced with a provision allowing both parties to veto implementation of the plan, was replicated verbatim in the Barents Sea context.[93]

Thus, whereas normative contributions have tended to echo commitments already established elsewhere, on the programmatic side the sub-regional instruments have generated a wide range of fresh activities designed to assess the state of the marine environment and to develop practical measures to reduce risks associated with military and industrial activity. Especially with regard to nuclear waste, those activities have been coordinated with efforts involving participants beyond the Barents Region.

CONCLUSIONS: EFFECTIVENESS OF SUB-REGIONAL COOPERATION IN THE BARENTS SEA

International cooperation at the sub-regional level – the bilateral Russian–Norwegian Environmental Commission and the multilateral Barents

[91] 'Overenskomst mellom Regjeringen i Kongeriket Norge og Regjeringen i Den russiske føderasjon angående samarbeid om bekjempelse av oljeforurensning i Barentshavet' (1994), Art. XII.

[92] A draft agreement was negotiated as early as 1991, when problems associated with the boundary dispute blocked further progress for several years; see D. Brubaker, *Marine Pollution and International Law: Principles and Practice* (London: Belhaven Press, 1993), p. 193.

[93] See Agreement Between the Government of the United States of America and the Government of the Union of Soviet Socialist Republics Concerning Cooperation in Combatting Pollution in the Bering and Chukchi Seas in Emergency Situations, available from the US Department of State, Washington DC; on file with author. See also 'Overenskomst mellom Regjeringen i Kongeriket Norge og Regjeringen i Den russiske føderasjon angående samarbeid om bekjempelse av oljeforurensning i Barentshavet' (1994), Arts. II and IX, respectively.

Region – has made notable contributions towards solving marine pollution problems in the Barents Sea. Efforts at the sub-regional level have differed from global processes by their clearer *programmatic* profile: relatively more resources, in terms of both expertise and funding, have been invested in order to enhance the knowledge-base for management decisions in the region, as well as the administrative and technical capacity to avoid behaviour liable to threaten the marine environment. Many of the programmatic activities encouraged at other levels, as for instance under the global dumping regime, have been planned, financed and organised at the sub-regional level. Comparatively less attention, however, has been given to establishing new *regulative* norms for environmental protection from either industrial or military activity in the region – a task which tends to be left to broader fora.

Whereas the bilateral Environmental Commission is a fairly standard bilateral environmental arrangement, certain *structural* features set BEAR apart from many other international arrangements addressing environmental problems on a regional scale. The BEAR Regional Council ensures that both county-level decision-makers and representatives of the indigenous population are involved. This has probably greater import in areas other than protection of the marine environment, however, as the latter type of problems emanate from activities of considerable national concern, such as military operations and offshore petroleum engagement, which tend to inhibit the leeway for sub-governmental action. Another feature of the BEAR structure is its comprehensive scope, which may enable productive linkages of issues where priorities differ among participant states. In the BEAR context, the general balance between the environmental and the economic component is a case in point. Moreover, the inclusiveness of the Barents Council, as demonstrated by the participation of the European Commission and non-regional states such as the USA, is meant to provide linkages to potential partners in development found beyond the Barents Sea area. This is highly relevant regarding marine pollution, where abatement costs can be high and may involve technologies not available in the region itself. While some of these institutional means have appeared elsewhere as well, it is the BEAR machinery which provides them in this particular context.

In capsule form, the sub-regional level has served to relate environmental protection to broader foreign policy issues and has strengthened environmental networks across the Nordic–Russian divide. In turn, this has generated the financial resources and expertise necessary for assessing environmental problems in the region and enhancing the capacity to cope with them. This is clearly a supportive linkage. At the general level, the main reason for the higher fund-raising capacity of sub-regional processes is that geographic proximity ensures denser networks of interdependence.

In the Barents Sea context, this affects the incentives of regional actors, and in particular those at the receiving end of transborder pollution, in at least two important ways. First, from a purely environmental point of view, geographic proximity renders highly visible the fact that the Nordic neighbours have a clear self-interest in financing environmental projects in Russia, especially those addressing industrial pollution from the border areas and those designed to prevent dumping of radioactive waste. This is all the more so as trouble-ridden Russia cannot be assumed to give to environmental problems in the Barents Sea region the same high priority as its wealthy Western neighbours do, especially Norway. Secondly, from a more general political point of view, geographic proximity ensures that environmental projects may serve broader purposes associated with national security. Security concerns were already emphasised with the establishment and early operation of the bilateral Environmental Commission, in that both Norway and the Soviet Union stressed the need to forge close cooperative ties in the north. This became even clearer with the Norwegian initiative to create the Barents Euro–Arctic Region, which from the very beginning was presented as a broader construct to promote stability in a region traditionally marked by tension and military rivalry. Thus, the willingness on the part of Norway and other Nordic states to flex their financial muscles for problem-solving purposes in the Barents Sea area is closely related to the sub-regional nature of the cooperation – which allows linkage to overarching goals such as national security and the integration of Russia into the larger cooperative structure of Europe.

7 Domestic perspectives and regulations in protecting the polar marine environment: Australia, Canada and the United States

DONALD R. ROTHWELL AND CHRISTOPHER C. JOYNER*

The protection of the marine polar environment has increasingly become a matter of concern for Antarctic Treaty Consultative Parties and Arctic states alike, as evident in the developments in the legal regimes which apply to the polar regions. In the case of the Antarctic Treaty System, attention has focused on the protection of the marine environment of the Southern Ocean through a combination of measures adopted at Antarctic Treaty Consultative Meetings and also through the 1991 Protocol on Environmental Protection to the Antarctic Treaty.[1] In the Arctic, concerns over the protection of the marine environment have been driven by the present and possible environmental consequences of land-based marine pollution, nuclear waste and the potentials for increased oil and gas exploitation as well as navigation through Arctic waters, especially by the Northwest Passage and the Northern Sea Route.[2] Particular emphasis has been placed on Arctic marine environmental protection in the process of development and implementation of the Arctic Environmental Protection Strategy (AEPS),[3] since 1991, and, since 1996, within the Arctic Council.[4]

This chapter addresses these issues by considering how prominent polar states have influenced developments on the international scene while also seeking to implement through domestic policy and law a range of international responses aimed at protecting the polar marine environment. The aim is thus to demonstrate the importance of the domestic level, in both initiative-giving and in implementing commitments agreed through international cooperative fora or processes.

Three states have been selected for this comparative assessment: Australia, Canada and the United States. In selecting these countries, we were led

* The authors acknowledge the assistance of William Bush and David VanderZwaag in the preparation of this chapter; any errors and omissions, however, remain the responsibility of the authors.

[1] ILM, Vol. 30, 1991, pp. 1,455ff. For an overview of Antarctic regional environmental cooperation, see Vidas, Chapter 4 in this book; and specifically on the Environmental Protocol, see Joyner, Chapter 5 in this book.

[2] See, respectively, the discussions by VanderZwaag, Chapter 8; Stokke, Chapter 9; Brubaker, Chapter 10; and Brigham, Chapter 11 in this book.

[3] ILM, Vol. 30, 1991, pp. 1,624ff.

[4] For an overview, see Vidas, Chapter 4 in this book.

by several criteria: leadership or importance in recent initiatives for the protection of the polar marine environment, elaboration of the relevant domestic implementing legislation, and geographic placement. We have thus included one 'unipolar' case each: Australia, with its interests and legislation linked primarily to the Antarctic and, correspondingly, Canada for the Arctic. There is also the 'bi-polar' case of the United States, with interests present in, and legislation adopted for, both polar regions. All three have substantial territorial and maritime claims in either of the polar regions, and have been leaders in the development of Arctic and Antarctic law and policy. Australia, the chief initiator of the Antarctic Environmental Protocol, asserts the largest territorial claim in Antarctica.[5] Canada, the original advocate of the Arctic Council, has one of the longest-standing and largest Arctic claims, with Canadian territory extending beyond the Arctic Circle as far as 83° North. The United States, current chair country of the Arctic Council and simultaneously an original Antarctic Treaty Consultative Party, has multiple polar interests. These interests apply appropriately, yet differently, to the Arctic region in the near north of the continental USA, but also in Antarctica, where the USA has long-standing foreign policy interests and has been present in the region ever since the expeditions of the nineteenth century.

CANADA AND THE ARCTIC MARINE ENVIRONMENT

Canada's initiatives to protect the Arctic Ocean are an interesting mix of unilateral action (as in the case of the response to the voyage of the *SS Manhattan*), bilateral and regional initiatives with neighbouring states and other Arctic nations, and global initiatives like the campaign to have Article 234 adopted as part of the LOS Convention. This trend, initiated in the 1970s, has continued throughout the 1980s and 1990s, and demonstrates Canada's commitment to the protection of the Arctic marine environment.

The Manhattan incident

The catalyst for the eventual development of the first specific provision for the polar marine environment in the international law of the sea was the 1969 voyage of the *Manhattan* through the Northwest Passage. The incident arose when the *Manhattan*, carrying a small cargo of oil, was intentionally sent through the Northwest Passage by its US owners to demonstrate that an icebreaking bulk carrier was capable of year-round sailings between Alaska and the east coast of the United States. The voyage was only the eleventh complete transit of the Northwest Passage, and the first since the end of World War II by a non-government vessel.[6]

[5] While this claim is not conclusive, some authors argue that it is probably stronger in international law than those which any other state may assert: see the discussion in G. D. Triggs, *International Law and Australian Sovereignty in Antarctica* (Sydney: Legal Books, 1986).

[6] T. C. Pullen, 'What Price Canadian Sovereignty?', *Proceedings of the US Naval Institute*, Vol. 113, 1987, pp. 69–71.

Although the voyage was innocent enough, it had great implications in Canada. Even though a representative of the Canadian government was on board the tanker during the passage, and the Canadian Coast Guard vessel *J. A. Macdonald* accompanied the *Manhattan*,[7] the voyage created considerable public controversy in Canada. However, any Canadian response to the voyage was hampered by the fact that at that time Canada had claimed only a three-mile territorial sea around the islands of the Canadian Arctic Archipelago: as a consequence, apart from where Canadian territorial waters overlapped in the narrow McClure Strait, the *Manhattan* was passing through high seas during its navigation of the Passage.[8]

Canada responded to the voyage of the *Manhattan* in 1970. First, the Arctic Waters Pollution Prevention Act was adopted, extending Canadian jurisdiction to proclaimed 'Arctic waters' 100 miles out into the Beaufort Sea and Arctic Ocean along the coastlines of the Yukon and Northwest Territories, including the islands of the Arctic Archipelago.[9] Under this Act, pollution control regulations – including standards for vessel construction, navigation and operation – were imposed on all ships passing through these Canadian waters. Failure to comply would result in passage by such vessels being prohibited. A Shipping Safety Control Zones Order was also issued under the Act which established sixteen zones in the 100 nautical mile offshore area within which navigational restrictions applied, including earliest and latest navigation dates for each particular type or class of ship.[10]

Secondly, Canada extended its territorial sea from three to twelve nautical miles,[11] including the waters around the Canadian Arctic mainland and the islands. This extension of the territorial sea resulted in a great deal of the Northwest Passage becoming enclosed within Canadian territorial sea, so that any vessels in transit would come more frequently under Canadian jurisdiction.

Thirdly, Canada varied its acceptance of the compulsory jurisdiction of the International Court of Justice in regard to matters dealing with Canadian jurisdiction in the Arctic.[12] The effect was to ensure that no challenge could be brought before the Court as to the validity in international law of the Arctic Waters Pollution Prevention Act without Canada accepting the Court's jurisdiction in the matter. In justifying these initiatives, Canada relied upon the growing concern for

[7] *Ibid.*, p. 71.

[8] D. Pharand, *The Law of the Sea of the Arctic with Special Reference to Canada* (Ottawa: University of Ottawa Press, 1973), p. 57.

[9] Section 3(2) of the Arctic Waters Pollution Prevention Act, *Statutes of Canada*, Vol. I, Chapter 47, 1970.

[10] R. S. Reid, 'The Canadian Claim to Sovereignty over the Waters of the Arctic', *Canadian Yearbook of International Law*, Vol. 12, 1974, pp. 117–29; and D. A. VanderZwaag, *Canada and Marine Environmental Protection: Charting a Legal Course Towards Sustainable Development* (London: Kluwer Law International, 1995), pp. 340–1.

[11] Act to Amend the Territorial Sea and Fishing Zones Act, *Statutes of Canada*, Vol. II, Chapter 68, 1970.

[12] Canadian Declaration Concerning the Compulsory Jurisdiction of the International Court of Justice, ILM, Vol. 9, 1970, pp. 598ff.

Arctic environmental protection. It argued that as international law had yet to develop sufficient measures to protect the area from the dangers of pollution, it was appropriate for Canada to take unilateral action.[13] The USA did not welcome these initiatives, noting that: 'International law provides no basis for these proposed unilateral extensions of jurisdiction on the high seas, and the United States can neither accept nor acquiesce in the assertion of such jurisdiction.'[14]

The *Manhattan* incident was an important milestone in polar state initiatives for the protection of the marine environment. Regardless of its underlying motives for enacting the Arctic Waters Pollution Prevention Act, Canada justified its action to the international community with reference to protection of the fragile Arctic marine environment from pollution impacts. This was the first occasion that any polar state had strongly asserted a claim to exercise a sovereign right over polar waters on environmental or conservation grounds (to be contrasted with initiatives to assert jurisdiction on resource management or resource conservation grounds). Canada's action was sufficiently influential to eventually have an impact upon the negotiations at the Third UN Conference on the Law of the Sea.

Responses to the Law of the Sea Convention

Article 234 of the LOS Convention provides multilateral recognition of the special features of the Arctic Ocean and the interests of the adjacent coastal states in protecting the marine environment. However, while Article 234 represents a significant advance in recognising the need for marine environmental protection in the Arctic, its limitations should also be recognised.[15] It does provide coastal states with the ability to implement unilaterally laws and regulations for the 'prevention, reduction and control of marine pollution from vessels in ice-covered areas within the limits of the exclusive economic zone'. It cannot, however, be read as a provision conferring upon states the ability to implement extensive marine pollution provisions for all polar waters, as it applies only within the EEZ and to ice-covered areas within the zone. Any provisions adopted under Article 234 must also be 'non-discriminatory' and have due regard for navigation.

For Canada, Article 234 represented international acceptance of the action it had taken in enacting the Arctic Waters Pollution Prevention Act and a basis upon which to expand legislative and policy initiatives to protect the Canadian Arctic. As noted by leading Canadian commentators on the Arctic, such as VanderZwaag:

[13] P. E. Trudeau, 'Canadian Prime Minister's Remarks on the Proposed Legislation', ILM, Vol. 9, 1970, pp. 600–4.

[14] J. A. Beesley and C. B. Bourne (eds.), 'Canadian Practice in International Law During 1970 as Reflected Mainly in Public Correspondence and Statements of the Department of External Affairs', *Canadian Yearbook of International Law*, Vol. 9, 1971, p. 288.

[15] C. Lamson and D. VanderZwaag, 'Arctic Waters: Needs and Options for Canadian–American Cooperation', *Ocean Development and International Law*, Vol. 18, 1987, p. 81; W. E. Westermeyer, 'Jurisdiction and Management of Arctic Marine Transportation', *Arctic*, Vol. 39, 1986, pp. 346–7.

Arguably, the provision grants Arctic coastal states the right to unilaterally regulate vessel design, construction, equipment and manning. These powers exceed coastal state control in the territorial sea and the general economic zone, and Article 234 was drafted with this in mind.[16]

The Polar Sea incident

In 1985, Canada became concerned once again over navigation through the Northwest Passage. This followed an announcement that the US icebreaker, *Polar Sea*, intended to sail through the Northwest Passage. Canada responded by announcing a review of Canadian Arctic policy; and on 10 September 1985, the Minister for External Affairs, Joe Clark, made a comprehensive statement on Canadian Arctic sovereignty to Parliament.[17] The statement included six major policy initiatives:

1. the establishment of straight baselines around the Canadian Arctic Archipelago, effective from 1 January 1986;
2. the adoption of new legislation to enforce Canadian civil and criminal laws in the offshore areas enclosed by the straight baselines;
3. talks with the United States on cooperation in Arctic waters on the basis of full respect for Canadian sovereignty;
4. increased aircraft surveillance and naval activity in the eastern Arctic;
5. the withdrawal of Canada's reservation to the International Court of Justice;
6. the construction of a Polar Class 8 icebreaker to operate in the enclosed waters and a review of the other means available through which effective control could be exercised over Canadian Arctic waters.[18]

The statement removed some doubts about Canada's intentions in the Arctic and clarified Canada's legal position over the region. By proclaiming straight baselines around its Arctic Archipelago, all the waters that fell within the baselines were 'internal waters' of Canada over which it now claimed complete sovereignty. The measures which accompanied the proclamation of the baselines were also designed to ensure that the Canadian action was not hollow but would be supported by positive evidence of Canadian sovereignty over the waters.[19]

The importance of the Canadian baseline declaration should not be underestimated. The declaration may be controversial, but there has been no formal legal challenge made against it. Canada's capacity to legislate over the waters on the landward side of the baselines appears to be unchallenged, subject,

[16] D. VanderZwaag, 'Canada and Marine Environmental Protection: The Changing Tides of Law and Policy', in D. McRae and G. Munro (eds.), *Canadian Oceans Policy* (Vancouver: University of British Columbia Press, 1989), p. 103. See also the discussion by Brubaker, Chapter 10 in this book.

[17] House of Commons Debates (Canada), 10 September 1985, Vol. 5, p. 6,463.

[18] *Ibid.*, p. 6,464.

[19] For an assessment of the Canadian action, see T. L. McDorman, 'In the Wake of the Polar Sea: Canadian Jurisdiction and the Northwest Passage', *Marine Policy*, Vol. 10, 1986, pp. 243–57.

however, to the contentious right of international navigation through the Northwest Passage.[20] This has substantial implications for Canada's capacity to enact laws and adopt polices for the protection of its Arctic waters and the marine environment which fall within the baselines.

Canada's response to the AEPS

While the AEPS places emphasis on the development of enhanced cooperative measures amongst the Arctic states on matters of common environmental concern, ultimately many of the commitments made under the AEPS relate to actions of individual states. Canada has perhaps been the best placed to meet its domestic obligations under the Strategy. In 1991 the government announced US$100 million in funding to support Canadian Arctic environmental research and clean-up operations,[21] and also released an 'Arctic Environmental Strategy' for the Canadian Arctic.[22] There has been continuing debate within Canada over the need for enhanced marine environmental protection in the Canadian Arctic;[23] however, notwithstanding some of the developments discussed below, a comprehensive response is still pending.

Relevant Canadian legislation

Canada has adopted a range of legislative and policy initiatives for the protection of its Arctic waters. The most notable remains the Arctic Waters Pollution Prevention Act, which is the principal Canadian legislative regime for the area. Among other initiatives is the Arctic Marine Conservation Strategy, developed in the late 1980s. It identified the following key principles:

1. Canada will exercise its sovereign rights and responsibilities in Arctic maritime areas;
2. Canada will conserve and protect Arctic marine waters and renewable resources for the benefit and enjoyment of present and future generations;
3. essential ecological components, processes and systems, and genetic diversity will be maintained in the Arctic marine environment;

[20] See the discussion in D. Pharand, *The Northwest Passage Arctic Straits* (Dordrecht: Martinus Nijhoff, 1984); R. R. Roth, 'Sovereignty and Jurisdiction over Arctic Waters', *Alberta Law Review*, Vol. 28, 1990, pp. 845–72; and D. R. Rothwell, 'The Canadian–US Northwest Passage Dispute: A Reassessment', *Cornell International Law Journal*, Vol. 26, 1993, pp. 331–72.

[21] R. Howard, 'Arctic Protection Program Unveiled', *Globe and Mail* (Toronto), 30 April 1991, p. A4; and R. F. Keith, 'Canada's Arctic Environmental Strategy: Critique and Prospect', *Northern Review*, Vol. 8/9, 1992, pp. 83–113.

[22] Department of Indian and Northern Affairs (Canada), *The Arctic Environmental Strategy* (Hull: Department of Indian and Northern Affairs, 1991).

[23] L. Beckman, 'Marine Conservation in the Canadian Arctic', *Northern Perspective*, Vol. 22, 1994, pp. 33–9.

4. conservation requires an ecosystem approach and integrated management of renewable and non-renewable resources; and
5. all users of Arctic marine resources will be recognised.[24]

Moreover, as already mentioned, in 1991 Canada developed an Arctic Environmental Strategy as part of the development of Canada's *Green Plan*. This Strategy has five main objectives:

1. ensure the health and well-being of northern ecosystems;
2. protect and enhance environmental quality and sustainable use of resources, including their use by indigenous peoples;
3. ensure that indigenous peoples' perspectives, values and practices are accommodated in the planning, development, conservation and protection of the north;
4. improve decision-making by integrating local, regional, national and international interests as part of new legal, constitutional and cooperative arrangements; and
5. develop international agreements to use, conserve and manage resources and protect the circumpolar environment.[25]

The main components of this strategy dealt with environment–economy integration, waste, water and contaminants. While none was specifically directed towards the marine environment, several did deal directly with developing response strategies to marine pollutants. This has especially been the case with the programme dealing with contaminants, which has focused on developing a more advanced understanding of the way in which contaminants enter the Canadian Arctic environment via a number of sources.[26] This Arctic Environmental Strategy has enhanced Canada's capacity to give effect to its commitments under the AEPS; however, it does not seem to have resulted in any substantive changes to the Canadian legislative regime dealing with the Arctic marine environment.[27]

Canada also adopted the Oceans Act in 1996, which became operative from 31 January 1997.[28] The Preamble to this Act specifically notes that the Arctic Ocean is a part of the common heritage of all Canadians, and that Canada wishes to promote the precautionary approach to the 'conservation, management and exploitation of marine resources in order to protect these resources and preserve the marine environment'. The Act, in its section 30, provides for the development of an 'Oceans Management Strategy' to be based upon the following principles:

[24] L. R. Kriwoken and R. P. Côté, 'Developments in Australian and Canadian Environmental Management', in L. R. Kriwoken, M. Haward, D. VanderZwaag and B. Davis (eds.), *Oceans Law and Policy in the Post-UNCED Era: Australian and Canadian Perspectives* (London: Kluwer Law International, 1996), p. 231.

[25] Department of Indian and Northern Affairs (Canada), *The Arctic Environmental Strategy: Five Years of Progress* (Hull: Department of Indian and Northern Affairs, 1996), p. 6.

[26] *Ibid.*, pp. 28–33.

[27] See the discussion in D. VanderZwaag, R. Huebert and O. Hurtzman, 'The Arctic Marine Environment: Not a Pristine Pole Apart', in Kriwoken *et al.* (eds.), *Oceans Law and Policy*, pp. 351–88.

[28] *Statutes of Canada*, Chapter 31, 1996.

1. sustainable development, i.e., development that meets the needs of the present without compromising the ability of future generations to meet their own needs;
2. the integrated management of activities in estuaries, coastal waters and marine waters that form part of Canada or in which Canada has sovereign rights under international law; and
3. the precautionary approach, i.e. erring on the side of caution.

The Act seeks to provide a framework for the overall management of Canada's oceans; with the exception of the provisions dealing with marine protected areas (i.e., sections 35–36), it does not specifically provide for environmental management mechanisms.

THE UNITED STATES AND THE ARCTIC MARINE ENVIRONMENT

As defined in US legislation, the Arctic encompasses 'all United States and foreign territory north of the Arctic Circle and all United States territory north and west of the boundary formed by the Porcupine, Yukon, and Kuskokwim Rivers; all contiguous seas, including the Arctic Ocean and the Beaufort, Bering, and Chukchi Seas; and the Aleutian chain'.[29] As recently as 1970, US attention to the Arctic was close to non-existent, although this is not to imply that activities important for the United States were not occurring in the polar north.[30] Rather, it suggests that, relative to other parts of the country and the world, US policy interests toward the Arctic received low priority. Unlike Russia and Canada, which possess extensive Arctic territory, the US Arctic remains isolated from most Americans. Even so, over the past three decades, there has been a significant re-evaluation of US national priorities in the Arctic.

US Arctic policy

Current US policy toward the Arctic was first articulated in National Security Decision Memorandum 144 of 1971.[31] Although this policy statement was not comprehensive, it did furnish the basic framework within which US Arctic policy, and national legislation implementing it, could be developed. National Security Decision Memorandum 144 asserts that:

> the President has decided that the United States will support the sound and rational development of the Arctic, guided by the principle of minimizing any adverse effects to the environment; will promote mutually beneficial

[29] Arctic Research and Policy Act of 1984, Public Law 98-373, Title I, section III, of 31 July 1984, 98 Stat. 1248 (codified in USCA, Vol. 15, section 4111).

[30] See, for instance, above in this chapter on the *Manhattan* incident.

[31] Done on 22 December 1971. See the discussion in B. D. Smith, 'United States Arctic Policy', *Ocean Policy Studies 1:1* (Charlottsville, VA: Center for Oceans Law and Policy, University of Virginia, 1978), pp. 38–40.

international cooperation in the Arctic; and will at the same time provide for the protection of essential security interests in the Arctic, including preservation of the principle of freedom of the seas and superjacent airspace.

This Memorandum also established the Interagency Arctic Policy Group (IAPG), which has become a key body for overseeing implementation of US Arctic policy and for reviewing and coordinating US activities in the Arctic.

The management framework for US ocean policy in the Arctic is established by national legislation. This has meant a functional, piecemeal approach, rather than that of unified, comprehensive management. There is no US 'Grand Plan' for the Arctic, no master scheme for a management authority to coordinate US ocean law or policy there. This is attributable to the fact that, as US territory, Alaska falls under federal jurisdiction; thus, all laws pertaining to the United States perforce pertain to the Alaskan Arctic. While beyond the scope of this analysis, the state laws of Alaska also generate important impacts upon the activities of US nationals in the Arctic region, which is not surprising given the *ad hoc* pattern of regulatory development in the United States.

US Arctic interests

The United States security interests in Arctic marine areas, and national legislation protecting those interests, have taken four principal themes: military security, scientific security, economic security and environmental security. Since the end of the Cold War, military concerns have waned, while the priority of other interests has tended to escalate.

During the Cold War, the United States perceived serious threats from Soviet maritime activities in the Arctic. The deployment into Arctic waters of Soviet ballistic missile submarines capable of firing nuclear missiles at US targets was a grave concern. Three-quarters of the most advanced Soviet submarines were based on the Kola Peninsula and operated in Soviet Arctic waters. The Soviet military build-up at the eastern end of the Northern Sea Route also gave the northern front new prominence.[32] Finally, the threat of bombers and land-based intercontinental ballistic missile attacks over the Arctic Circle remained a constant strategic concern, creating the need for early warning systems on both sides of the Arctic Ocean.[33] The disintegration of the Soviet Union and the end of the Cold War diminished these threats.

[32] On the Northern Sea Route (NSR) in the context of vessel-source pollution, see Brubaker, Chapter 10 in this book. For a general review of the NSR, see D. Brubaker and W. Østreng, 'The Military Impact on Regime Formation for the Northern Sea Route', in D. Vidas and W. Østreng (eds.), *Order for the Oceans at the Turn of the Century* (The Hague: Kluwer Law International, 1999), pp. 261–90; and W. Østreng (ed.), *National Security and International Environmental Cooperation in the Arctic – The Case of the Northern Sea Route* (Dordrecht: Kluwer Academic Publishers, 1999).

[33] See G. L. Johnson, D. Bradley and S. Winokur, 'United States Security Interests in the Arctic', in W. E. Westermeyer and K. E. Shusterich, *United States Arctic Interests: The 1980s and 1990s* (New York: Springer Verlag, 1984), pp. 268–94.

As to the case of the USA and Canada, security-related disagreements in the Arctic have strained the relationship since 1970. First, while the United States and Canada share a common boundary in the Arctic – the 141st meridian – their maritime boundary in the Beaufort Sea remains unresolved.[34] The two governments have 'agreed to disagree' on the issue, which remains non-problematic so long as the overlapping ocean area appears to have little strategic value. The hydrocarbon potential of the region, however, is thought to be high. That could in the future raise the economic stakes, and complicate the resource claims, for both governments. Secondly, there is ambiguity concerning jurisdiction over offshore areas in the Arctic. This is potentially most problematic, as it relates to transit rights through the Northwest Passage. As already described in this chapter, the complexities were demonstrated during the voyages of the *Manhattan* and the *Polar Sea*. US–Canadian relations over the Northwest Passage have also been strained by the assertion of the right of submarines to pass through it submerged. While Canada maintains that these Arctic waters are subject to Canadian jurisdiction, the United States regards the Northwest Passage as an international strait, subject to international rights and regulations.

Access to and control of living and non-living marine resources in the Arctic remains a US security concern of considerable importance. Since the 1950s, the US government has sought to protect these interests in its coastal waters, including those in the polar north. As technology for exploiting underwater reserves of crude oil and natural gas developed after World War II, interest arose in producing hydrocarbons from the Outer Continental Shelf (OCS), that being the federal portion of the continental shelf which extends outward beyond the three nautical mile line in most cases. While submerged lands within three miles of the coast belonged to the states, the Outer Continental Shelf Lands Act 1953 (OCSLA) established federal jurisdiction over submerged lands on the outer continental shelf seaward of Alaska's state boundary.[35] The OCSLA provided for orderly leasing of these lands, while ensuring protection of the environment and that the federal government received fair market value for the land and for mineral production. The outer continental shelf is the source of 15 per cent of US crude oil production and 25 per cent of natural gas output.[36]

Under the OCSLA, the US Secretary of the Interior is responsible for the administration of mineral exploration and development of Alaska's outer continental shelf. The Act empowers the Secretary to grant leases to the highest qualified

[34] See generally K. L. Lawson, 'Delimiting Continental Shelf Boundaries in the Arctic: The United States–Canadian Beaufort Sea Boundary', *Virginia Journal of International Law*, Vol. 22, 1981, p. 221; and D. R. Rothwell, *Maritime Boundaries and Resource Development: Options for the Beaufort Sea* (Calgary: Canadian Institute of Resources Law, 1988).

[35] Done 7 August 1953, 67 Stat. 462, as amended, USCA, Vol. 43, 1988, section 1331ff.

[36] L. C. Kumins, '95115: Outer Continental Shelf Leasing for Oil and Gas Development' (Washington, DC: Congressional Research Service, Library of Congress, CRS Issue Brief updated, 1 November 1996), p. 1.

responsible bidder(s) and to formulate such regulations as necessary to carry out provisions of the Act. In general, the Act provides guidelines for implementing the OCS oil and gas exploration and development programme. The basic goals of the OCSLA in the offshore US Arctic region are threefold:

1. to establish policies and procedures for managing the oil and natural gas resources of Alaska's OCS that are intended to come from development of the OCS;
2. to preserve, protect and develop oil and natural gas resources of Alaska's OCS in a manner that is consistent with the need: (a) to make such resources available to meet the nation's energy requirements as rapidly as possible; (b) to balance orderly resources development on the continental shelf with protection of the human, marine and coastal environments offshore Alaska; (c) to ensure the public a fair and equitable return on the resources of the Alaskan OCS; and (d) to preserve and maintain free enterprise competition; and
3. to encourage the development of new and improved technology for energy resources production, which will eliminate or minimise risk of damage to Alaska's human, marine and coastal environments.[37]

The US Minerals Management Service (MMS), which collects royalties for petroleum and natural gas production, is responsible for administering mineral leasing of submerged OCS lands and for supervising offshore operations after leases are issued. Regulations administered by the MMS govern the leasing of oil, gas and sulphur mineral deposits on the OCS.[38] The Secretary of the Interior is responsible for monitoring the human, marine and coastal environments of any area or region in order to obtain data for determining whether any significant impacts are being made on the quality of productivity of the environment.

Opposition by environmental groups to drilling on the outer continental shelf since 1970 has seriously deterred leasing prospects offshore Alaska. This is notwithstanding the enactment of the Outer Continental Shelf Deep Water Royalty Relief Act 1995,[39] which provides for a 'royalty holiday' (i.e., rate reduction) for deep-water drilling operations in order to encourage hydrocarbon prospectors to go further offshore on the OCS. In response to such opposition to drilling in Alaska's offshore areas, Congress approves OCS moratoria annually and bans expenditure of appropriated funds for any leasing activity on environmentally sensitive areas of the OCS. While a moratorium is in place preventing leasing activities along most coastal areas of the United States, five Alaskan planning areas have been included in the five-year leasing plan for 1997–2002 adopted by the MMS.

[37] See USCA, Vol. 43, 1988, section 1332.

[38] See CFR, Vol. 30, section 256. Regulations for the conduct of mineral operations are contained in CFR, Vol. 30, sections 250 and 251.

[39] Public Law 104-58, Title III, section 301, of 28 November 1995, 109 Stat. 563, USCA, Vol. 43, sections 1301 and 1337.

US Arctic conservation and protection laws

Protection of the marine environment through conservation and protection laws has assumed increasing importance over the past decade for the US Arctic. US legislation affecting the Arctic ocean environment has focused on managing coastal resources, preserving areas offshore the Alaska wildlife refuge, and sustaining living marine resources in the Arctic Ocean. A prominent example is the Coastal Zone Management Act (CZMA), enacted by Congress in 1972 to check increasing pressures of over-development of coastal resources and land-use conflicts in US coastal areas.[40] The CZMA encourages US states (including Alaska) to preserve, protect and where possible restore valuable natural coastal resources such as wetlands, floodplains, estuaries, barrier islands and coral reefs, as well as fish and wildlife using those habitats. An interesting feature of the CZMA is that participation by states is voluntary, and that Alaska is included among the participating states.

The CZMA sets an important precedent as it establishes the role of state and local governments in developing coastal planning and management programmes. It also encourages state governments to participate by providing federal assistance to any US coastal state willing to develop and implement a comprehensive programme of coastal management. In addition to resource protection, the CZMA specifies that coastal states, including Alaska, may manage development offshore. The CZMA requires that Alaska's programme management anticipate impacts from energy development facilities and that they plan for such impacts.[41] Among the facilities affecting US Arctic waters are petroleum refineries; gasification facilities, used for transport, treatment, conversion, transfer or storage of liquefied natural gas; oil and gas facilities, including platforms, assembly plants, storage depots and refining facilities; and transfer facilities, deepwater ports, pipelines and related terminals – all of which are active in coastal areas offshore Alaska.

Throughout the 1980s, the US federal government considered the need for greater scientific research in the Arctic. Debate within the Congress over the merits of a special Arctic science policy drew attention to the growing importance of the Arctic for US interests. Policy-makers came to realise that the Arctic contains vital resources, both onshore and offshore, that can reduce US dependence upon imported foreign oil. They also came to appreciate that the Arctic is critical to US national defence and that the renewable resources of the Arctic – inclusive of fisheries – constitute one of the country's greatest commercial assets. Consequently, a comprehensive national policy to organise hitherto-neglected research on the region was deemed necessary in order to fulfil the objectives of national resource, strategic, environmental and foreign policy.

[40] Coastal Zone Management Act of 1972, Public Law No. 92-583, *Statutes*, Vol. 86, section 1280, of 1972 (codified as amended at USCA, Vol. 16, sections 1451–1465).

[41] *Ibid.*, section 306(d)(2)(H).

Thus, in 1984 Congress enacted the Arctic Research and Policy Act (ARPA) which provides for a comprehensive national policy dealing with US research needs and objectives in the Arctic.[42] The ARPA establishes an interagency Arctic Research Policy Committee and an Arctic Research Commission to implement the Act. The Committee helps to set federal Arctic research priorities and works with the Commission to develop an integrated Arctic research policy to guide federal agencies in implementing their research programmes in the Arctic. The Committee thus develops a five-year plan to implement the national policy, and updates the plan biennially. Marine science remains a principal focus of US Arctic research activities, so the Committee includes representatives from federal agencies with ocean interests, among them the National Science Foundation, the Department of Commerce (especially the National Oceanic and Atmospheric Administration), the Department of Defense, the Department of Energy, the Department of the Interior, the Department of State, the Department of Transportation and the Environmental Protection Agency.[43]

Canada and the United States in protecting the Arctic marine environment: final remarks

Both Canadian and US responses in protecting the Arctic marine environment are driven by individual national interests in having extensive Arctic territorial and maritime claims. While these interests have not always been identical, and at times have even been in conflict, in recent years there has been greater complementarity in approaches. This may partly be a reflection of the impact the AEPS and the Arctic Council have had upon Arctic environmental protection; on the other hand, the impact of major Arctic maritime incidents such as the *Exxon Valdez* in Alaska should not be ignored.

For the United States, moreover, the state of Alaska remains the dominant political, economic and legal concern in the Arctic. The situation here is quite different from that affecting US policy in the Antarctic: the Arctic is largely a matter of the US *domestic* rather than foreign policy.

AUSTRALIA AND THE ANTARCTIC MARINE ENVIRONMENT

To assess Australia's initiatives in protecting the Antarctic marine environment it is necessary to survey Australia's Antarctic maritime claims and the laws enacted to apply within those areas. In some instances, these claims and the

[42] Arctic Research and Policy Act of 1984, Public Law 98-373, Title I, *Statutes*, Vol. 98, section 1242, of 1984 (codified as amended at USCA, Vol. 15, sections 4101–4111).

[43] See US Senate, *Arctic Oceans Research: Hearing Before the National Ocean Policy Study of the Committee on Commerce, Science, and Transportation, 102 Cong., 1st Sess., April 24, 1991* (Washington, DC: US Government Printing Office, 1991).

application of these laws raise issues as to their consistency with the Antarctic Treaty.[44]

Australia's Antarctic maritime claims

In November 1990 Australia extended its territorial sea from three to twelve nautical miles, applicable also to the waters adjacent to the Australian Antarctic Territory (AAT).[45] Despite this enlargement of Australia's AAT territorial sea claim, there is no evidence that Australia has sought to exercise any more extensive jurisdiction over activities within that area than it did previously. This apparent reluctance by Australia to assert jurisdiction more vigorously within its AAT territorial sea claim may be partly explained by the restriction that Article IV of the Treaty places upon the assertion of new claims,[46] and by the limitations which Article VIII places on the exercise of jurisdiction. In addition comes the great difficulty in enforcing laws of any type in Antarctica.[47] As a result, it is in Australia's best interests not to adopt an overly assertive approach towards sovereignty and jurisdiction within the territorial sea.

With respect to the continental shelf in the Southern Ocean, while Australia has now adopted the LOS Convention's definition, it has yet formally to proclaim the limits of the new area which will also include the sub-Antarctic waters offshore Heard and McDonald Islands and Macquarie Island. Australia has not to date had occasion to apply any of its laws to activities taking place within its Southern Ocean continental shelf adjacent to the AAT.[48] However, in 1991, in a response to concern over mining activities occurring in Antarctica, the Antarctic Mining Prohibition Act was adopted to prohibit mining in the AAT. The Act extended to the continental shelf of the AAT (section 3), and applied not only to Australian nationals but also to nationals of other contracting parties. This Act has since been replaced by the amended Antarctic Treaty (Environment Protection) Act 1980 upon the entry into force of the 1991 Environmental Protocol, which prohibits mining in Antarctica.[49]

Antarctic exclusive economic zones or fisheries zones have been claimed only by Argentina, Australia and Chile. However, varying practices have been

[44] For a general review of Australian law in Antarctica, see D. R. Rothwell and R. Davis, *Antarctic Environmental Protection* (Annandale: Federation Press, 1997), pp. 147–295.

[45] See the discussion in J. Brown (ed.), 'Australian Practice in International Law 1990 and 1991', *Australian Yearbook of International Law*, Vol. 13, 1992, p. 277; and B. R. Opeskin and D. R. Rothwell, 'Australia's Territorial Sea: International and Federal Implications of its Extension to 12 Miles', *Ocean Development and International Law*, Vol. 22, 1991, pp. 395–431.

[46] For a discussion of how this may apply to maritime claims, see J. Crawford and D. R. Rothwell, 'Legal Issues Confronting Australia's Antarctica', *Australian Yearbook of International Law*, Vol. 13, 1992, pp. 82–3.

[47] See the discussion in S. Blay and J. Green, 'The Practicalities of Domestic Legislation to Prohibit Mining Activity in Antarctica: A Comment on the Australian Perspective', *Polar Record*, Vol. 30, 1994, p. 29.

[48] Crawford and Rothwell, 'Legal Issues Confronting Australia's Antarctica', p. 81.

[49] See the discussion in Blay and Green, 'The Practicalities of Domestic Legislation'.

adopted by the claimant states towards their Antarctic claims. The status of these claims and the enforcement of fisheries laws in Southern Ocean EEZs has become the subject of extensive debate in the second half of the 1990s due to illegal fishing activities for Patagonian toothfish (*Dissostichus eleginoides*).[50] In addition, the Australian approach towards asserting a fisheries claim and EEZ shows a particular sensitivity to the Antarctic Treaty's limitations upon the assertion of new sovereignty claims and the exercise of coastal state jurisdiction. This also reflects upon Australia's capacity to implement marine environmental protection laws and regulations in parts of the Southern Ocean.

In September 1979 Australia claimed a 200 nautical mile 'Australian Fishing Zone' (AFZ) offshore the mainland and external territories.[51] However, a little over one month later, a new proclamation was made which excepted the waters around the AAT so they were no longer part of the AFZ.[52] The effect of this action was to exempt foreign vessels from the reach of Australian law so that the waters of the AAT beyond the territorial sea remained open to foreign fishing. Australian nationals and vessels were, however, still caught by the reach of Australian law, because even though the waters were not part of the AFZ they were still 'proclaimed waters' for the purposes of the Fisheries Act 1952.[53] This legislative regime was eventually replaced by the Fisheries Management Act 1991; however, it adopted a similar exemption, so that the waters of the AAT were not considered part of the AFZ.[54] Australian nationals and vessels were nevertheless still bound by the provisions of the Act within waters offshore the AAT.

Australia's declaration in 1994 of an EEZ offshore the AAT was consistent with a policy designed to ensure that Australia had claimed the range of maritime zones allowed under the new law of the sea; it also put in place a regime which would eventually allow it to ratify the LOS Convention. However, various difficulties attend the EEZ claim. First, the Australian EEZ proclamation provides that in the case of Australia's external territories (which include the AAT) the outer limits of the EEZ comprise 'the lines that are 200 international nautical miles seaward of the baselines established under international law'.[55] However, as there are currently no proclaimed baselines around the coastline of the AAT, it is impossible to determine accurately the outer limits of the EEZ.[56] An official chart showing Australia's maritime claims adjacent to the AAT has been published; however, this

[50] See the discussion in S. Bateman and D. R. Rothwell (eds.), *Southern Ocean Fishing: Policy Challenges for Australia* (Wollongong: Centre for Maritime Policy, 1998).

[51] See *Commonwealth of Australia Gazette*, 26 September 1979; and W. M. Bush, *Antarctica and International Law: A Collection of Inter-State and National Documents*, Vol. II (London: Oceana Publications, 1982), pp. 202–3.

[52] See *Commonwealth of Australia Gazette*, 31 October 1979; and Bush, *Antarctica and International Law*, Vol. II, p. 208.

[53] Bush, *Antarctica and International Law*, Vol. II, pp. 205 and 209.

[54] Fisheries Management Act 1991 (Commonwealth of Australia), sections 7, 8 and 11.

[55] *Commonwealth of Australia Gazette*, No. S290, 29 July 1994.

[56] For a review of how baselines could be proclaimed around the Antarctic coastline, see C. C. Joyner, *Antarctica and the Law of the Sea* (Dordrecht: Martinus Nijhoff, 1992), pp. 81–7; and S. Kaye, *Australia's Maritime Boundaries* (Wollongong: Centre for Maritime Policy, 1995), pp. 191–211.

chart does not purport to indicate the extent of the EEZ claim. Secondly, the declaration of an EEZ seems to run counter to Article IV of the Antarctic Treaty. This follows because, unlike both the territorial sea and continental shelf, the EEZ was not recognised in international law prior to 1961 and was not therefore an inherent sovereign right of a coastal state at the time the Antarctic Treaty entered into force.[57] If then Article IV(2) of the Antarctic Treaty applies to maritime claims, the conclusion seems inescapable that the declaration of an Antarctic EEZ is either an enlargement of an existing claim or an assertion of a new claim and thereby infringes Article IV(2).[58] However, while Australia has now asserted an EEZ offshore the AAT, what legal content has it given to such a claim? The outer limits of the claim have not yet been fixed, due to the uncertainty of the baselines. The legal regime which applies in the new EEZ is not dissimilar to that prevailing previously as the AFZ regime still remains in place. This action has been taken to enable the Fisheries Management Act 1991, the principal Australian legislation dealing with fisheries, to remain in place without the need for substantial amendments. The consequence of retaining the AFZ for the AAT is that the exception which previously applied has remained in place. Australia has therefore declared an EEZ offshore the AAT, but has in place only a fisheries management regime that applies to Australian nationals and vessels. Foreign nationals and vessels are exempt.[59]

One change in Australian law offshore the AAT is that the Whale Protection Act 1980 now applies in the area.[60] This has important implications for any whaling activities offshore the AAT, as the Act applies to foreign persons, vessels and aircraft (section 6(2)(b)). However, the Act is 'subject to the obligations of Australia under international law, including obligations under any agreement between Australia and another country or countries' (section 6(3)). Irrespective then of whatever limitations may be imposed upon Australia by the terms of the Antarctic Treaty in regard to the assertion of jurisdiction within the AAT, as a party to the International Convention for the Regulation of Whaling, Australia is constrained from adopting legislation that may be contrary to the Convention's accepted rights and duties.

While Australia has yet actively to enforce the Whale Protection Act offshore Antarctica, in 1997 and 1998 Australia arrested vessels operating in EEZ waters off Heard and McDonald Islands. On both occasions, the arrests related to illegal fishing for Patagonian toothfish. This is the first time Australia has sought to enforce its fishery laws offshore its Antarctic territories; however, it should be noted that, while the Heard and McDonald Islands are truly sub-Antarctic and fall within the area of operation of the 1980 Convention on the Conservation of Antarctic

57 D. P. O'Connell, *The International Law of the Sea*, Vol. 1 (Oxford: Clarendon Press, 1982), pp. 553–8; and B. Kwiatkowska, *The 200 Mile Exclusive Economic Zone and the New Law of the Sea* (Dordrecht: Martinus Nijhoff, 1989), pp. 7–9.

58 Crawford and Rothwell, 'Legal Issues Confronting Australia's Antarctica', p. 81.

59 I. Shearer, 'Australia's New Maritime Zones', *Australian Law Journal*, Vol. 69, 1995, p. 29.

60 Maritime Legislation Amendment Act 1994 (Commonwealth of Australia); see the discussion in Shearer, 'Australia's New Maritime Zones'.

Marine Living Resources (CCAMLR),[61] Australian sovereignty over these islands is uncontested, so the enforcement of Australian law in these waters is not questionable.[62]

Australia and the 1991 Environmental Protocol to the Antarctic Treaty

The 1991 Environmental Protocol was a result of a campaign commenced by Australia in 1989 to reject CRAMRA in favour of a comprehensive environmental protection regime for Antarctica. This campaign, which also had the strong support of France, developed from long-standing concerns by environmentalists in Australia over the effects of mining on Antarctica and domestic political factors which influenced the Australian Government at that time to support a pro-environment, anti-mining campaign. The effect of this campaign for Antarctic environmental protection has been considerable and, if not for the initial leadership shown by Australia and France, the Antarctic Treaty Parties through the sheer force of the momentum generated by the CRAMRA negotiations may have ratified the minerals regime. Instead, the Treaty Parties did an about-turn, and rejected the possibility of mining in Antarctica while at the same time creating a comprehensive environmental protection regime.

Australia ratified the Environmental Protocol in 1994. At that time the Antarctic Treaty (Environmental Protection) Legislation Amendment Act was enacted to give effect to Australia's international obligations under the Protocol, and these amendments became operative with the entry into force of the Protocol in 1998.[63] Included in these new provisions is the power to declare Specially Protected Areas or Specially Managed Areas within the maritime areas of the AAT. New categories of environmental offences are also created to reflect the new regime. Mining is now also prohibited on any part of the continental shelf of the AAT or the continental shelf of Australia's sub-Antarctic islands, including Heard and McDonald Islands, which fall north of the Antarctic Treaty area but are within the CCAMLR area of application.[64] New legislation has also been introduced to give effect to the provisions of Annex IV of the Environmental Protocol dealing with marine pollution. Under the Protection of the Sea (Prevention of Pollution from Ships) Act 1983 amendments have been introduced which make it an offence under Australian law to discharge oil, sewage or garbage from a ship within the Antarctic Treaty area. The prohibition, however, is not absolute: some exceptions do exist depending on necessity or the rate of the discharge.

[61] ILM, Vol. 19, 1980, pp. 841ff.

[62] See Art. IV of CCAMLR; see also Final Act, Conference on the Conservation of Antarctic Marine Living Resources, Canberra, 7–20 May 1980, paras. 1–5; text reproduced in J. A. Heap (ed.), *Handbook of the Antarctic Treaty System*, 8th edn (Washington, DC: United States Department of State, 1994), pp. 175ff.

[63] See the discussion in Rothwell and Davis, *Antarctic Environmental Protection*, p. 161.

[64] Antarctic Treaty (Environment Protection) Legislation Amendment Act 1992 (Commonwealth of Australia), section 18.

THE UNITED STATES AND THE ANTARCTIC MARINE ENVIRONMENT

The United States has made no territorial claim to Antarctica, although it reserves the right to do so. While the Antarctic lies some 8,000 miles distant of the continental USA, the area has been of prominent scientific interest to the USA since the 1930s, and of serious geopolitical concern from the late 1940s. Considerations of national interests in circumpolar Antarctic seas have since figured in the calculus of US foreign policy, although the polar south became less salient in the 1990s because of the successful legal regime and the redirection of government interest in the Antarctic toward policies aimed at resource conservation and environmental protection, rather then political rivalry and commercial exploitation in the region.[65]

US Antarctic security interests

Although not a territorial claimant in Antarctica, the United States has long-standing security interests in Antarctic seas. These have been protected mainly by provisions of the Antarctic Treaty, by which the United States is pledged to non-militarisation, non-nuclearisation and peaceful uses of the Treaty area (Articles I, V and VII), which includes the circumpolar ocean area out to 60° South. In fact, as treaty commitments, these obligations are rendered part of the 'supreme law of the land', as codified by Article III, section 2 of the US Constitution.

Concern over US economic interests in the polar south emerged in the 1980s as international attention focused on the issue of minerals development in and around Antarctica, resulting in the adoption of the 1988 Convention on the Regulation of Antarctic Mineral Resource Activities (CRAMRA).[66] Negotiation of CRAMRA aroused considerable controversy, both internationally and domestically. The environmental community had serious concerns that CRAMRA would actually promote minerals development, not discourage it, and that minerals development would produce unacceptable ecological impacts.[67] Within the United States, environmentalists pressured Congress to pass legislation that would prohibit US citizens from participating in any minerals resource development activities in the Antarctic. Congress eventually responded with the tersely worded Antarctic Protection and Conservation Act of 1990.[68] This legislation aimed to

[65] For a general assessment of US interests and policy in the polar south, see C. C. Joyner and E. Theis, *Eagle Over the Ice: The US in the Antarctic* (Hanover, NH: University Press of New England, 1997).

[66] ILM, Vol. 27, 1988, pp. 868ff.

[67] See C. C. Joyner, 'CRAMRA: The Ugly Duckling of the Antarctic Treaty System?', in A. Jorgensen-Dahl and W. Østreng (eds.), *The Antarctic Treaty System in World Politics* (London: Macmillan, 1991), pp. 161–85.

[68] Antarctic Protection Act of 1990, Public Law No. 101-594, section 2, *Statutes*, Vol. 104, sections 2975–2976, of 1990 (codified at USCA, Vol. 16, section 2461). This Act was subsequently repealed in part by the Antarctic Science, Tourism, and Conservation Act of 1996, Public Law 104-227, section 202(b), *Statutes*, Vol. 110, sections 3034 and 3044, of 1996 (repealing Antarctic Protection Act of 1990, sections 5 and 7, *Statutes*, Vol. 104, sections 2977–2978).

strengthen overall environmental protection of Antarctica by prohibiting the prospecting, exploration or development of Antarctic mineral resources by any US citizen or other persons under US jurisdiction. As the operative clause of the Act asserts: 'It is unlawful for any person to engage in, finance, or otherwise knowingly provide assistance to any Antarctic mineral resource activity.'[69] Authority to enforce the Antarctic Protection Act was allocated to the Secretary of Commerce, as any prohibited activities would be deemed violations of the Antarctic Marine Living Resources Act 1984.

Thus, US security interests in the Antarctic marine ecosystem have become mainly environmentally oriented and conservation-based. The twin themes of environmental protection and resource conservation have overtaken prospects for the economic development of Antarctic resources, however. Albeit somewhat reluctantly, US foreign policy objectives and domestic law have shifted to follow suit. The principal interests in the polar south as codified in US legislation have been specifically enacted to implement US legal commitments to international agreements and measures adopted to augment the Antarctic Treaty.

Environmental protection

The demise of CRAMRA redirected the Antarctic Treaty parties away from minerals development and towards more serious efforts at environmental protection. Consistent with this development, the most recent US legislation affecting Antarctica is environmentally the most comprehensive. On 2 October 1996 the Antarctic Science, Tourism, and Conservation Act was signed into law.[70] This Act implements and integrates into US law the specific provisions of the Antarctic Environmental Protocol, mainly by amending in substantial part the Antarctic Conservation Act 1978 and by repealing the Antarctic Protection Act 1990.

The Antarctic Science, Tourism, and Conservation Act 1996 makes it unlawful for any person subject to US jurisdiction to introduce prohibited products (under Annex III) onto the land or ice shelves, or waters in Antarctica; to dispose of or openly burn waste in Antarctica; to damage or remove a historic site or monument, or to refuse permission to an authorised US officer to inspect a vessel in connection with enforcement of the Act. Certain other actions are prohibited unless authorised by permit. Among these are the disposal of wastes in Antarctica, disposing wastes from land into the sea, incinerating wastes on land or ice shelves.[71]

US federal agencies operating in Antarctica are also bound by the National Environmental Policy Act 1969,[72] which requires environmental impact assessments to be undertaken for development activities. This essentially means that the same legal standards applied to assess the environmental impacts of US federal activities within United States territory will be applied to US federal agency

[69] *Ibid.* [70] Public Law 104-227, 110 Stat. 3034, USCA, Vol. 16, section 2401.
[71] See CFR, Vol. 45, section 103. [72] USCA, Vol. 42, sections 4321–4347.

activities in Antarctica. The key threshold of having 'more than a minor or transitory impact' for environmental impact assessment in Annex I is retained in the US legislation.

The 1996 Antarctic Science, Tourism, and Conservation Act also strives to deter US ships from polluting Antarctic seas. Provisions in Annex IV to the Protocol are activated for US law by amending the Act to Prevent Pollution from Ships,[73] which implements the 1978 Protocol of the 1973 International Convention for the Prevention of Pollution from Ships (MARPOL 73/78).[74] In effect, the Antarctic Science, Tourism, and Conservation Act of 1996, which implements the Antarctic Environmental Protocol into US domestic law, also specifically links the legal obligations for the United States under MARPOL 73/78 to the obligations set out in Annex IV of the Antarctic Environmental Protocol, including enforcement actions and penalties for violations.[75] As a result, US vessels are not allowed 'any discharge of oil or oily mixture', save in circumstances permitted under MARPOL 73/78. US vessels are also forbidden to discharge 'any noxious liquid substance, and any other chemicals or other substances, in quantities or concentrations that are harmful to the marine environment'.[76] US ships are further prohibited from disposing of plastics and garbage within the Antarctic Treaty area.[77] Finally, US vessels must have 'sufficient capacity' on board for the retention of garbage while within the Antarctic Treaty area and have 'adequate facilities' provided for the reception of all sludge, dirty ballast, tank washing water, oily residues, and garbage from those ships.[78]

US scientific interests and environmental protection

US activities and government-sponsored research in Antarctica are managed under a single integrated programme, the United States Antarctic Research Program (USAP). The National Science Foundation (NSF) funds and manages this programme to support the range of US interests and the government's adherence to the Antarctic Treaty. Overall responsibility for these activities was transferred to the NSF in 1970, as formalised in National Security Council Memorandum 71.[79] In 1976 the NSF was assigned government-wide management of the entire Antarctic programme.[80] In 1982 President Reagan reaffirmed in Presidential Memorandum 6646 the prior national policy underlying the USAP and directed that the United States maintain 'an active and influential presence in

[73] Public Law No. 96-478, *Statutes*, Vol. 94, section 2297, of 1980 (codified at USCA, Vol. 33, sections 1901–1907).

[74] Protocol of 1978 relating to the International Convention for the Prevention of Pollution from Ships, 1973; entered into force 2 October 1983.

[75] See Public Law 104-227, Title, II, section 201(e) and (f).

[76] Annex IV to the Protocol on Environmental Protection to the Antarctic Treaty: Prevention of Marine Pollution, Arts. 3 and 4. [77] *Ibid.*, Art. 5. [78] *Ibid.*, Art. 9.

[79] National Security Decision Memorandum 71, of 10 July 1971.

[80] See National Security Decision Memorandum 318, of 26 February 1976.

Antarctica, which shall include the conduct of scientific activities in major disciplines [and] year-round occupation at the South Pole and two coastal stations'.[81] More recently, in June 1994, Presidential Decision Directive NSC 26 asserted that US policy towards the Antarctic has four fundamental objectives, namely:

1. to protect the relatively unspoiled environment of Antarctica and its associated ecosystems;
2. to preserve and pursue unique opportunities for scientific research to understand Antarctica and the global physical and environmental system;
3. to maintain Antarctica as an area of international cooperation reserved exclusively for peaceful purposes; and
4. to assure the conservation and sustainable management of the living resources in the oceans surrounding Antarctica.[82]

Each year the USAP sends 2,500 scientists and support personnel to the Antarctic region to conduct research in a variety of disciplines. The United States operates three principal facilities to support research activities on the continent: McMurdo Station, the main US facility, on Ross Island on the coast of Antarctica; Amundsen-Scott Station, at the geographic South Pole; and Palmer Station, on Anvers Island due west of the Antarctic Peninsula. In addition, two US ice-strengthened research vessels operate in the Southern Ocean: the icebreaker *Nathaniel B. Palmer* and the *R/V Laurence M. Gould*.

The principal legislation now governing US scientific activities is the Antarctic Science, Tourism, and Conservation Act 1996. As persons under the jurisdiction of the United States, scientists – like tourists and other visitors – are legally obligated to conform to the requirements of environmental protection and conservation as they perform their research and related activities, whether in station or in the field.

Australia and the United States in protecting the Antarctic marine environment: final remarks

Protection of the Antarctic marine environment for both Australia and the United States has in recent years been driven by the entry into force of the 1991 Environmental Protocol. Both countries have adopted wide-ranging initiatives to give effect to the provisions of the Protocol in their domestic legislation; however, neither has adopted a comprehensive legislative response to the Protocol. As key parties to the Antarctic Treaty, Australia and the USA have also been mindful of the limitations which the Treaty imposes upon their exercise of jurisdiction, although Australia has from time to time taken a more assertive approach because of its territorial interests.

[81] Presidential Memorandum 6646, of 5 February 1982. This remains US policy today.

[82] Presidential Decision Directive NSC 26, 'United States Policy on the Arctic and Antarctic Regions', of 9 June 1994.

As to the United States, the principal legislation relevant to the Antarctic is federal law designed to implement multilateral agreements to which the USA has become party. There is no commingling of state and federal powers; though the only polar state of the USA, Alaska's jurisdictional reach stops at the edge of its own state boundaries. The US law that reaches into the polar south is that which sets out federal jurisdiction over activities of US nationals visiting there, mainly as scientific researchers, support staff or tourists.

POLAR STATES AND MARINE ENVIRONMENTAL PROTECTION: CONCLUSIONS

This chapter has reviewed the laws and policies of three polar states – one with bi-polar interests and the other two with specific interest in each of the polar regions. The survey does not claim to be complete. In Antarctica, there are six other claimant states which could have been reviewed, as well as another nineteen Consultative Parties to the Antarctic Treaty (and twenty parties to the Environmental Protocol) which have a direct interest in the Southern Ocean. Likewise, in the north, there are six other Arctic states plus numerous other states which have interests in the region. Nevertheless, the chapter provides an appreciation of the national legislation and perspectives, and the constraints which operate upon polar states seeking to protect the polar marine environment.

The first of these is the common element of the international legal regime which provides the framework within which polar states operate. In Antarctica, the Antarctic Treaty is a constraining factor for both claimant and non-claimant states in how they either assert sovereignty or enforce jurisdiction. On the other hand, the Environmental Protocol now provides a basis for implementing legal obligations for marine environmental protection. In the Arctic, the law of the sea has provided a basis for various responses in addition to other international legal regimes such as those dealing with marine pollution. Another factor is the assertion of sovereign rights, recently expressed in concern for marine environmental protection. Canada set the precedent for this with its response to the voyage of the *Manhattan*; however, both Australia in the Antarctic and, though perhaps to a lesser extent, the USA in the Arctic have also taken various initiatives which clearly demonstrate their concerns as sovereign states for the marine environment.

The US case reveals differences between its Arctic and Antarctic perspectives and regulations. In the Arctic, the national link is clearly fixed by having the US state of Alaska as a vested federal interest. Consequently US Arctic activities are more intimately domestic and legally apparent. Moreover, there are viable national security and economic interests of the USA in the Arctic. The presence of Russian ballistic missiles and submarine activity there remain national security concerns for the United States, though less so today than during the Cold War era. There are also US concerns over the rights of indigenous peoples in the Arctic; the US Arctic is permanently populated with more than 610,000 citizens. In the

Antarctic, there is no native population, only visitors; and some 1,500 scientists and supporting logistical staff are stationed temporarily. Rather then foster economic development and exploitation of natural resources, as regularly occurs in the US Arctic, conservation and environmental protection appear as more salient US Antarctic concerns. Science, not commercial development, is the main activity in the Antarctic. The national security threat to the USA from foreign military activities in and around the Antarctic is slight.

The role of individual states in polar marine environmental protection should not be underestimated. Even setting aside the role these states play at the political level in having regional initiatives adopted to protect the Arctic and Antarctic, much international law ultimately depends upon state implementation. This chapter has shown how Australia, Canada and the USA have sought to implement their international obligations incurred under both global legal regimes (such as the law of the sea) and regional regimes (such as the regime based on the Antarctic Treaty). Moreover, they have responded with a range of domestic policies which supplement and enhance their legal responses. This is not to suggest that these responses have been perfect. This is far from the case, as demonstrated by the limitations upon the exercise of jurisdiction, from both a legal and a practical perspective. However, by the exercise of their state sovereignty and jurisdiction, these polar states have managed to make a practical impact in polar marine environmental protection in ways which international and regional legal regimes alone would have been unable to achieve.

II Current trends and issues in protecting the polar marine environment

8 Land-based marine pollution and the Arctic: polarities between principles and practice

DAVID VANDERZWAAG*

The concept of sustainable development calls for the application of many principles – including public participation, polluter pays, intergenerational equity, community-based management, indigenous rights and environmental impact assessment.[1] Fundamental in combating land-based pollution is the *precautionary principle*, also known as the precautionary 'approach'.[2] It urges a shift away from the traditional belief in the assimilative capacity of the oceans to absorb wastes and faith in end-of-pipe standards to achieve acceptable environmental quality standards. The precautionary principle is torn between competing philosophies towards nature and natural resources. In its strictest form, it calls for quite extreme law and policy reforms that emphasise pollution prevention and the need to develop clean technologies and products.

Extreme control measures may include the establishment of zero discharge (or virtual elimination) standards for synthetic chemicals, a 'reverse listing' approach to chemicals management where only 'safe' chemicals are listed for use, and a shift in the burden of proof to those proposing development activities to show some standard of safety.[3] Best available technology without regard to costs is a further direct measure often advocated.[4]

* The author wishes to acknowledge the research assistance of Justin Mellor, third-year law student at Dalhousie Law School, and research funding from the Social Sciences and Humanities Research Council of Canada.

1 See J. M. Van Dyke, 'The Rio Principles, and Our Responsibilities of Ocean Stewardship', *Ocean and Coastal Management*, Vol. 31, 1996, pp. 1–23; and P. Sands, 'International Law in the Field of Sustainable Development: Emerging Legal Principles', in W. Lang (ed.), *Sustainable Development and International Law* (London and Dordrecht: Graham & Trotman and Martinus Nijhoff, 1995), pp. 53–72.

2 For reviews, see D. Freestone and E. Hey, 'Origins and Development of the Precautionary Principle', in D. Freestone and H. Hey (eds.), *The Precautionary Principle and International Law: The Challenge of Implementation* (The Hague: Kluwer Law International, 1996), pp. 3–12; and E. Hey, 'The Precautionary Principle', *Marine Pollution Bulletin*, Vol. 26, 1993, p. 53.

3 J. Moffett, 'Legislative Options for Implementing the Precautionary Principle', *Journal of Environmental Law and Practice*, Vol. 7, 1997, pp. 157, 166–7; A. Deville and R. Harding, *Applying the Precautionary Principle* (Sydney: Federation Press, 1997), p. 71; and T. Lundmark, 'Systemizing Environmental Law on a German Model', *Dickinson Journal of Environmental Law & Policy*, Vol. 7, 1998, pp. 1, 17.

4 O. McIntyre and T. Mosedale, 'The Precautionary Principle as a Norm of Customary International Law', *Journal of Environmental Law*, Vol. 9, 1997, pp. 221, 237.

A more utilitarian view of precaution, as articulated in Principle 15 of the Rio Declaration on Environment and Development,[5] supports an array of less extreme management measures. Reliance on cost–benefit and risk–benefit assessments is viewed as critical in reaching rational decisions. Flexible trade-offs between environmental and economic values are encouraged through application of best available techniques (with consideration given to costs).[6]

The principles of pollution prevention and precaution are closely related, as pollution prevention measures are one of the best ways of being 'precautionary'. There is also a distinction, however, since the precautionary principle or approach applies where there is scientific uncertainty regarding the cause–effect links of an activity.[7]

The twin principles of precaution and pollution prevention raise various questions which will be explored in this chapter. These include:

1. How far are the abstract principles heralded by politicians and bureaucrats being translated into concrete action in the Arctic?
2. To what extent do ethical, economic, scientific and cultural perspectives influence interpretations?
3. How effectively are concerns over toxic contamination of traditional foods of Arctic inhabitants and calls for preventing transboundary pollution being articulated and heeded at the regional and global levels?

Practical questions have also arisen over the adequacy of financial and human resource commitments to environmental protection, and the effectiveness of compliance and enforcement.

This chapter examines the tensions between the principles of precaution and pollution prevention, on the one hand, and actual practices relating to land-based pollution control in the Arctic, on the other. Following an introductory summary of the sources of land-based marine pollution in and into the Arctic, further sections review global efforts to address land-based pollution in terms of the precautionary principle or approach, as well as examining extra-regional efforts to address land-based sources of pollution. The focus is then shifted to 'regional seas' agreements tangential to Arctic land-based pollution control, specifically the two 1992 conventions for the protection of the marine environment of the northeast Atlantic[8] and the Baltic Sea area, respectively.[9] Finally, efforts

[5] UN doc. A/CONF.151/26, Vol. 1, text reprinted in ILM, Vol. 31, 1992, pp. 874ff.

[6] F. B. Cross, 'Paradoxical Perils of the Precautionary Principle', *Washington & Lee Law Review*, Vol. 53, 1996, p. 851.

[7] See J. Cameron and W. Wade-Gery, 'Addressing Uncertainty: Law, Policy and the Development of the Precautionary Principle', *CSERGE Working Paper GEC 92-43* (London: Centre for Social and Economic Research on the Global Environment, 1992), pp. 6–8.

[8] Reprinted in ILM, Vol. 32, 1993, pp. 1,069ff.

[9] Reprinted in K. R. Simmonds (ed.), *New Directions in the Law of the Sea (New Series)*, Vol. 3, J. 47, Release 92–2 (London: Oceana Publications, 1993), pp. 1–53.

undertaken at the Arctic regional level to address land-based sources of marine pollution are reviewed.[10]

CRISES IN ARCTIC SEAS

Although the Arctic Ocean has been described as less polluted than other marine regions,[11] the Arctic seas are coming under increasing stress from various sources: from land-based activities in the Arctic, particularly in the Russian Federation; from long-range movements of hazardous substances from areas outside the Arctic; and from global emissions of greenhouse gases and ozone-depleting substances.

Land-based activities in the Arctic

From one perspective, land-based marine pollution in the Arctic would not appear a major regional problem. Since population levels are relatively low, with small communities dotting the coastline, quantities of human sewage are low as well, and tend to have only local effects. For example, the population of Greenland is about 57,000, out of which about 80 per cent live in villages. In the Canadian Far North, approximately fifty settlements fringe the coast, with the average community population being 742.[12]

However, major local sources of marine and coastal pollution do exist – among them urban settlements, mining wastes, oil and gas operations, nuclear activities, industrial complexes (particularly smelters), and pulp and paper mills. Sewage from urban settlements is a special concern in the Russian Federation, where some two million people live in its Arctic part.[13] The most serious discharges of untreated sewage and wastes come from Murmansk, Arkhangelsk, Severodvinsk, Naryan-Mar, Anderma, Igarka, Dudinka, Tiksi and Pevek.[14] The annual discharge of mining wastes into watercourses in the Murmansk region of the Russian Federation is reported to be over two billion m^3.[15] In addition to sources in the Russian Arctic, mining wastes also enter the Arctic marine environment from two

[10] A discussion of global agreements addressing land-based *nuclear* pollution is beyond the scope of this chapter. On dumping of radioactive wastes in parts of the Arctic Ocean, see Stokke, Chapter 9 in this book.

[11] *Arctic Pollution Issues: A State of the Arctic Environment Report* (Oslo: Arctic Monitoring and Assessment Programme, 1997).

[12] Working Group on the Protection of the Arctic Marine Environment, *Report to the Third Ministerial Conference on the Protection of the Arctic Environment, 20–21 March 1996, Inuvik, Canada* (Oslo: Norwegian Ministry of the Environment, 1996), p. 31 (hereinafter PAME 1996 Report).

[13] *AMAP Assessment Report: Arctic Pollution Issues* (Oslo: Arctic Monitoring and Assessment Programme, 1998), p. 168.

[14] *Report of the Russian Arctic Group of the Advisory Committee on Protection of the Sea (ACOPS): Identification and Assessment of Land-Based Sources Which Lead to the Degradation of the Arctic Marine Environment*, Executive Summary (London: Advisory Committee on Protection of the Sea, 1996) (hereinafter ACOPS Report).

[15] PAME 1996 Report, pp. 34–7.

lead-zinc mines in Canada that discharge tailing effluents – the Polaris mine on Little Cornwallis Island and the Nanisivik mine on the north coast of Baffin Island – and three former mines in Greenland known to be sources of heavy metals entering the sea.

Considerable amounts of hydrocarbons enter Arctic waters by means of rivers and the atmosphere. Atmospheric transport is estimated to add about 40,000 tons of hydrocarbons to the Arctic marine environment annually. Input of petroleum hydrocarbons into the European Arctic is thought to stem mainly from river transport, with Russian measurements indicating a concentration of petroleum hydrocarbons four to twenty times higher in the Ob Gulf than in the Rhine or Elbe rivers.[16] A major land-based oil pollution concern remains the Russian Federation's extensive pipeline network, which is often in poor condition and experiences frequent leaks. Six trunk oil pipelines stretch over 10,000 km of western Siberia; the network is capable of carrying 400 million tons of oil per year. Estimates of losses from oil pipelines for western Siberia and Timan-Pechora, the two main petroleum provinces of Russia, are 1–1.2 per cent. The notorious 1994 oil pipeline leaks north of the town of Urinsk in the Komi Republic spilled over 100,000 tons across a 60 km^2 area, with some oil entering the Usa and Pechora rivers.[17]

In addition to the contamination hazards associated with some 130 decommissioned former Soviet nuclear submarines, most of which remain afloat in coastal areas and have spent nuclear fuel aboard, two nuclear power plants operate in the Russian north. The Kola nuclear power plant, estimated to produce over 1,000 m^3 of radioactive solid and liquid wastes per year, has four nuclear reactors. The Bilibino nuclear power plant in the Chuckchi autonomous district also has four blocks; more than 3,600 spent nuclear assemblies are stored at the site.[18]

The largest point-source contributors of land-based marine pollution in the Arctic appear to be major mining-metallurgical complexes. Hot spots of heavy metal emissions include the Pechenganikel industrial complex and the Severonickel smelter complex on the Kola Peninsula of the Russian Federation. Severonickel, the largest nickel-copper smelter in the world, emits an estimated 3,000 tons of copper and 2,700 tons of nickel annually to the atmosphere. The Norilsk mining and metallurgical combine is also a major polluter, responsible each year for up to 1,300 tons of nickel emissions, 3,000 tons of copper, 44 tons of lead and almost 31,000 tons of sulphuric acid emissions.[19] Human health effects from pollution may be severe, with overall child mortality in the Kola Peninsula exceeding the Russian average by 39 per cent.[20] Also outside Russia there are problems. In northeastern Sweden, airborne emissions from the primary smelter at

[16] *A State of the Arctic Environment Report*, p. 151; and J. R. Hansen, R. Hansson and S. Norris (eds.), 'The State of the European Arctic Environment', *EEA Environmental Monograph No. 3* (Copenhagen: European Environment Agency, 1996), p. 97 (hereinafter State of the European Arctic Report).

[17] *A State of the Arctic Environment Report*, p. 150; and ACOPS Report, pp. 31–3.

[18] ACOPS Report, pp. 38–42.

[19] *A State of the Arctic Environment Report*, p. 98; and ACOPS Report, pp. 9–10.

[20] State of the European Arctic Report, p. 102.

Rönnskär carry arsenic, copper, lead, mercury, zinc as well as sulphur and nitrogen compounds to the Arctic region.[21]

Finally, pulp and paper mills are also among major land-based sources of marine pollution on the Arctic. The volume of waste discharged from the Russian Federation's Arkhangelsk pulp mill between 1985 and 1990 ranged between 309 and 345 million m^3 per year, while the Solombalsky pulp mill contributed an annual 82–90 million m^3. Both mills dispose of wastes into the North Dvina river, whose estuarine area has been found to have dioxin levels exceeding industrial areas of central Europe.[22]

Long-range transport of hazardous substances from land-based sources outside the Arctic

Three main categories of hazardous pollutants are transported from outside the region. *Radionuclides* are carried by ocean currents into the Arctic from three nuclear reprocessing plants in western Europe: Sellafield on the northwest coast of England, La Hague near Cherbourg in France and Dounreay in northeast Scotland.[23] *Heavy metals*, including mercury, lead, nickel, cadmium and copper, are mainly transported from sources in Eurasia and North America.[24] *Persistent organic pollutants* (POPs), typically semi-volatile and enabling long-distance movements, include many persistent pesticides such as dieldrin, DDT, toxaphene, chlordane and hexachlorocyclohexane; also several industrial compounds, among them the polychlorinated biphenyls (PCBs); and various combustion by-products such as polycyclic aromatic hydrocarbons (PAHs) as well as dioxins and furans.[25]

Substantial amounts of POPs may be reaching the Arctic, transported on air masses from Europe, Russia, North America and Asia, and are a special concern. Concentrating in the fatty tissues of wildlife, POPs raise health risks to indigenous communities that rely on traditional diets high in lipid content. For example, concentrations of organochlorines in mothers' milk of the Inuit from Nunavik in northern Quebec are two to ten times higher than levels in southern non-aboriginal populations. While a whole range of socio-cultural, nutritional and spiritual benefits accompany traditional foods, various threats are posed by toxic contaminants, including neurological effects, reproductive problems, immune suppression and cancer.

Global climate change and ozone depletion threats

Global warming, fuelled by greenhouse gas emissions, may have particularly severe impacts on the Arctic. Temperatures in the Arctic are predicted to rise approximately twice the global average, and melting ice contributing to sea-level

[21] PAME 1996 Report, p. 36. [22] ACOPS Report, pp. 36–7.
[23] *A State of the Arctic Environment Report*, p. 114; and PAME 1996 Report, pp. 61–3.
[24] *A State of the Arctic Environment Report*, p. 98. [25] *Ibid.*, pp. 72–91.

rise may increase coastal erosion and inundate low-lying areas.[26] Thawing of the permafrost could damage vegetation, while changes in sea ice could shift the migration routes of marine mammals and reduce feeding areas for polar bears.[27] Higher temperatures and lower salinity (from increased snow melting) could also affect global ocean circulation, including the warm North Atlantic current, and might result in colder climates, especially in Scandinavia and northwest Russia.[28]

The thinning of the ozone layer over the Arctic is also a growing concern, with its environmental impacts still uncertain. A general trend of ozone depletion greater than 8 per cent per decade has been reported for the Arctic. Numerous ozone holes, likened to Swiss cheese, occur over the Arctic in late winter and early spring. At such times, depletion may be severe indeed – up to 40 per cent. Snow, with its highly reflective surface, can double ultraviolet radiation exposure. Polar plants and plankton that have become adapted to low light and radiation conditions might be more susceptible to damage than organisms from other regions. Zooplankton and fish eggs and larvae might also be threatened. Human health risks include increases in skin cancer, cataracts and immune system suppression.[29]

GLOBAL INSTRUMENTS ADDRESSING LAND-BASED MARINE POLLUTION

Since no global convention exists on land-based pollution control, management efforts here have been fragmented. The three main initiatives to date have shunned a strict precautionary approach to pollution control: the 1998 Convention on the Prior Informed Consent Procedure for Certain Hazardous Chemicals and Pesticides in International Trade,[30] the proposed global convention on POPs, and the Global Programme of Action for the Protection of the Marine Environment from Land-Based Activities (GPA).[31] The provisions of the 1982 Law of the Sea Convention relating to land-based pollution will not be discussed here, as they are analysed elsewhere in this book.[32]

The 1998 Convention on Prior Informed Consent

This Convention promises to address – partially and indirectly – the problem of long-range chemical transport. The Convention will make mandatory the previously voluntary 'prior informed consent' procedures for banned and severely restricted chemicals promoted by the FAO Code of Conduct on the

[26] State of the European Arctic Report, p. 32; and *A State of the Arctic Environment Report*, p. 162.

[27] P. Prestrud and I. Stirling, 'The International Polar Bear Agreement and the Current Status of the Polar Bear Conservation', *Aquatic Mammals*, Vol. 20, 1994, pp. 119–20.

[28] *A State of the Arctic Environment Report*, p. 161.

[29] *Ibid.*, pp. 165–8, 180.

[30] Text reprinted in ILM, Vol. 38, 1999, pp. 1ff.

[31] UNEP (OCA)/LBA/IG.2/7, of 5 December 1995.

[32] See Boyle, Chapter 1; and Vukas, Chapter 2 in this book.

Distribution and Use of Pesticides, and by the London Guidelines for the Exchange of Information on Chemicals in International Trade.[33]

The Convention requires each party to inform other parties of national bans or severe restrictions on chemicals; it further makes chemicals listed under its Annex III subject to the 'prior informed consent' procedure whereby the importing state can withhold consent. The Convention initially covers twenty-two pesticides and five industrial chemicals, of which many are POPs.

However, the Convention does not adopt a precautionary approach. Its Preamble, while recalling pertinent provisions of the Rio Declaration in a general way, does not explicitly call for a precautionary approach. Nor does the Convention promote pollution prevention and toxic chemical reduction. It does not ban trade in hazardous chemicals but is aimed at preventing illegal international traffic in dangerous chemicals. The Convention does not establish a proactive chemical management regime, but envisages a reactive chemical-by-chemical addition to the 'prior informed consent' procedure. Before additional listings are possible, notifications are required from two different regions that a chemical is banned or severely restricted. A Chemical Review Committee must recommend listing and prepare a draft decision guidance document; and the Conference of the Parties must approve the listing.

The Convention's provisions on technical assistance are also slanted towards promoting chemical use. No financial mechanism is established and no binding financial commitments are made to assist developing countries or countries with economies in transition. General funding commitments are made to promote technical assistance so that countries can implement the Convention and manage chemicals throughout their life-cycle.

Global POPs Convention Initiative

In early 1997 the Governing Council of UNEP launched negotiation efforts to convene, together with relevant international organisations, an intergovernmental negotiating committee to prepare an international legally binding instrument for POPs. The initial focus was to be on the so-called 'dirty dozen' persistent organic pollutants (aldrin, chlordane, DDT, dieldrin, dioxins and furans, endrin, heptachlor, hexachlorobenzene, mirex, PCBs, and toxaphene).

The negotiation process commenced in Montreal, in mid-1998, at which time the Intergovernmental Negotiating Committee (INC) for an International Legally Binding Instrument for Implementing International Action on Certain Persistent Organic Pollutants considered some of the key sections necessary for inclusion in a future convention and suggested the completion of an international

[33] The documents are reprinted in H. Hohmann (ed.), *Basic Documents of International Environmental Law*, Vol. 1 (London: Graham & Trotman, 1992), pp. 173–86 and 157–72, respectively.

convention by the year 2000. The Committee decided to establish two groups to forge approaches in two critical areas. An expert group on POPs has been mandated to develop science-based criteria and a procedure for identifying other substances for management actions. Another group, on implementation aspects, was to consider convention provisions on technical and financial assistance to developing countries and countries with economies in transition.[34]

At the second negotiation session held in Nairobi, in January 1999, numerous issues remained unresolved. One contentious point was the need for a financial mechanism, like the multilateral fund of the Montreal Protocol. Another open question was whether those who had produced or exported POPs should be responsible for the removal and destruction of unused stockpiles in developing countries. Details on technology transfer and non-compliance procedures also needed to be worked out.[35]

Adoption of a strict precautionary approach seems unlikely. A slow chemical-by-chemical regulatory approach is what is being endorsed. Even for chemicals understood as being 'super nasty', gradual phase-outs are likely to be recommended for some, in order to allow the development of economically viable alternatives and to continue the battle against diseases such as malaria. Moreover, the Group of 77 and China pressed for the inclusion of various principles in the Convention, such as the right to development and the need for differential obligations for developing countries, which may run counter to precautionary measures.[36] The group of African countries emphasised the numerous obstacles to effective phasing out of POPs, such as the lack of national inventories, and the lack of financial resources for research, monitoring and management of chemicals.[37]

The Global Programme of Action

The Global Programme of Action for the Protection of the Marine Environment from Land-Based Activities, adopted in November 1995 by over 100 countries at a conference held in Washington DC, marked a further step in building international consensus on the need for a global POPs convention. Paragraph 88 of the GPA endorses the need for a global, legally binding instrument on POPs and highlights the need to address the technical and financial needs of developing countries.

The GPA also promises to promote national actions to control land-based

[34] *Report of the Intergovernmental Negotiating Committee for an International Legally Binding Instrument for Implementing International Action on Certain Persistent Organic Pollutants on the Work of Its First Session*, UNEP/POPS/INC.1/7, of 3 July 1998.

[35] *Report of the Intergovernmental Negotiating Committee for an International Legally Binding Instrument for Implementing International Action on Certain Persistent Organic Pollutants on the Work of Its Second Session*, UNEP/POPS/INC.2/6, of 29 January 1999.

[36] *Ibid.*, Annex V.

[37] *Ibid.*, Annex III.

pollution. Its Chapter 2 calls on states to develop, within a few years' time, national programmes of action. Such national programmes should follow a six-step process: identifying and assessing problems; establishing priorities; setting management objectives; identifying and selecting management strategies and measures; developing criteria for evaluating the effectiveness of strategies and programmes; and ensuring programme support, such as financial mechanisms and new legislation.

The GPA also encourages strengthening of regional cooperation to address land-based pollution and activities. Its Chapter 3 calls for building institutional arrangements for regional cooperation and negotiating new regional conventions, as appropriate. Regional programmes of action are advocated; these should include the harmonisation of environmental standards, the protection of critical habitats and endangered species, the exploration of innovative financing mechanisms, and the identification of regional centres of excellence in research and management training.

Chapter 5 of the GPA urges national, regional and international action for nine specific problem areas. Targets and actions are set for sewage, POPs, radioactive substances, heavy metals, oil (hydrocarbons), nutrients, sediments, litter and physical alterations/destruction of habitats.

The GPA entrenches a utilitarian version of the precautionary approach, emphasising the validity of weighing environmental values and interests against economic costs and benefits. Paragraph 24 of the GPA states:

> The precautionary approach should be applied through preventive and corrective measures based on existing knowledge, impact assessments, resources and capacities at national level, drawing on pertinent information and analyses at the subregional, regional and global levels. Where there are threats of serious or irreversible damage, lack of full scientific certainty should not be used as a reason for postponing cost-effective measures to prevent the degradation of the marine environment.

The effectiveness of the GPA remains uncertain. Besides being full of undefined buzzwords and often stating the obvious, the GPA is weak on the financial side. It leaves implementation to the happenstance of available national funding and private sector contributions. The GPA lists in an annex likely sources of financing, such as pollution fees, taxes and loans from international financial institutions, but does not include a trust fund or firm financial commitments to assist developing countries and countries with economies in transition.[38]

The prospects for a global convention on land-based marine pollution do not look good. 'Treaty fatigue' is evident: many developing countries in particular are averse to any further binding commitments until technical and financial

[38] D. L. VanderZwaag, P. Wells and J. Karau, 'The Global Programme of Action for the Protection of the Marine Environment from Land-Based Activities: A Myriad of Sounds, Will the World Listen?', *Ocean Yearbook*, Vol. 13, 1998, pp. 183, 208.

assistance is assured. State sovereignty concerns remain paramount, and calls for differential standards between developed and developing countries constrain the finding of common ground.[39] The broad array of activities needing to be controlled also complicates the picture.[40]

EXTRA-REGIONAL EFFORTS ADDRESSING LAND-BASED POLLUTION

Various regional agreements and actions developed outside the Arctic region proper promise to assist in controlling long-range pollutants, but they are also limited in advocating a precautionary approach. Key agreements adopted under the auspices of the UN Economic Commission for Europe include the Protocols on POPs and heavy metals adopted pursuant to the 1979 Convention on Long-Range Transboundary Air Pollution (LRTAP)[41] and the Convention on Environmental Impact Assessment in a Transboundary Context.[42] Also, four regional action plans (covering DDT, chlordane, mercury and PCBs) have been adopted pursuant to the North American Sound Management of Chemicals Initiative.

Protocol on Persistent Organic Pollutants (POPs)

The 1998 POPs Protocol to LRTAP,[43] adopted at Aarhus, Denmark, in June 1998, does include various provisions encouraging precautionary and pollution-prevention approaches. The Preamble to the Protocol articulates the parties' resolve 'to take measures to anticipate, prevent or minimise emissions of persistent organic pollutants, taking into account the application of the precautionary approach, as set forth in principle 15 of the Rio Declaration on Environment and Development'. Twelve substances are targeted for elimination: aldrin, chlordane, chlordecone, DDT, dieldrin, endrin, heptachlor, hexabromobiphenyl, hexachlorbenzene, mirex, PCBs and toxaphene. Parties also pledge to facilitate the exchange of information and technology for reducing the generation and emissions of POPs and to develop alternatives.

In many ways, however, the Protocol is weak and not strictly precautionary. For the twelve substances scheduled for elimination, some major exceptions exist. For example, production of DDT is to be eliminated only within one year of consensus by the parties that suitable alternatives are available for public health

[39] D. B. Magraw, 'Legal Treatment of Developing Countries: Differential, Contextual, and Absolute Norms', *Colorado Journal of International Environmental Law & Policy*, Vol. 1, 1990, p. 69.

[40] Qing-nan Meng, *Land-Based Marine Pollution: International Law Development* (London: Graham & Trotman and Martinus Nijhoff, 1987).

[41] ILM, Vol. 18, 1979, pp. 1,442ff. Additional protocols adopted but not discussed in this chapter include those on the European Monitoring and Evaluation Program (1984), Reduction of Sulphur Emissions (1985), Control of Nitrogen Oxide Emissions (1988), Control of Volatile Organic Compounds (1991) and Further Reduction of Sulphur Emissions (1994).

[42] ILM, Vol. 30, 1992, pp. 800ff.

[43] Text available at www.unece.org/env.

protection from diseases such as malaria and encephalitis. PCB production by countries with economies in transition is to be eliminated as soon as possible, but those countries are allowed to continue production until 31 December 2005. While countries are required to reduce emissions of PAHs, dioxins and furans and hexachlorobenzene from a chosen reference year, no overall percentage reductions are required, and parties are mandated to apply best available techniques – a highly flexible approach that permits consideration of economic, technical and practical factors. Emission limit values for dioxins and furans are established for major stationary sources – specifically, municipal and medical solid waste incinerators and hazardous waste incinerators – but the emission standards are diluted for existing stationary sources with the qualification 'insofar as . . . technically and economically feasible'. The Protocol also allows eight years from the time of its entry into force for parties to apply the emission limits to existing sources.

The Protocol does not establish a comprehensive and proactive chemical management framework. Only sixteen substances are initially designated for control actions. Parties can add substances one by one, according to agreed procedures.

Heavy Metals Protocol

The Protocol on Heavy Metals[44] to LRTAP adopts precautionary and pollution-prevention approaches in a rather limited fashion. Its Preamble states the parties' resolve to take anticipatory and prevention measures to minimise emissions of heavy metals and to take into account the precautionary approach as set forth in the Rio Declaration. Parties are encouraged to develop alternatives to the use of heavy metals in various products.

Several provisions of the Protocol run counter to strong precautionary or pollution-prevention approaches. Only three heavy metals are subject to initial controls: cadmium, lead and mercury. While parties are required to reduce total annual emissions into the atmosphere from a reference year, no specific percentage reductions are required. Moreover, parties are mandated to apply best available techniques to major stationary sources, leaving wide discretion as to control measures. Although emission limits are established for selected stationary sources, such as municipal, medical and hazardous waste incinerators, the limits for existing sources are weakened by the loophole 'insofar as . . . technically and economically feasible' as well as the allowable eight-year delay in applicability from the Protocol's entry into force.

Product-control measures for lead in gasoline and for mercury content in batteries are also not stringent in terms of precaution or prevention. While the lead content of marketed petrol for on-road vehicles is not to exceed 0.013 g/l, states are

[44] Also adopted at Aarhus, June 1998; text available at *ibid.*

given the option of extending the cutback time period by up to ten years in case of serious socio-economic or technical problems. Although parties are required to limit the content of mercury in alkaline manganese batteries, they are given up to five years (ten years in the case of countries with economies in transition) to achieve these concentration levels.

Convention on Environmental Impact Assessment in a Transboundary Context

The 1991 UN/ECE Convention on Environmental Impact Assessment in a Transboundary Context, requires parties to undertake environmental impact assessments (EIAs) for any proposed activities, listed in its Appendix I, likely to cause significant adverse transboundary impacts. The Preamble notes the need for anticipatory policies and to prevent and mitigate significant adverse environmental effects in a transboundary context. For activities subject to the EIA requirement, the party of origin and the affected party must hold consultations where possible alternatives, including the no-action alternative, are to be considered and mitigation measures discussed.

However, the Convention also displays several weaknesses, raising questions as to how precautionary and anticipatory actual state practice will be. The Convention, negotiated before the 1992 UN Conference on Environment and Development, does not adopt the precautionary principle or approach. The Convention's list of activities subject to EIA is not comprehensive and could allow some activities with substantial transboundary environmental risks to escape assessment, such as major port developments.[45] The Convention does not require parties to apply the EIA provisions to proposed policies, plans or programmes that may have significant transboundary impacts. It leaves the final 'go–no go' decision with the party proposing the development activity. And, finally, no explicit provision is made for carrying out joint environmental assessment reviews.

North American Sound Management of Chemicals Initiative

A further extra-regional effort for addressing hazardous substances is the North American Sound Management of Chemicals Initiative, launched in 1995 under the auspices of the North American Commission for Environmental Cooperation,[46] whose Council of Ministers established a Working Group on the Sound Management of Chemicals. To be composed of two senior environmental

[45] Arctic Environmental Protection Strategy, *Guidelines for Environmental Impact Assessment (EIA) in the Arctic* (Helsinki: Finnish Ministry of the Environment, 1997), p. 41.

[46] The Commission was established pursuant to the North American Agreement on Environmental Cooperation; see ILM, Vol. 32, 1993, pp. 1,480ff.

officials from each party, this working group is intended to promote cooperation in studying and managing chemicals of mutual concern with priority to be given to chemicals that are persistent, bio-accumulative and toxic.[47] Through the use of task forces, it has developed four regional action plans: on PCBs, DDT, chlordane and mercury.[48]

The Sound Management of Chemicals Initiative does include some consideration of the precautionary and pollution-prevention approaches. Resolution 95-5, establishing the Initiative, directed the Working Group to 'incorporate, as appropriate, pollution prevention principles and precautionary approaches in making recommendations to reduce risk associated with toxic substances'.[49] Pursuant to a document recently adopted by the Working Group, the selection of any additional substances for regional action is to be based on various principles, including the precautionary approach, in keeping with Principle 15 of the Rio Declaration.[50] The PCB Regional Action Plan specifically calls for a pollution-prevention approach and the promotion of PCB waste reduction and recycling.

The North American Initiative does not represent a strict precautionary approach. What is envisaged is a process of regional management response on a chemical-by-chemical basis. The Initiative requires consideration of the differing economic, political and regulatory circumstances of the parties.[51] The Regional Action Plan on DDT, recognising the importance of DDT use in Mexico for malaria control, provides for the gradual reduction of DDT with a target of 80 per cent (volume) reduction in five years. The Regional Action Plan on Mercury, mainly calling for workshops and database development, leaves the setting of specific targets and time-frames for mercury reductions to the future.[52]

REGIONAL SEA AGREEMENTS RELEVANT TO LAND-BASED POLLUTION OF THE ARCTIC

Two regional agreements, already mentioned in the introduction to this chapter, focusing on regional sea management may serve as further avenues for controlling land-based pollution relevant to the Arctic: the 1992 Paris Convention for the Protection of the Marine Environment of the North-East Atlantic and the 1992 Convention on the Protection of the Marine Environment of the Baltic Sea Area (the latter often referred to as the Helsinki Convention of 1992).

[47] A copy of the relevant resolution and a description of subsequent initiatives are available in 'Overview and Update on the Sound Management of Chemicals Initiative under the North American Agreement on Environmental Co-operation Council Resolution 95-5', available on the Commission's website www.cec.org.

[48] The Action Plans are available on the Commission's website, *ibid.*

[49] 'Overview and Update', available on *ibid.*

[50] 'Process of Identifying Candidate Substances for Regional Action under the Sound Management of Chemicals Initiative', available on *ibid.* [51] 'Overview and Update', available on *ibid.*

[52] Pursuant to section 5(5)(1), the parties agreed to forward to the Commission's Council by June 1999 any proposed additional specific actions.

The 1992 Paris Convention

Applying to parts of the northeast Atlantic and Arctic Ocean, this Convention includes several provisions supportive of the precautionary and pollution-prevention approaches. Parties are required by Article 2(2)(a) to apply the precautionary principle. Article 2(1)(a) requires parties to 'take all possible steps to prevent and eliminate pollution' and to 'take the necessary measures to protect the maritime area against the adverse effects of human activities'. Annex I, detailing commitments on the prevention and elimination of land-based sources, calls for consideration of using clean technologies.

However, the Convention is disappointing when it comes to addressing land-based pollution and activities. It limits application of the precautionary principle to pollution activities. This means that precautionary approaches are not mandated for other activities such as coastal forestry operations or sand/gravel extraction. The Convention fails to establish specific control standards or set specific targets and timetables for regulatory actions.[53] Moreover, it leaves wide discretion to states by requiring application of best available techniques (BAT) for point sources of land-based pollution and best environmental practices (BEP) for point and diffuse sources. Appendix I to the Convention, seeking to clarify the meaning of BAT and BEP, grants considerable leeway for parties to weigh economic and social factors. Management decisions by the Commission, established under the Convention, require acceptance by individual parties. Majority vote standard-setting, which is one way of implementing the precautionary principle, is not strictly followed.[54]

The 1992 Helsinki Convention

The Baltic Sea, surrounded by nine countries (Denmark, Estonia, Finland, Germany, Latvia, Lithuania, Poland, Russia and Sweden) has a population of approximately 66 million living around its drainage basin area.[55] Thus, there is considerable potential for long-range transboundary atmospheric pollution from the region as well as the potential sources linked to the Arctic.

The 1992 Helsinki Convention promotes precautionary and pollution-prevention approaches in various ways. The parties are required to apply the

[53] E. Hey, T. IJlstra and A. Noelkaemper, 'The 1992 Paris Convention for the Protection of the Marine Environment of the North-East Atlantic: A Critical Analysis', *International Journal of Marine and Coastal Law*, Vol. 8, 1993, especially pp. 13 and 48.

[54] E. Hey, 'The Precautionary Concept in Environmental Policy and Law: Institutionalizing Caution', *Georgetown International Environmental Law Review*, Vol. 4, 1992, pp. 257 and 304.

[55] J. M. Broadus, S. Demisch, K. Gjerda, P. Haas, Y. Karou, G. Peet, S. Repeto and A. Roginko, *Comparative Assessment of Regional International Programs to Control Land-Based Marine Pollution: The Baltic, North Sea, and Mediterranean* (Woods Hole, MA: Marine Policy Center, Woods Hole Oceanographic Institution, 1993), p. 11.

precautionary principle where introduced substances or energy may create environmental hazards. The parties are further required to take all appropriate legislative and administrative measures to prevent and eliminate pollution, so as to promote ecological restoration and to preserve the ecological balance of the Baltic Sea area. Annex I, dealing with harmful substances, prohibits the use of certain substances, such as DDT and its derivatives and PCBs (except in closed system equipment). The Annex also bans (or at least requires reduction of) various pesticides, among them many types of POPs.

The Baltic Marine Environment Protection Commission (Helsinki Commission) has adopted numerous recommendations on land-based pollution measures. For example, at its meeting in March 1996, the Commission recommended reduction of atmospheric emissions from pulp and paper mills, wastewater discharge standards of chemical industries, emission limits from incineration of household wastes and discharge reductions from textile plants.[56]

However, the Convention has not guaranteed strict precautionary and pollution-prevention measures. The Helsinki Convention embraces best available technology for point-sources of land-based pollution and BEP for all land-based sources. This allows parties wide discretion to consider social and economic factors. The Convention, being of a programmatic nature, does not set detailed standards for industries, but rather leaves details to national permit limits.[57] Out of forty-seven land-based recommendations issued by the Commission in the period 1980–91, only twelve (or about one-quarter of all) were reported as implemented by the parties.[58] A proactive chemical management system is not ensured; reliance is placed on a limited number of chemical bans set out in Annex I of the Convention.

ARCTIC REGIONAL INITIATIVES ADDRESSING LAND-BASED POLLUTION AND ACTIVITIES

Regional efforts to address land-based pollution in the Arctic have involved at least five main steps: the establishment of the Arctic Environmental Protection Strategy (AEPS); the designation of the Working Group on Protection of the Arctic Marine Environment (PAME); the development of Arctic Environmental Assessment Guidelines; the drafting of a Regional Programme of Action for the Protection of the Arctic Marine Environment from Land-Based Activities (RPA); and the formation of the Arctic Council.

[56] The recommendations are reprinted in W. E. Bruhenne and N. A. Robinson (eds.), *International Protection of the Environment: Conservation in Sustainable Development*, Vol. 3, Booklet 17-03-96/1 (Dobbs Ferry, NY: Oceana, 1997), pp. 1–27.

[57] P. Ehlers, 'The Helsinki Convention, 1992: Improving the Baltic Sea Environment', *International Journal of Marine and Coastal Law*, Vol. 8, 1993, pp. 191 and 212–13.

[58] Broadus *et al.*, *Comparative Assessment*, pp. 74–6.

Arctic Environmental Protection Strategy

A non-treaty document and process initiated in June 1991, the Arctic Environmental Protection Strategy[59] commits the eight Arctic states to undertake research and to develop strategies for six priority environmental problems, five of which are at least partly land-based. Besides investigating underwater noise from ships, the states agreed to address persistent organic contaminants, oil pollution, heavy metals, radioactivity and acidification.

National commitments for cooperative actions under the AEPS have tended to be quite general. For example, regarding POPs the Arctic states agreed to implement measures to reduce or control the use of chlordane, DDT, toxaphene and PCBs, but no specific targets were established. Countries have also agreed to implement measures to control heavy metal releases from industrial activities including, as appropriate, the use of best available technology.

The AEPS pledged the holding of further ministerial meetings, which led to the issuance of additional declarations in part relevant to land-based pollution activities. The 1993 Nuuk Declaration pledged precautionary approaches to developments in the Arctic, including prior assessments of environmental impacts.[60] As discussed in further detail in Chapter 4 of this book, the Nuuk Declaration also endorsed the establishment of the Protection of the Arctic Marine Environment (PAME) Working Group to assess the need for further actions or instruments to prevent pollution of the Arctic marine environment. The Inuvik Declaration directed PAME to develop a Regional Programme of Action for the Protection of the Arctic Marine Environment from Land-Based Activities.[61] The 1997 Alta Declaration reiterated support for the principles of the Rio Declaration and Agenda 21, and pledged determined efforts to secure support for international action to reduce Arctic contamination.[62]

The AEPS has greatly advanced efforts to monitor and assess the effects of anthropogenic pollutants. In particular, AMAP supported the development of two major reports. The first of these, *Arctic Pollution Issues: A State of the Arctic Environment Report*, which became available in mid-1997, provides an overview of pollution pathways in and into the Arctic and reviews the sources and potential impacts of POPs, heavy metals, radioactivity, acidification and Arctic haze, petroleum hydrocarbons, climate change and ozone depletion. A companion report, published in 1998, *The AMAP Assessment Report: Arctic Pollution Issues*, is

[59] ILM, Vol. 30, 1991, pp. 1,624ff. See Vidas, Chapter 4 in this book.

[60] The Nuuk Declaration on Environment and Development in the Arctic, in *Report of the Second Ministerial Conference, 16 September 1993, Nuuk, Greenland* (Copenhagen: Ministry of Foreign Affairs, 1993), p. 5, section 8.

[61] Inuvik Declaration on Environmental Protection and Sustainable Development in the Arctic, in *Report of the Third Ministerial Conference on the Protection of the Arctic Environment, 20–21 March 1996, Inuvik, Canada* (Ottawa: Indian and Northern Affairs Canada, 1996), p. 4, section 6.

[62] The Declaration is available at the Arctic Council website at http://arctic-council.usgs.gov.

a fully referenced version, containing the scientific basis behind the AMAP assessment.[63]

PAME Working Group

Besides developing a Regional Programme of Action on Land-Based Activities (discussed below), PAME undertook a study of land-based activities contributing to the degradation of the Arctic marine environment and an analysis of existing international instruments relating to land-based activities. The 1996 PAME Report to the Third AEPS Ministerial Meeting in Inuvik designated the four main substances of concern in the Arctic: POPs, heavy metals, radionuclides and oil. The report estimated that land-based activities are responsible for as much as 80 per cent of pollutants entering the Arctic marine environment.[64]

Moreover, the report reviewed the fragmented array of legal instruments touching on land-based activities and explored various options for strengthening management actions. Possible actions included amending the Convention for the Protection of the Marine Environment of the North-East Atlantic (OSPAR Convention) to cover the entire Arctic, encouraging the Russian Federation to become a party to the OSPAR Convention, and developing a new Arctic Seas Convention for Land-Based Sources.[65]

In conclusion, the report did not recommend a new legal agreement on land-based pollution. In the view of PAME, sufficient additional protection should be afforded by new instruments already being developed for POPs, heavy metals and radioactive wastes. The ratification and implementation of existing legal and other instruments was also viewed as a key challenge. PAME left open the possibility of reconsidering the need for an Arctic agreement on land-based activities in the future.

Arctic Environmental Impact Assessment Guidelines

At the 1996 AEPS meeting in Inuvik, Arctic environmental ministers requested the preparation of EIA guidelines. Finland subsequently led this work, which was completed in 1997 with suggested good practices.[66] Arctic EIA Guidelines have the potential to influence national, provincial, land claim or international EIA procedures applying to proposed land-based activities that may have significant adverse impacts. The Guidelines urge the application of the precautionary approach in carrying out EIAs in the Arctic and encourage the full involvement of indigenous communities and the use of traditional knowledge at all stages of EIA work.

[63] See Preface to *A State of the Arctic Environment Report*, p. v.

[64] PAME 1996 Report, p. 75.

[65] *Ibid.*, pp. 79–81.

[66] Arctic Environmental Protection Strategy, *Guidelines for Environment Impact Assessment (EIA) in the Arctic* (Helsinki: Finnish Ministry of the Environment, 1997).

The Arctic EIA Guidelines are critical of the existing UN/ECE Convention on EIA in a Transboundary Context. Because of the particularly vulnerable nature of the Arctic environment, the Guidelines urge a much broader inclusion of projects than the major types of developments listed in Appendix I of the Convention on EIA in a Transboundary Context. For example, migratory species may be impacted by land-based developments far from border areas, such as land drainage and road building. Harmonisation of project lists is recommended through further bilateral or multilateral agreements.

Regional Programme of Action for the Protection of the Arctic Marine Environment from Land-Based Activities

Adopted on 18 September 1998 at the first meeting of the Arctic Council in Iqaluit, the Regional Programme of Action[67] (RPA) might be described as small, soft steps in addressing land-based pollution in the Arctic. Indeed, in its introductory section the RPA emphasises the 'first step' nature by recognising the need for a 'phased and stepwise approach' to programme development. While adopting various principles of sustainable development, including the precautionary approach, it does not elaborate on measures necessary to implement the principles. Mention is made of the need for integrated coastal area management, harmonised with river basin and land-use planning, but coastal zone definition and consideration of integrated management efforts are left to future work.

Regional priorities for action on sources of pollution have been set according to three criteria: severity of risk to human health, to the environment or to socio-economic uses (including cultural values); transboundary pollution effects or habitat degradation; and issues which benefit from common approaches. On the basis of these criteria, POPs and heavy metals have been ranked as high priorities; radionuclides and petroleum hydrocarbons as medium; and sewage, nutrients and litter as low.

Disposal of sewage and solid wastes is identified by the RPA as a local concern for virtually all coastal communities in the Arctic, partly due to the cold climate. However, the RPA does not suggest any actions. Detailed actions are recommended only for POPs and heavy metals, and, even here, the actions are hardly pioneering. For POPs, Arctic states are urged to sign, ratify and implement the 1998 UN/ECE Protocol on POPs, to participate actively in negotiations of a global POPs convention and to give financial and technical support to the negotiation process, as well as to distribute information on POPs pollution to Arctic communities. Moreover, as to heavy metals, Arctic states are encouraged to sign, ratify and implement the 1998 UN/ECE Heavy Metals Protocol and to put pressure on international financial institutions towards financing heavy metal management efforts. Finally,

[67] Regional Programme of Action for the Protection of the Arctic Marine Environment from Land-Based Activities (Ottawa: Ministry of Public Works and Government Services Canada, 1999), text available at the Arctic Council website at http://arctic-council.usgs.gov.

Arctic states are in the longer term called to consider the need to set dates for phasing out and providing substitutes for POPs not covered by the UN/ECE Protocol on POPs, and to assess the need for global action on mercury reduction.

Rather than ensuring programme support, the RPA leaves financial and technical commitments uncertain. Specific budgetary commitments are lacking, and Arctic states are merely ‘encouraged’ to explore innovative financing approaches. While the need for secretariat support is emphasised, no specific obligations are spelled out. The need to assist the Russian Federation in taking pollution prevention and remedial actions is highlighted, but the financial arrangements remain to be worked out.

The Advisory Committee on the Protection of the Sea has been assisting the Russian Federation in preparing a National Plan of Action for the Protection of the Marine Environment from Anthropogenic Pollution in the Arctic Region of the Russian Federation (NPA-Arctic),[68] but funding support is still a critical question. In view of the severe budget constraints in the Russian Federation, a Partnership Conference has been proposed for 2001, involving donor organisations and international financial institutions. The main purpose of the conference would be to seek financial support for activities and projects implementing the Russian NPA-Arctic.[69]

A further weakness of the initial RPA relates to reporting on implementation. The RPA urges progress reports on regional programme implementation to Arctic Council ministers and other inter-governmental bodies (such as UNEP, UN/ECE and the Commission on Sustainable Development), but details on reporting procedures and format are left to be worked out. Countries are not specifically required to report on the efforts made or the effectiveness of their national programmes of action.

Limited implementation of the Global Programme of Action at the national level may be a further weak link in addressing land-based pollution in the Arctic. For example, the United States does not have a national plan of action specifically based on the GPA, because the US non-point source programme was already in place through an amendment to the Coastal Zone Management Act before the GPA was adopted.[70]

In March 1999, Canada issued a Draft National Programme of Action (NPA) for the Protection of the Marine Environment from Land-Based Activities, which included an Arctic chapter. However, this chapter was more descriptive of existing environmental problems and very general when describing actions to be taken. Suggested strategies and actions included using a community-based

[68] See ACOPS and Goskomsever, *Report of the Conference of Official Representatives of the State Duma and the Government of the Russian Federation Dedicated to Approval of the National Plan of Action for the Protection of the Marine Environment from Anthropogenic Pollution in the Arctic Region of the Russian Federation (NPA-Arctic)* (Moscow, 2 October 1998).

[69] *PAME Working Group Meeting Report, 15-18 February, 1999* (Ottawa: Environment Canada, 1999).

[70] Personal communication, Kathryn Ries, National Ocean Service, National Oceanic and Atmospheric Administration, 26 April 1999.

approach to identify and assess sewage-related problems and treatment requirements, promoting integrated coastal planning with watershed development plans, and helping to build capacity within land-claims organisations to address issues defined by the NPA. Federal government financial commitments also seemed limited, as the draft document emphasised 'the NPA will be based on existing resources and an approach of increasing cost-effectiveness, efficiency and cooperation among existing policies, programmes, resources and legislation'.[71]

The Arctic Council

The Arctic Council assumed responsibility for the AEPS and its four working groups, and thus became a further forum for addressing issues of land-based marine pollution. In the 1998 Iqaluit Declaration,[72] the Council welcomed the US proposal for a Technology Transfer Project to Improve Arctic Sanitation Systems, and endorsed the Regional Programme of Action for the Protection of the Arctic Marine Environment from Land-Based Activities. The Council urged support for the Russian Federation Partnership Conference and called for a further assessment of the adequacy of existing international agreements and arrangements related to the protection of the Arctic marine environment. The Council also instructed Senior Arctic Officials to continue development of the Arctic Council Action Plan to Eliminate Pollution in the Arctic (ACAP), to complement existing legal arrangements and the RPA.

At the PAME meeting in February 1999, the PAME Working Group decided upon a two-phased approach to reviewing international instruments, including those covering land-based activities. An initial updating of factual and legal information in the 1996 PAME report and a description of instruments covering habitat protection is to be prepared and reported at the Arctic Council meeting in 2000. The second phase is to include an assessment of the adequacy of the instruments and recommendations to the Arctic Council in 2002.

The Arctic Council has also launched a multinational initiative to identify and phase out sources of PCBs in the Russian Federation. This initiative, commenced in February 1999, is to consist of three phases involving the collection of information on PCB uses and wastes, an analysis of the information, followed by demonstration projects aimed at reducing the sources of PCBs causing transboundary impacts.[73]

The US-led Technology Transfer Project could be a useful step. In Alaska alone, approximately 40 per cent of rural households do not have access to sanitary means of sewage disposal, and 'honey buckets' or privies are the only ways of

[71] Federal/Provincial/Territorial Advisory Committee on Developing Canada's National Programme of Action for the Protection of the Marine Environment from Land-Based Activities, *Draft Canada's National Programme of Action (NPA) for the Protection of the Marine Environment from Land-Based Activities* (March 1999), p. 3; available at www.ec.gc.ca/nat_action.

[72] Available at http://arctic-council.usgs.gov.

[73] US Department of State, Office of the Spokesman, Press Statement, 26 February 1999.

disposing of human waste. The poor capacity of local governments to provide public services due to lack of training or on-site technical assistance is a common problem throughout the Arctic. The project proposal, besides suggesting a first-phase demonstration project to assist a candidate community in addressing sewage treatment, proposes to convene a conference to compare Arctic experiences in rural sanitation.[74]

How effective the Arctic Council will be in addressing land-based pollution/activity issues remains to be seen, but a major constraint is likely to be funding. Currently, Arctic Council projects and programmes are financed on a voluntary basis. Whether funding of programme secretariats and permanent participants should be made mandatory continues to be debated.[75] Iceland has offered to host the PAME Secretariat on a voluntary funding basis for a two-year trial period beginning in 1999.[76]

CHALLENGES AND FUTURE DIRECTIONS

Management of land-based pollution in and into the Arctic still has a long way to go from the limited regulatory and institutional responses to date. There are many challenges to effective management, including the proliferation of global and regional agreements and arrangements, weak acceptance of the precautionary principle, giving priority to scientific assessment over management critiques, the complexity of North–South relations, the preference for preparations over action, the lack of firm financial commitments, and not least a tendency to separate Arctic and Antarctic environmental threats.

The *proliferation* of international environmental agreements and arrangements threatens to overwhelm national and local management capabilities and to foster fragmentation. For example, controls over POPs are spread across the 1998 Prior Informed Consent Convention, the UN/ECE Protocol on POPs, and the Basel Convention (for shipments of hazardous wastes). Complexity is certain to increase with the addition of a global POPs convention. Coordination has been limited. However, the FAO, UNEP, UNIDO and the Secretariat of the Basel Convention have agreed to join forces in developing national inventories of obsolete stocks of chemicals, as well as disposal and capacity-building programmes.[77]

Global and regional practices for managing chemicals have drifted towards a weak version of the precautionary principle. Global trade in banned or severely restricted chemicals is allowed to continue, subject to prior informed consent by the country of import. The global community, while seeking to phase

[74] Information provided by Jutta Paczulla, Arctic Council Sustainable Development Working Group, 5 May 1999.

[75] *Report of Senior Arctic Officials to the Arctic Council*, Iqaluit, Northwest Territories (Nunavut), Canada, 17 September 1998; available at http://arctic-council.usgs.gov. See also D. Scrivener, 'Arctic Environmental Cooperation in Transition', *Polar Record*, Vol. 35, 1999, pp. 51–8.

[76] *PAME Working Group Meeting Report*, 15–18 February 1999.

[77] 'Further Measures to Reduce the Risks from a Limited Number of Hazardous Chemicals', *Report of the Executive Director*, UNEP/GC. 20/37, 9 January 1999.

out the 'dirty dozen', has yet to develop a proactive approach to chemical management. A scientific, rational approach to risk assessment and risk management predominates, as demonstrated by the search for science-based procedures for listing additional POPs under a global convention. No serious consideration is being given to a broad 'reverse listing' approach, whereby only chemicals specifically listed as 'safe' would be allowed on the market.[78] The UN/ECE Protocol on POPs also adopts a chemical-by-chemical listing approach.

A further challenge is the lack of information on and attention to national- and local-level management capabilities, including the effectiveness of compliance and enforcement. Priority in Arctic research has been given to assessing the sources, pathways and effects of pollutants, and not on investigating the political, legal and socio-economic constraints to effective pollution control. There have been few specific case studies on the difficulties of Arctic pollution control.[79]

Differing North–South interests, with concomitantly differing perspectives and capabilities towards environmental protection, represent a particular challenge. Negotiations towards a global POPs convention remain complicated, with debates between developed and developing countries over whether differential environmental standards should be allowed, and the extent of financial and technical assistance to be made available. UNEP did convene eight regional awareness-raising workshops on POPs between July 1997 and June 1998, but long-term educational and capacity-building efforts will be needed. Meanwhile, the necessary global standards and focused assistance are not in place to address the emissions of heavy metals from facilities such as incinerators and coal-fired plants in developing countries.

A further challenging phenomenon is to move from conference discussions, working group meetings and the growing number of well-meant programmes and agreements to specific management actions. The 1998 report of the UN Secretary-General on 'Oceans and the Law of the Sea' recognised the problem in these words:

> Important developments have occurred in relation to the reduction and control of different sources of pollution . . . The challenge now lies in implementing all those agreements, protocols and programmes of action.[80]

Perhaps one of the greatest challenges to addressing land-based marine pollution in the Arctic is to ensure adequate financial support for Arctic environmental and sustainable development programmes. The Arctic Council and the Arctic Regional

[78] D. VanderZwaag, 'International Law and Arctic Marine Conservation and Protection: A Slushy, Shifting Seascape', *Georgetown International Environmental Law Review*, Vol. 9, 1997, pp. 303 and 343–4.

[79] See, e.g., V. Kotov and E. Nikitina, 'Regime and Enterprise: Norilsk Nickel and Transboundary Air Pollution', in D. G. Victor, K. Raustiala and E. B. Skolnikoff (eds.), *The Implementation and Effectiveness of International Environmental Commitments: Theory and Practice* (Cambridge, MA: MIT Press, 1998), pp. 549–74.

[80] UN doc. A/53/456, of November 1998; available at www.un.org/Depts/los.

Programme of Action continue to depend on voluntary financial contributions, and funding for the Russian National Plan of Action on Land-Based Pollution remains uncertain.

Funding of international environmental initiatives relevant to the Arctic also continues to be problematic. For example, UNEP has relied on voluntary financial contributions to support negotiation efforts for a global POPs convention, and a shortfall of approximately US$3 million has been announced for negotiations to be held between 1999 and 2001. UNEP has resorted to encouraging contributions through a 'POPs Club' where contributors to the trust fund for global negotiations on POPs receive recognition through yearly certificates, and substantial contributors receive silver or gold pins.[81]

A further aspect of lack of progress in addressing long-range and global atmospheric pollution problems is the failure to have a strong 'united polar voice' in international fora. While wildlife in the Antarctic also displays elevated levels of various organochlorines transported from outside the region,[82] there has been no strong South Pole lobby for stringent controls on POPs. The lack of indigenous peoples in Antarctica at least partly explains this difference. Environmental effects on polar regions have also tended to be treated separately: for example, the disturbing news of major retreats in the Larsen B and Wilkens ice shelves by nearly 3,000 km^2 has not been effectively linked to Arctic ice variabilities such as the reported thinning of year-round sea-ice over the Arctic Ocean by about 25 centimetres over the past decade.[83]

Future directions in initiatives to address land-based marine pollution related to the Arctic remain uncertain. Several main scenarios can be identified. One is *fragmented incrementalism*, which in the near term appears most likely, with continuing implementation efforts under the UN/ECE Protocols on POPs and heavy metals and the Arctic Regional Programme of Action. A global POPs convention is likely to mobilise efforts in countries outside the region to develop alternatives to POPs and strengthen toxic chemical controls. Global governance innovations, such as a reformed UN Environment Organisation with powers to impose global environmental standards,[84] or a comprehensive international convention on land-based marine pollution, appear to be distant prospects.

[81] 'International Action to Protect Health and the Environment Through Measures Which Will Reduce or Eliminate Emissions and Discharges of Persistent Organic Pollutants, Including the Development of an International Legally Binding Instrument', *Report of the Executive Director*, UNEP/GC.20/41, 7 January 1999.

[82] See T. E. Bidleman, M. D. Walla, R. Roura, E. Carr and S. Schmidt, 'Organochlorine Pesticides in the Atmosphere of the Southern Ocean and Antarctica, January–March 1990', *Marine Pollution Bulletin*, Vol. 26, 1993, pp. 258–62; and M. Oehme, M. Schlabach and I. Boyd, 'Polychlorinated Dibenzo-p-dioxins, Dibenzofurans and Coplanar Biphenyls in Antarctic Fur Seal Blubber', *Ambio*, Vol. 24, 1995, pp. 41–6.

[83] 'Melting of Antarctic Ice Shelves Accelerates', and 'Scientists Ask Why Ice Cap is Melting', *Climate News*, text available at www.listproc.mbnet.mb.ca:8080guest/guest/archives.climate-l/climate-I.9904/msg00013.html.

[84] G. Palmer, 'New Ways to Make International Environmental Law', *American Journal of International Law*, Vol. 86, 1992, pp. 259, 280–2.

The other scenario would envisage conclusion of *Arctic-specific environmental agreements*, which could take various forms, including: the UNEP regional seas model, i.e. a framework agreement with protocols addressing particular priorities, such as land-based pollution, contingency planning and biodiversity protection; a single, comprehensive environmental agreement addressing various pollution problems including land-based and vessel-source (models include agreements for the Baltic, the northeast Atlantic and the Antarctic[85]); a stand-alone convention on Arctic land-based pollution; and a broader sustainable development and environmental protection convention for the Arctic covering both marine and terrestrial areas.[86]

Various arguments can be made in favour of a treaty approach. Public allocation and political profile could be raised, compliance and enforcement obligations could be strengthened and financial commitments might be enhanced.[87] More important than the form of a future agreement, or agreements, are the substantive approaches of future cooperative efforts. If sustainable development is to be taken seriously, a strong precautionary approach will have to be guaranteed through such means as making sure that development proponents bear the burden of demonstrating no serious or irreversible harm (or some other standard) to Arctic ecosystems. Waste minimisation planning and clean technologies need to be strongly advocated. The next major decision point for Arctic marine management arrangements, including land-based pollution controls, is likely to be in 2002. It remains to be seen whether the PAME Working Group's updated report on international instruments will remain the same in terms of recommending against new agreements.

At least two realities may hinder the formalisation of regional arrangements in treaty form. First, Arctic states are likely to be hesitant to skate in new directions as long as the Arctic Council is still 'learning to stand'. Secondly, the practicality of a 'regional sea' treaty approach has increasingly come under question, in view of the evolving regulatory efforts at global and extra-regional levels.

Several common environmental features stand out for both the Arctic

[85] For a discussion of the Protocol on Environmental Protection to the Antarctic Treaty, see F. Orrego Vicuña, 'The Effectiveness of the Protocol on Environmental Protection to the Antarctic Treaty', and 'The Legitimacy of the Protocol on Environmental Protection to the Antarctic Treaty' in O. S. Stokke and D. Vidas (eds.), *Governing the Antarctic: The Effectiveness and Legitimacy of the Antarctic Treaty System* (Cambridge University Press, 1996), pp. 174–202 and 268–93. See also C. C. Joyner, *Governing the Frozen Commons: The Antarctic Regime and Environmental Protection* (Columbia, SC: University of South Carolina Press, 1998), pp. 147–80; and D. Vidas (ed.), *Implementing the Environmental Protection Regime for the Antarctic* (Dordrecht: Kluwer Academic Publishers, 2000).

[86] D. VanderZwaag, 'Regionalism and Arctic Marine Environmental Protection: Drifting between Blurry Boundaries and Hazy Horizons', in D. Vidas and W. Østreng (eds.), *Order for the Oceans at the Turn of the Century* (The Hague: Kluwer Law International, 1999), p. 247.

[87] D. Scrivener, 'Environmental Cooperation in the Arctic: From Strategy to Council', *Security Policy Library, No. 1/1996* (Oslo: Norwegian Atlantic Committee, 1996), p. 27.

and the Antarctic. The greatest threats are arguably not from within but from outside. Global warming and ozone depletion, followed by long-range transport of toxic substances, appear to be the most pressing concerns for both polar regions and their marine environments.

9 Radioactive waste in the Barents and Kara Seas: Russian implementation of the global dumping regime

OLAV SCHRAM STOKKE*

During the 1990s, protection of the Arctic marine environment has attracted intense political attention, engaging diplomats, parliamentarians, researchers and non-governmental organisations across the Arctic rim – and well beyond. The disclosure of Soviet dumping of radioactive waste in the Barents and Kara Seas is among the main reasons for this. It is now clear that such dumping has been conducted for decades – by the Northern Fleet as well as by the civilian Murmansk Shipping Company, the operator of nuclear-run icebreakers in the Northern Sea Route. Measured at the time of disposal, the total radioactivity dumped into Arctic seas by the Soviet Union is twice as high as that of all previously known dumping worldwide.[1] The most intensely radioactive type of waste stems from nuclear vessel reactors which still contain high-level spent fuel.

Parts of this dumping occurred in violation of Soviet commitments to the 1972 Convention on the Prevention of Marine Pollution by Dumping of Wastes and Other Matter[2] (London Convention); this forms the point of departure for this chapter. In particular, we will focus on how international regimes may affect domestic implementation in member states.[3] The core of the argument is that Soviet and later Russian management of nuclear waste in the north has been significantly influenced by regulations and programmes generated under international dumping instruments.

* I would like to thank Davor Vidas for his very helpful comments. Part of the material in this chapter draws upon O. S. Stokke, 'Nuclear Dumping in Arctic Seas: Russian Implementation of the London Convention', in D. G. Victor, K. Raustiala and E. B. Skolnikoff (eds.), *The Implementation and Effectiveness of International Environmental Commitments: Theory and Practice* (Cambridge, MA: MIT Press, 1998), pp. 475–517.

[1] Of a total of 136,682 TBq, Soviet dumping in Arctic seas from 1960 to 1991 accounted for 90,152 TBq; see K.-L. Sjoeblom and G. Linsley, 'Sea Disposal of Radioactive Wastes: The London Convention 1972', *IAEA Bulletin*, Vol. 37, 1995, p. 14.

[2] ILM, Vol. 11, 1972, pp. 1,291ff. The Convention was adopted in London, on 13 November 1972 and entered into force on 30 August 1975; the Soviet Union ratified it in 1976.

[3] 'Implementation' is here understood as the process of converting international agreements into behavioural adaptation on the part of target groups.

THE PROBLEM OF RADIOACTIVE WASTE

More than five decades after the first controlled nuclear fission, no one has come up with a widely accepted solution to the problem of how to deal with the most highly radioactive products – high-level waste and spent nuclear fuel. According to the International Atomic Energy Agency (IAEA), 'high-level waste' comprises irradiated reactor fuel, liquid or solidified wastes from the first solvent extraction cycle of chemical reprocessing (or equivalent processes) of such fuel, or any other matter of activity concentration exceeding certain limits specified for alpha, beta/gamma, and tritium emitters.[4] Globally, the spent fuel produced by the military sector is modest compared to that from the civilian sector, but in the north, the nuclear waste dumped by the Soviet Union in Arctic seas is chiefly of military origin. As documented in the Yablokov Report, a Russian governmental White Paper published in 1993, as many as sixteen nuclear reactors have been dumped in the Kara Sea since 1965; seven of these are especially dangerous because of a failure to remove spent fuel prior to disposal.[5] In addition, large amounts of low- and medium-level solid waste have been dumped by the Northern Fleet in flimsy metal containers that are highly liable to corrosion. And liquid waste – like water used in cooling, incineration or deactivation of radioactive installations – has been disposed of in the Barents Sea since the mid-1960s. This past dumping is a matter of substantial concern in Russia and its neighbouring states as well. Various remedial measures have been considered, including sealing, capping and retrieval for storage on land.[6] Such action, however, may itself involve great hazards and would definitely be very costly. Measurements at several sites in the Barents and Kara Seas, including the dump-sites for hot reactors in some bays of Novaya Zemlya, indicate that so far there has not been significant release of radioactivity into the marine environment.[7] Indeed, levels in these seas are comparatively low, and certainly much lower than in the Black Sea or the Baltic.[8] Simulation models suggest that even a worst-case scenario of rapid release of all the dumped activity would not result in considerable danger to marine food-chains, although local-scale

[4] IAEA Safety Series No. 78, reproduced in *The London Dumping Convention: The First Decade and Beyond* (London: International Maritime Organisation, 1991).

[5] A. V. Yablokov, V. K. Karasev, V. M. Ruyantsev, M. Y. Kokeyev, O. I. Petrov, V. N. Lystsov, A. F. Yemelyanenkov and P. M. Rubtsov, *Facts and Problems Related to Radioactive Waste Disposal in Seas Adjacent to the Territory of the Russian Federation* (Albaquerque: Small World Publishers, 1993).

[6] See Office of Technology Assessment (OTA), *Nuclear Wastes in the Arctic: An Analysis of Arctic and Other Regional Impacts from Soviet Nuclear Contamination* (Washington, DC: Office of Technology Assessment, Congress of the United States, 1995), pp. 68–9.

[7] Joint Russian–Norwegian Expert Group for Investigation of Radioactive Contamination in the Northern Areas, *Dumping of Radioactive Waste and Investigation of Radioactive Contamination in the Kara Sea: Results from 3 Years of Investigations (1992–1994) in the Kara Sea* (Østerås: Norwegian Radiation Control Authority, 1996), pp. 42–9.

[8] North Atlantic Treaty Organisation (NATO), *Cross-Border Environmental Problems Emanating from Defence-Related Installations and Activities: Volume 1, Radioactive Contamination (Final Report)* (Brussels: North Atlantic Treaty Organisation, 1995), p. 287.

effects would need to be studied more.[9] These conclusions should be seen as preliminary, as considerable uncertainty attends both the rate of release and the transport models underlying them.[10]

Even more alarming than past dumping is the current imbalance between the steady generation of new waste and Russia's capacity to deal with it. First, the 100-odd nuclear-powered vessels currently operated by the Northern Fleet regularly generate large amounts of both solid and liquid waste, yet adequate storage or treatment facilities are lacking. As for spent nuclear fuel, the highly deficient temporary storage facilities for removed fuel assemblies are already full to capacity. Secondly, the compilation of waste will accelerate further in the years to come, as submarines are taken out of operation due to old age or to comply with commitments under the Strategic Arms Reduction Treaty regime.[11] Sixty Northern Fleet vessels were laid up in the period from 1989 to 1993, and it is expected that another thirty will be scrapped within the next few years.[12] Only a fraction of the vessels taken out so far have been properly decommissioned by removal of reactor fuels and the reactor section. According to Western sources, in 1994 the dismantlement capacity of the Northern Fleet was one submarine a year[13] – partly due to lack of storage facilities for the reactor cores and an inadequate system of transporting the waste out of the region,[14] but also because of a tendency to allocate scarce docking facilities to the reloading of operative vessels rather than the unloading of laid-up ones.

Hence, the backbone of radioactive waste management, a key problem addressed by the London Convention, is *adequate storage*. This involves interim storage on the site where waste is generated, as well as a satisfactory system for transporting high-level waste and spent fuel for final deposition or, in the case of spent fuel, reprocessing.[15] In practice, it also involves treatment capacity for concentrating or solidifying liquid waste and for compacting solid waste to facilitate storage. Ever since the 1960s the Northern Fleet in particular, but the Murmansk Shipping Company as well, have experienced a widening gap between actual and needed capacity along those dimensions; and this is the basic reason why both

[9] See A. Baklanov, R. Bergman and B. Segerståhl, *Radioactive Sources in the Kola Region: Actual and Potential Radiological Consequences for Man. Final Report of the Kola Assessment Study of the RAD Project* (Laxenburg: International Institute for Applied Systems Analysis, 1996).

[10] OTA, *Nuclear Wastes in the Arctic*, pp. 89, 108.

[11] See, respectively, the Treaty on the Reduction and Limitation of Strategic Offensive Arms, Moscow, 31 July 1991; in force 5 December 1994 (START I Treaty); and the Treaty on Further Reduction and Limitation of Strategic Offensive Arms, Moscow, 3 January 1993 (START II Treaty).

[12] Report to the Storting, *St.meld. 34 (1993–94), Atomvirksomhet og kjemiske våpen i våre nordlige nærområder*, p. 20. For Russia as a whole, the total number is 170 by the year 2000; the comparative figure for the United States is 120; see NATO, *Cross-Border Environmental Problems*, p. 276.

[13] NATO, *Cross-Border Environmental Problems*, p. 283.

[14] N. N. Yegorov, 'Plenary Address', *International Cooperation on Nuclear Waste Management in the Russian Federation* (Vienna: International Atomic Energy Agency, 1995), pp. 15–26.

[15] While several are working on programmes for final disposal, mostly opting for deep underground sites in stable geological strata, the first operative repository is still at least twenty years away; see *IAEA Yearbook 1995* (Vienna: International Atomic Energy Agency, 1995), p. C83.

have resorted to the dumping of some of the waste generated in the nuclear complex in Russia's northwest.

THE GLOBAL DUMPING REGIME

The basic principle of the regime based on the 1972 London Convention is that the disposal at sea of hazardous waste – defined in terms of toxicity, persistence and tendency to bioaccumulate in marine organisms – must be forbidden, save in cases where all other options are deemed more harmful.[16] Putting this into practice involves at least three types of activities: (1) generating the knowledge necessary to enable informed choices; (2) adopting regulative measures which give life to the principles and take heed of existing knowledge; and (3) sustaining a collective system to further compliance, including reporting and verification of whether international commitments are matched by behavioural adaptation. While radioactive waste is only one of the substances dealt with by this Convention, it has been the single most politicised issue.

The main decision-making body is the Consultative Meeting of the Parties, usually held every year. A 'black' and 'grey' list system is applied, in which 'black' items may not be dumped at all, whereas 'grey' ones require special permits from a designated national authority to be reported to the secretariat of the Convention,[17] located with the International Maritime Organisation (IMO). Members are obliged to monitor and keep a record of the nature and quantities of matter permitted to be dumped as well as when, where and how such dumping occurred and the condition of the seas where it took place.[18] When the 1996 Protocol enters into force, a reverse listing will be introduced: all dumping will be prohibited unless explicitly permitted; the impact of this is further enhanced by a strong statement of the precautionary principle.[19] Unlike many other international arrangements, the London Convention permits regulative decisions to be taken without unanimity: amendments to the lists may be passed by a two-thirds majority, balanced however by an opt-out clause allowing states to avoid being legally

[16] See *Report of the Fourth Consultative Meeting of Contracting Parties to the Convention on the Prevention of Marine Pollution by Dumping of Wastes and Other Matter*, IMO doc. LDC 4/12, Annex 2; see also the discussion in P. W. Birnie and A. E. Boyle, *International Law and the Environment* (Oxford: Clarendon Press, 1992), p. 321. Those main criteria also guide regulative decisions under regional conventions such as the 1992 OSPAR and 1974 Helsinki Conventions; see, respectively, Convention on the Protection of the Marine Environment of the Baltic Sea Area, Helsinki, 22 March 1974, reproduced in ILM, Vol. 13, 1974, pp. 546–84, and Convention for the Protection of the Marine Environment of the North-East Atlantic, Paris, 22 September 1992, reproduced in ILM, Vol. 32, 1993, pp. 1,069ff. See further VanderZwaag, Chapter 8 in this book.

[17] Art. IV(1) and (2), and Art. VI, respectively. [18] Art. VI(1).

[19] Compare Arts. 3 and 4 of the 1996 Protocol to the Convention on the Prevention of Marine Pollution by Dumping of Wastes and Other Matter, 1972 and Resolutions Adopted by Special Meetings, London, 7 November 1996, reproduced in ILM, Vol. 36, 1997, pp. 7ff, with Art. IV of the 1972 London Convention. On the emergence of the precautionary principle, see in general Birnie and Boyle, *International Law and the Environment*, pp. 97ff; see also VanderZwaag, Chapter 8 in this book.

bound by provisions they do not wish to adhere to.[20] A tacit consent procedure, whereby amendments become binding on the parties after 100 days unless they file a reservation, adds speed to the implementation process.[21] In addition, the meeting may adopt non-binding resolutions by simple majority. As to enforcement, the London Convention sets out a broad range of provisions for the prevention, discovery and punishment of violations, obliging members to enforce rules in their capacities as, respectively, flag states, port states and coastal states; the latter can apply the Convention not only to their territorial waters but to their exclusive economic zones and continental shelves as well.[22] A dispute settlement arrangement (adopted in 1978, but yet to enter into force) provides for arbitration or submission to the International Court of Justice.[23]

While the London Convention forms the core of the international dumping regime, other global and regional processes complement it. The obligation to control dumping is confirmed by the 1982 Law of the Sea Convention, which in Article 210 refers implicitly to the London Convention and its annexes when requiring that national regulation shall be no less effective than the rules and standards set globally.[24] As to radioactive waste, the Helsinki Convention targeting the Baltic Sea banned dumping of radioactive waste in 1974;[25] and, in 1992, the OSPAR Convention elicited commitments to this effect from two of the most outspoken recalcitrants in the London process, the United Kingdom and France.[26]

Since the late 1980s, various cooperative political vehicles have been set in motion in the Arctic realm. Those processes, including their interaction with activities under the London Convention, are also important to the current management of marine disposal of nuclear waste. At the bilateral level, several Russo-Norwegian research cruises into the Barents and Kara Seas were launched in the 1990s, endorsed rather than initiated by London Consultative Meetings, for the purpose of gauging nuclear contamination in water masses and subsoil sediments

[20] Art. XV(1) and (2).

[21] Art. XV(2); see also A. Kiss and D. Shelton, *International Environmental Law* (Ardsley-on-Hudson, NY and London: Transnational Publishers and Graham & Trotman, 1991), p. 102; a more general discussion of procedural mechanisms designed to get around the 'slowest-boat' problem in international regimes is provided by P. H. Sand, 'Lessons Learned in Global Environmental Governance', *Environmental Affairs Law Review*, Vol. 18, 1991, pp. 213–77.

[22] See IMO doc. LDC 11/14, p. 32.

[23] IMO doc. LDC 3/12, p. 11; see also Annex 4.

[24] Birnie and Boyle, *International Law and the Environment*, p. 320; UN Convention on the Law of the Sea, Montego Bay, 10 December 1982, UN doc. A/CONF.62/122, reproduced in ILM, Vol. 21, 1982, pp. 1,261ff. For a condensed analysis of this relationship between the London Convention and the Law of the Sea Convention, see J. L. Canfield, 'Soviet and Russian Nuclear Waste Dumping in the Arctic Marine Environment: Legal, Historical, and Political Implications', *Georgetown International Environmental Law Review*, Vol. 6, 1994, pp. 353–444, especially pp. 358–60.

[25] Art. 9 of the Helsinki Convention.

[26] Annex 2, Art. 3(3) of the OSPAR Convention. The OSPAR prohibition would expire after fifteen years; France and the United Kingdom also unsuccessfully opted for this solution in the London Convention; see IMO doc. LC 16/14, p. 16. The International North Sea Conference had already agreed in 1990 that the North Sea was unsuitable for the dumping of radioactive waste; see Birnie and Boyle, *International Law and the Environment*, p. 324.

in the areas close to dumping sites.[27] For its part, the trilateral Declaration on Arctic Military Environmental Cooperation (AMEC), involving the defence ministries of Russia, Norway and the United States, has framed several projects aimed at enhancing nuclear safety practices in northwest Russia.[28] And the fairly ambitious Arctic Monitoring and Assessment Programme (AMAP) under the 1991 Arctic Environmental Protection Strategy, which has singled out radionuclides as a priority area, submitted its major reports on the state of the Arctic environment in 1997 and 1998.[29] Thus, on both the regulative and the programmatic side, the London Convention interlocks with a range of other cooperative processes, largely on a regional and sometimes bilateral level.

Since the adoption of the London Convention, a system of *scientific advice* has been elaborated, with three strands. The broadest advisory mechanism is the Scientific Group on Dumping, comprising experts nominated by the parties, which achieved permanent status in 1984.[30] Secondly, a range of *ad hoc* groups of experts has been set up to compile information and further recommendations on particularly vital or controversial matters, such as the Panels on Sea Disposal of Radioactive Waste formed in 1983 and 1985.[31] Similarly, in 1987 the Inter-Governmental Panel of Experts on Radioactive Waste Disposal at Sea (IGPRAD) began addressing the wider political, legal, economic and social aspects of radioactive waste dumping, the comparative costs and risks of dumping as compared to land-based disposal, and whether it can be proven that radioactive dumping is not harmful to human life or the marine environment.[32] IGPRAD's final report in 1993 paved the way for the subsequent global prohibition of all dumping of radioactive waste at sea.[33]

A third strand of the information-related activities generated by the London Convention is the work conducted by external organisations at the request of the Consultative Meetings. The significance of being able to trigger or forward investigations conducted by others becomes clear when we note that in 1990, the budget of the London Convention was a mere US$0.76 million, and the IMO staff allocated to it consisted of five persons.[34] The International Atomic Energy Agency (IAEA), with a budget of roughly US$225 million and a staff of some

[27] For a more detailed analysis, see Stokke, Chapter 6 in this book.

[28] Text available at www.denix.osd.mil/denix/Public/Intl/AMEC/declar.html. See an analysis by S. G. Sawhill, 'Cleaning-Up the Arctic's Cold War Legacy: Nuclear Waste and Arctic Military Environmental Cooperation', *Cooperation and Conflict*, Vol. 35, 2000, pp. 5–36.

[29] The two reports by the Arctic Monitoring and Assessment Programme, *AMAP Assessment Report: Arctic Pollution Issues* (Oslo: Arctic Monitoring and Assessment Programme, 1998); and *Arctic Pollution Issues: A State of the Arctic Environment Report* (Oslo: Arctic Monitoring and Assessment Programme, 1997). See Vidas, Chapter 4 in this book.

[30] IMO, *The London Dumping Convention*, p. 117.

[31] See, respectively, IMO doc. LDC 7/12, pp. 19–30 and Annex 6; IMO doc. LDC 8/10, pp. 19–20, and Annex 4; and IMO doc. LDC 9/12, pp. 19–29.

[32] IMO doc. LDC 10/15, Annex 11.

[33] IMO doc. LC 16/14, pp. 19–20.

[34] P. H. Sand (ed.), *The Effectiveness of International Environmental Agreements: A Survey of Existing Legal Instruments* (Cambridge: Grotius Publications, 1992), p. 16.

2,000,[35] has been vital to the work of IGPRAD by conducting several specialised technical and scientific studies.[36]

In terms of *regulative provisions* pertaining to radioactive waste, high-level radioactive waste was placed on the original black list in 1972 – and state parties are thus obliged to abstain from any dumping of such material.[37] While that prohibition had been highly controversial, at first strongly opposed by the United Kingdom and the United States,[38] subsequent regulative discussion on nuclear matters revolved around extending it to low- and medium-level waste as well. The parties to the London Convention had designated the IAEA as the competent international advisory authority on whether given nuclear materials are unsuitable for dumping. Accordingly, the IAEA set up geographic criteria for the localisation of such dumping,[39] including requirements that it should occur only in the belt between 50° North and 50° South latitude, beyond the continental shelf and at depths greater than 4,000 metres. The Barents and Kara Seas are located roughly between 70° and 80° North; moreover, most of the area is on a continental shelf with depths rarely exceeding a few hundred metres.

In 1983 a proposed ban failed to gain sufficient support, but Spain, strongly backed by South Pacific and Nordic countries, successfully sponsored a resolution on a *voluntary moratorium* on all dumping of radioactive materials until an expert meeting had presented their final report to the contracting parties.[40] The Soviet Union abstained from voting,[41] as it also did when the moratorium was prolonged in 1985; the reasons cited were that the moratorium lacked adequate scientific basis and violated the spirit of consensus underlying the Convention.[42] Four years later, the Soviet delegation officially declared that it had not dumped such materials in the past, and would not do so in the future.[43] But when in 1993 a binding prohibition on the dumping of low- and medium-level waste was established unanimously, Russia was among the five states abstaining from the vote.[44]

[35] *Yearbook of International Co-operation on Environment and Development 1999/2000* (London: Earthscan Publications, 1999), pp. 221–4; of those, more than 800 are professional scientists.

[36] IMO doc. LDC 13/15, p. 32.

[37] London Convention, Annex 1.

[38] The Soviet Union had favoured an even more comprehensive prohibition, including not only high-level but also low- and medium-level waste; see L. Ringius, *Radwaste Disposal and the Global Ocean Dumping Convention: The Politics of International Environmental Regimes* (Florence: Thesis towards the Degree of Doctor of the European University, Department of Political Science, 1992), pp. 9, 114. This view was repeated by Soviet delegations on later occasions; see for instance IMO doc. LDC 5/12, p. 12.

[39] IAEA doc. INF CIRC/205/Add.1/Rev 1, Convention on the Prevention of Marine Pollution by Dumping of Wastes and Other Matter: The Definition Required by Annex I, para. 6 to the Convention, and the Recommendations Required by Annex II, sec. D (Vienna: International Atomic Energy Agency, 1978).

[40] IMO doc. LDC 7/12, pp. 19–30. The voluntary moratorium was established by Resolution LDC 14 (7), reproduced in IMO doc. LDC 7/12, Annex 3.

[41] The states voting against were Japan, the Netherlands, South Africa, Switzerland, the United Kingdom and the United States; see IMO doc. LDC 7/12, p. 29.

[42] See IMO doc. LDC 9/12, p. 41 and Annex 5.

[43] Yablokov *et al.*, *Facts and Problems*, section 1.1.

[44] See Resolution LC 51 (16), reproduced in IMO doc. LC 16/14, Annex 5. IMO doc. LC 16/14, p. 17. The four other abstainees were the United Kingdom, Belgium, France and China.

Having tried in vain to obtain a two-year delay, Russia filed, as the only contracting party, a formal reservation to the amendment, so that it is currently not formally bound by this prohibition.[45]

The *compliance system* of the London Convention is the weak part of its implementation profile.[46] This system is based largely on self-reporting; in addition to a widespread inclination to ignore existing obligations to file reports, there is scant opportunity for the Secretariat or other members to subject reports to critical assessment. Nor can the regime, at least directly, provide significant positive incentives to induce compliance with its requirements. It should be noted here that relatively undeveloped compliance systems are quite common for environmental and resource management regimes.[47] To some extent and in some situations, the formal reporting system of the Convention is complemented by information made available to the meetings by non-governmental organisations with access to the deliberations. Thus, it was a document presented by Greenpeace International that triggered the animated discussion at the 1991 Consultative Meeting on Soviet dumping in Arctic seas, which in turn produced a Soviet pledge to submit more information on the matter to the Secretariat.[48]

IMPLEMENTING THE DUMPING REGIME: THE RUSSIAN CASE

The major source of the radioactive waste dumped into Arctic seas is the Soviet, later Russian, Northern Fleet, based on the Kola Peninsula. It has thus been the key target for regulations in this field. A second regional target is the Murmansk Shipping Company, which operates seven nuclear icebreakers engaged in keeping the Northern Sea Route open, especially the western part between Murmansk and Dudinka on the banks of the Yenisey. In addition, the nuclear icebreaker *Lenin* has been taken out of operation. The civilian nuclear power plant in Polyarnye Zori in Murmansk *oblast* has not engaged in dumping of waste in Arctic seas, so it is not among the relevant target groups in our context.

As to domestic regulative agencies, two sets of distinctions are particularly relevant. One is the classic differentiation between legislative, executive and judicial powers. In matters directly related to foreign affairs and international commitments, the normal situation in most countries is that the executive will be in charge unless the matter becomes politicised enough to engage one or both of the others. In the Soviet case, the judiciary has failed to play an independent role.

[45] IMO doc. LC 17/14, p. 6.

[46] See also M. Nauke and G. L. Holland, 'The Role and Development of Global Marine Conventions: Two Case Histories', *Marine Pollution Bulletin* (Special Issue on Progress and Trends in Marine Environmental Protection), Vol. 25, 1992, pp. 75–9.

[47] For an overview of a range of environmental agreements in this respect, see S. Andresen, 'International Verification in Practice: A Brief Account of Experiences from Relevant International Cooperative Measures', in E. Lykke (ed.), *Achieving Environmental Goals: The Concept and Practice of Environmental Performance Review* (London: Belhaven Press, 1992), pp. 101–21.

[48] IMO doc. LDC 14/16, pp. 36–7.

And, for most of its lifetime, the Soviet political system was marked by a strong executive: while the formal apex of power was placed in the legislative Supreme Soviet, real power resided in the Communist Party and was wielded primarily through the huge bureaucratic apparatus coordinated by the Council of Ministers. When a decree was issued in 1990 on measures to improve implementation of previous legislation to protect the northern environment, the relevant Supreme Soviet committee was not even consulted.[49] The introduction of presidential rule the same year implied some executive de-linking from the Communist Party;[50] the 1993 Constitution endowed the President of the Russian Federation with extensive powers, including the right to overrule legislative initiatives and to issue legally binding decrees. However, in the period from the dissolution of the Soviet Union to the 1993 assault on the Parliament by troops loyal to President Yeltsin, the legislature was very active on nuclear matters in the north, especially regarding nuclear testing at the Novaya Zemlya site.[51]

A second distinction regarding regulative agencies may be termed territorial. In the Soviet and later Russian context, it is generally helpful to scrutinise both federal and regional levels of government.[52] However, in the case of nuclear waste management, we do not lose much by blackboxing the latter because, while there have been a few recent attempts on the part of regional governments to regulate the nuclear safety practices of the military, they have been futile. In 1991, for instance, the governor of Murmansk set up operational rules for the removal of spent fuel from nuclear reactors in the naval bases;[53] those rules were stillborn, however, because physical access to the bases is up to the military to decide. The Northern Fleet flatly turned down a 1993 request from the environmental committee in the Murmansk *oblast* administration for information on nuclear waste management on the bases, although a visit was granted to one base two years later.[54] And when Yeltsin decreed in 1992 that the lands on which the Novaya Zemlya nuclear test site is located should be federalised, county authorities in Arkhangelsk were neither consulted nor informed prior to the decision.[55]

The politics of publicity

Throughout the 1960s and 1970s, Soviet handling of nuclear waste was a closed policy matter with few access points, and the pattern of inclusion clearly

49 Canfield, 'Soviet and Russian Nuclear Waste Dumping', p. 371.

50 Å. Egge, *Fra Alexander II til Boris Jeltsin. Russlands og Sovjetunionens moderne historie* (Oslo: Universitetsforlaget, 1993), p. 270.

51 On the role of the legislative Supreme Soviet in this matter, see Canfield, 'Soviet and Russian Nuclear Waste Dumping', pp. 375–9.

52 Under the 1993 Constitution, the Russian Federation has a total of eighty-nine subjects, which may be either republics, counties (*oblast*), territories (*kray*) or autonomous areas (*okrug*).

53 R. Castberg and O. S. Stokke, 'Environmental Problems in Northwest Russia: Regional Strategies', *International Challenges*, Vol. 12, 1992, pp. 33–45.

54 T. Nilsen, N. Bøhmer and A. Nikitin, 'Den russiske Nordflåten. Kilder til radioaktiv forurensning', *Bellona rapport*, No. 2 (Oslo: Bellona Foundation, 1996), p. 87.

55 Canfield, 'Soviet and Russian Nuclear Waste Dumping', p. 376.

biased in favour of the Navy. Twelve of the sixteen reactors disposed in the Kara Sea were dumped in this period, all of them before the entry into force of the London Convention.[56] In addition, the liquid and solid low- and medium-level waste dumped in this period fluctuated between close to zero in some years and some 300 TBq in a peak year.[57]

Largely because of their military significance, most aspects of the nuclear programmes of the former Soviet Union have been shrouded in a thick veil of secrecy. In the immediate post-World War II years, marked by a determined effort to catch up with the USA, the nuclear programme was placed under the Minister for State Security, Lavrenti Beria, who directed the establishment of several closed nuclear laboratories in secluded cities.[58] As of 1990, there were more than 100 such 'secret cities', some with tens of thousands of inhabitants, omitted from official maps and with strictly controlled access. Several of these, such as Arzamas-16 or Chelyabinsk-65, are key components of the Russian nuclear-military complex today. Except for a short period in the late 1950s, when Sakharov corresponded with Khrushchev on the matter and was allowed to publish several critical articles, the public nuclear discourse in the Soviet Union before Chernobyl was either non-existent or silent about problems and hazards involved. Yet another illustration of the traditional difficulty of gaining access to information about the Soviet nuclear complex is the way crises and accidents have been handled by Soviet officials at home and abroad. An explosion at the nuclear facility in Kyshtym in 1957, for instance, was denied by Soviet officials until 1989,[59] although details of the accident had been published in the West a decade earlier.[60]

On the other hand, there is nothing so uncommon about a general line of secrecy in nuclear affairs. Even in the United States, with a greater tradition of openness, organised opposition to nuclear waste management has largely been limited to the civilian sector, principally because access to information is confined to this sector.[61] Thus, a North Atlantic Cooperation Council report on cross-border environmental problems associated with military installation notes, before detailing the situation in Russia, that little is known about the temporary storage of spent nuclear fuel from Western naval vessels.[62]

In the late 1980s, however, a sea change occurred regarding both access rules and patterns of participation in Soviet environmental affairs; unlike in the

56 Yablokov *et al.*, *Facts and Problems*, section 2.3.1.

57 Due to very large discharges of liquid waste in the Kara Sea, 1976 was such a peak year. The total activity of low- and medium-level waste dumped by the Soviet Union in Arctic seas, measured at the time of disposal, was 1,342 TBq; see NATO, 'Cross-Border Environmental Problems', pp. 17 and 33. As noted, the total activity dumped by the Soviet Union, including the reactors with spent nuclear fuel, is about 90,000 TBq.

58 S. P. Weart, *Nuclear Fear: A History of Images* (Cambridge, MA: Harvard University Press, 1988), p. 122.

59 A. Blowers, D. Lowry and B. D. Salomon, *The International Politics of Nuclear Waste* (London: Macmillan Press, 1991), p. 40.

60 Z. Medvedev, *Disaster in the Urals* (London: Angus and Robertson, 1979).

61 Blowers *et al.*, *The International Politics of Nuclear Waste*, p. 240.

62 NATO, 'Cross-Border Environmental Problems', p. 283.

past, the nuclear area was no exception. While semi-official organisations for nature protection had thrived throughout the Soviet period, a critical environmental movement independent of state authorities did not emerge until the mid-1980s; and, when it did, nuclear fear was an important stimulus. These so-called 'informals' were able to organise large street demonstrations and public hearings on the ecological situation in the Russian northwest.[63] A survey carried out in 1990 in 850 cities throughout the Soviet Union indicated that more than half the respondents were unhappy with the environmental situation, and radiation was on the top-three list of worries cited.[64] And while today's economic hardships are, quite predictably, making it more difficult for the environmental movement to command the political attention of the average northwest Russian, such fluxes have been observed in the West as well. They need not mean that the Russian 'greens' are a thing of the past.

Assessing the hazards

When dumping of solid radioactive waste began in the early 1960s, the Northern Fleet was itself responsible for mapping the environmental situation around the sites used. This was conducted by four research institutions administered by the Defence Ministry, and all investigations proved very reassuring for the military.[65] The faintness of this praise is revealed by the fact that, after 1967, no water or sedimentary measurements were taken closer than 50 kilometres away from the solid waste disposal areas around Novaya Zemlya.[66] This respectful distance to the most interesting dumping sites was maintained even after *Goskomgidromet*, a civilian agency, was assigned responsibility for monitoring these areas, following a 1979 Council of Ministers resolution.[67] Moreover, *Goskomgidromet*'s mandate was never extended to military bases or repair yards.[68]

In 1987, following Gorbachev's reshuffling of the Soviet apparatus, a State Committee on Nature Protection (*Goskompriroda*) was established, and two years later it took over from *Goskomgidromet* the leadership of Soviet delegations meeting under the London Convention. The domestic influence of this agency, under shifting names, was to rise steadily to peak by around 1990, and then decline. One might have expected that the growing clout of *Goskompriroda* would strengthen those actors in the Soviet system who opposed the dumping practices of the Northern Fleet and encourage the formation of an effective coalition to counter the hitherto predominant resisters in the implementation game. Instead, *Goskomgidromet* and *Goskompriroda* reportedly clashed in a disruptive turf

[63] On the emerging environmental movement in this region, see O. A. Andreev and M.-O. Olsson, 'The Ecological Situation and Environmental Organizations in the Russian North-West', *CERUM Working Paper*, No. 15 (Umeå: CERUM, 1992).

[64] M. Feshbach and A. Friendly Jr, *Ecocide in the USSR: Health and Nature under Siege* (London: Arum Press, 1992), p. 238.

[65] Yablokov *et al.*, *Facts and Problems*, sections 3.1 and 3.3.

[66] *Ibid.*, section 3.3.

[67] Resolution 222 on Measures to Ensure Performance of the Soviet Side's Obligations Following from the 1972 [London] Convention, cited in Canfield, 'Soviet and Russian Nuclear Waste Dumping'.

[68] Yablokov *et al.*, *Facts and Problems*, section 3.1.

struggle regarding responsibility for assessment of the radiological situation in the north.[69] Those two institutions held a series of closed meetings between 1988 and 1990 regarding the flagrant disregard of the IAEA guidelines, without being able to generate any action at the level of government.[70]

By that time, military handling of radioactive waste was becoming an international issue. A former radiation safety engineer in the Murmansk Shipping Company, Andrey Zolotkov, who was also an activist in the non-governmental group 'To a New Earth', played an important role in the disclosure of Soviet dumping activities in the Arctic.[71] Zolotkov was also a delegate from Murmansk to the Congress of People's Deputies, an assembly set up in the Gorbachev era as part of the effort to vitalise the legislative branch of government.[72] Zolotkov's stature, and his previous employment in the northern nuclear complex, provided his detailed account of past and ongoing dumping activities with sufficient credibility to generate a huge scandal both domestically and internationally. When his allegations were neither withdrawn nor rejected by competent Soviet authorities, Greenpeace International compiled a report, primarily based on Zolotkov, tabled at a press conference in Moscow in September 1991 and circulated informally at the 1991 meeting under the London Convention, complementing a Soviet–Norwegian information paper on plans for cooperative investigations of the radiological impacts of the alleged dumping.[73] These environmental activists were much helped by the fact that nuclear dumping became politically linked to the even more salient issue of nuclear tests on Novaya Zemlya; indeed, the establishment of the Yablokov Commission, so important to the subsequent Russian implementation game, resulted from a struggle between Yeltsin and the Congress of People's Deputies regarding access to information on the dumping conducted by the test site authority.[74] An earlier accomplishment of the nuclear activists, Yeltsin's 1991 decision on a unilateral moratorium on nuclear tests had been widely interpreted as a strategic move to match Gorbachev's Soviet-level decision to the same effect[75] – an interpretation supported by certain post-Soviet decisions on the part of Yeltsin seen as favouring the nuclear industrial complex, especially the federalisation of the test site area in 1992.

The significance of the London Convention for the creation of the Yablokov Commission, and the subsequent leap in terms of information available on the Russian nuclear complex, should not be exaggerated. True, the 1991 Consultative Meeting had encouraged the compilation of information on past dumping operations,[76] but when the Consultative Meeting sharpened this to an actual request the year after, the Russian delegation responded by outlining the

69 Canfield, 'Soviet and Russian Nuclear Waste Dumping', pp. 371–2.
70 Yablokov *et al.*, *Facts and Problems*, section 2.2.
71 *Ibid.*, preface; in Russian, 'Novaya Zemlya' means 'New Land'.
72 Canfield, 'Soviet and Russian Nuclear Waste Dumping', p. 386; on the establishment of the Congress of People's Deputies, see Egge, *Fra Alexander II til Boris Jeltsin*, p. 268.
73 IMO doc. LDC 14/16, pp. 36–7.
74 Canfield, 'Soviet and Russian Nuclear Waste Dumping', p. 379.
75 *Ibid.*, p. 375.
76 IMO doc. LDC 14/16, p. 37.

broad composition and extensive tasks of an already established fact-finding commission headed by the distinguished scientist, Alexey Yablokov.[77] On balance, internal Russian dynamics were far more important than international requests for generation of support for the Yablokov Commission, especially the inter-institutional rivalries in the transition period which could be exploited by environmental activists and regulative agencies favouring adherence to the London Convention and greater openness on nuclear affairs.

Thus, it was not until the dumping issue became internationalised by the early 1990s that assessment obligations under the London Convention were taken seriously in the Russian northwest. Under the bilateral Russo-Norwegian Environmental Commission, three Russo-Norwegian cruises, involving participation from the IAEA Marine Environmental Laboratory, were conducted from 1991 onwards. They included measurements in the fjords where reactors with remaining spent fuel had been dumped.[78] Encouraged by the Consultative Meeting of the London Convention, the IAEA established an International Arctic Seas Assessment Programme. A combination of continuous political pressure in a range of international fora, including the Consultative Meetings under the London Convention,[79] and the provision from Western participants of equipment, expertise and funds for conferences and working group activities, have been decisive for the generation of adequate information about the hazards associated with the Soviet and Russian dumping of radioactive waste.

Regulating dumping

From the outset, domestic regulation of dumping had been largely left to the Northern Fleet itself and concerned safety precautions for the personnel involved in the operations. The first health requirements were established in 1960, with the Navy in the driver's seat, seconded by the Ministry of Medium Machine Building;[80] the latter used to be the hub of the Soviet nuclear military-industrial complex, operating the network of closed nuclear research cities.[81] In addition, while not having regulative authority, an agency under the Ministry of Health was included in the drafting of these health standards throughout the Soviet period.[82] Involving the same agencies, these regulations were made more specific in 1962

[77] IMO doc. LC 15/16, pp. 38–40.

[78] See Stokke, Chapter 6 in this book; a fourth cruise was conducted by the Norwegian Akvaplan-NIVA and the Murmansk Marine Biological Institute in 1992; see K.-L. Sjoeblom and G. Linsley, 'The International Arctic Seas Assessment Project: Progress Report', *IAEA Bulletin*, Vol. 37, 1995, pp. 25–30.

[79] See in particular IMO doc. LDC 14/16, pp. 36–7; IMO doc. LC 15/16, pp. 38–40; and IMO doc. LC 16/14, pp. 19, 23–4.

[80] Yablokov *et al., Facts and Problems*, section 2.1.

[81] OTA, *Nuclear Wastes in the Arctic*, p. 218; in 1989, this ministry was merged with the Ministry of Nuclear Power to the Ministry of Atomic Power and Industry, renamed in 1992 the Ministry for Atomic Power.

[82] *Ibid.*, p. 219; on the continuation of this situation, see O. I. Shamov, 'Ministry of Health Care and the Medical Industry of the Russian Federation', *International Cooperation on Nuclear Waste Management in the Russian Federation* (Vienna: International Atomic Energy Agency, 1995).

and 1966;[83] but it was the Navy itself that made the key decisions in 1965 and 1967 which permitted dumping of liquid waste beyond ten miles and the dumping of solid waste in thin metal containers, or even without containment, using the Barents Sea for liquid waste and the bays of Novaya Zemlya for solid waste.[84]

When in 1979, three years after Soviet ratification of the London Convention, the Council of Ministers passed domestic implementing legislation, *Goskomgidromet* was designated as responsible not only for monitoring, but also for granting permits regarding dumping of low- and medium-level waste, as well as reporting them to the IMO.[85] Radioactive waste was only one of a large number of compounds regulated by that Convention, and with its multisectoral nature and extensive environmental monitoring responsibilities, *Goskomgidromet* was the natural coordinating unit. For the first time, radioactive waste management was extended beyond the military-industrial complex and, at least formally, naval self-regulation was brought to an end.

The IAEA geographic criteria for site selection regarding radioactive waste promulgated under the London Convention, according to which the Barents and Kara Seas were particularly poorly suited for the purpose, acquired considerable significance at this stage. In accordance with the new access rules, *Goskomgidromet* had participated in the elaboration of new standards on dumping of radioactive waste in 1983. However, its endorsement of these regulations, which permitted continued dumping of low- and medium-level waste, had been given on the understanding that the Northern Fleet would realise plans to build installations for treatment – concentration and solidification – of that waste, in order to phase out its dumping operations.[86] In the meantime, the Murmansk Shipping Company, which had far smaller volumes of waste to handle in the first place, had built such an installation at its Atomflot base outside Murmansk and was able to discontinue dumping of liquid waste in 1984 and solid waste two years later.[87] When the Northern Fleet failed to build similar capacity, *Goskomgidromet* first expressed disagreement with the selection of dumping sites, citing the IAEA guidelines, and then in late 1987 withdrew its endorsement of the permit to dump radioactive waste in the sites used by the Northern Fleet.[88]

While this regulative controversy between *Goskomgidromet* and the Navy was clearly related to norms produced under the London Convention, it is not fully explained by them. We must recall that in 1985, Gorbachev had ascended to power in the Soviet Union, rapidly embarking upon his project of slackening restrictions on access to bureaucratic decision-making. The Chernobyl accident the following year had channelled much of the public disapproval into the environmental area, in particular of activities involving nuclear risks. Thus, while *Goskomgidromet* had

[83] Yablokov *et al.*, *Facts and Problems*, section 2.1. [84] *Ibid.*

[85] *Ibid.*, p. 22; where appropriate, *Goskomgidromet* was to consult with the Ministry of Fisheries.

[86] *Ibid.*, p. 25.

[87] A. E. Berkov, 'Ministry of Transport of the Russian Federation (Mintrans)', *International Cooperation on Nuclear Waste Management in the Russian Federation* (Vienna: International Atomic Energy Agency, 1995), p. 65. [88] Yablokov *et al.*, *Facts and Problems*, section 2.1.

voiced concern with the 1983 regulations because they deviated from the IAEA criteria, the boldness of its move four years later must be seen in the context of a rapidly changing society far more concerned with radioactive contamination and managed by a modernising leadership that encouraged criticism of bureaucratic malpractice.

The key target of regulation – the Northern Fleet – was at first less than impressed with this stricter policy line assumed by *Goskomgidromet*. In 1988, the year after *Goskomgidromet* had withdrawn its permission, the Northern Fleet dumped more low- and medium-level waste than it had in twelve years;[89] and, even more grave in terms of potential release into the environment, two reactors were dumped the same year in a bay off Novaya Zemlya.[90]

So again, effective measures were not taken before the dumping scandal became an international one. In terms of political influence, the establishment of the Yablokov Commission and the publication of its report in 1993 marked the highest point for the proponents of stricter controls regarding handling of radioactive waste. Even though leading representatives from the nuclear military complex took part in its preparation, the Yablokov Report was highly critical of both the dumping and the secrecy surrounding it. Indeed, the Report itself reveals a strong belief among its authors in the domestic political clout of the global dumping regime – because, in the Report, Soviet commitments under the London Convention are systematically exaggerated. No mention is made of the distinction between resolutions and amendments in the London Convention, nor of the opt-out clause pertaining to the latter. Thus, the Report does not bring out that the Soviet abstention from the votes on the voluntary moratorium in 1983 makes it very hard to argue that this country was legally or even politically bound by them in this period.[91] Likewise, while the Commission boldly states that the permission to conduct dumping of low- and medium-level waste in the Barents and Kara Seas was illegal,[92] in reality the IAEA guidelines have no more than quasi-legal status.[93]

This notwithstanding, harsh criticism from several parties to the London Convention, especially Japan, following a 1993 dumping operation of low-level liquid radioactive waste in the Sea of Japan, induced Russia to reverse a plan to conduct a second operation and to pledge to cease operations such as this completely within a few years.[94] Russia is not known to have dumped any radioactive materials since.

In Russian decision-making on nuclear waste, the scope of participation has levelled off since the Yablokov peak. There are several reasons for this. At the

[89] NATO, 'Cross-Border Environmental Problems', pp. 17 and 33.

[90] Yablokov *et al.*, *Facts and Problems*, section 2.3.2.

[91] See IMO doc. LDC 7/12, p. 29. The Soviet Union also abstained when the moratorium was prolonged in 1985; see IMO doc. LDC 9/12, p. 41. As noted above, when the 1993 prohibition was adopted, Russia abstained from the vote and was the only state which subsequently filed a reservation.

[92] Yablokov *et al.*, *Facts and Problems*, section 2.2.

[93] Birnie and Boyle, *International Law and the Environment*, p. 324.

[94] See IMO doc. LC 16/14, pp. 22–5.

level of societal organisation, there is currently less enthusiasm for environmental matters than in the late 1980s, in part due to disillusionment with the early experiments of political activism and direct democracy. Also, as noted, the economic hardships and the political turmoil of the 1990s have pushed environmental affairs down the agenda. Perhaps more significantly, along with the political consolidation of presidential power, including stronger authoritarian features in the governance style,[95] the number of access points for those still interested in affecting nuclear developments have become fewer. By the turn of the decade *Goskompriroda*, now named the State Committee for Environmental Protection, held a formally strong position in the Russian bureaucratic structure; the agency has subsequently lost several important battles for regulative competence. While, for a period, the mighty State Committee for Water and also those for Forestry and Cartography, were placed administratively under the environmental agency, they soon re-emerged as separate federal agencies.[96] In the nuclear safety area, the environmental agency is now seen as having very limited enforcement powers. Its regulative role is impeded by the fact that it is a new agency with very limited financial backing, inadequate informational basis for making environmental decisions and poorly defined internal structures.[97]

In contrast, the Ministry of Atomic Power appears gradually to have recovered much of its strength after the setbacks associated with Chernobyl. A merger with the Ministry of Medium Machine Building in 1989 brought both the military and the civilian parts of the nuclear complex into its portfolio.[98] In the following years, new reactors were put on line in the Russian nuclear programme, partly to compensate for the loss of control over nuclear plants in the Ukraine. On matters related to nuclear issues in general, the feeling grew in the environmental movement that Yeltsin was increasingly yielding to the demands of the Ministry of Atomic Power and the nuclear-industrial lobby.[99] Western observers have described the ministry as 'extremely large and powerful', noting also that the minister, Viktor Mikhailov, in July 1995 was appointed to the Russian Security Council.[100] Along with the failure of the State Committee for Environmental Protection to assert its authority in areas formally placed under it and the gradual recuperation of the Ministry of Atomic Power, secrecy is returning to the nuclear waste arena in Russia. Already in 1992, the latter ministry and the nuclear industry had managed to convince the Supreme Soviet to extend the secret status of

[95] See V. Baranovsky, 'Russia and its Neighborhood: Conflict Developments and Settlement Efforts', *SIPRI Yearbook 1995: Armaments, Disarmaments and International Security* (Stockholm: International Peace Research Institute, 1995), pp. 231–64.

[96] Compare *Vsya Moskva: Informatsionno-reklamny ezhegodnik 1992/93* (All Moscow: information and Advertisement Yearbook 1992/93) (Moscow: Vsya Moskva, 1992), p. 16 and *Novaya Rossiya: Informatsionno-statistichesky almanakh '94* (New Russia: Information and Statistical Almanac '94) (Moscow: Vsya Moskva/Mezhdunarodnaya akademiya informatsii, 1994), pp. 53–5. The environmental agency had ministerial status between 1992 and 1996.

[97] OTA, *Nuclear Wastes in the Arctic*, p. 217. [98] *Ibid.*, p. 218.

[99] Canfield, 'Soviet and Russian Nuclear Waste Dumping', p. 425.

[100] OTA, *Nuclear Wastes in the Arctic*, p. 218.

governmental information on nuclear programmes.[101] Earlier that year, Yeltsin had reversed a decision to open up the nuclear city of Severodvinsk, home to one of the major military shipbuilding complexes in the Soviet era. Another indication of this trend towards less openness on nuclear matters is that *Gosatomnadzor*, the Federal Nuclear and Radiation Safety Authority of Russia, which in 1991 had been assigned the task of regulating and inspecting safety practices both at civilian and military facilities, lost the military part of its portfolio by a presidential decree of July 1995 after a very critical inspection report.[102] Even clearer evidence that access rules are being sharpened is the new and tougher policy pursued towards environmental organisations in the nuclear field. In 1992, a representative of 'To a New Earth' had been included as senior expert and author in the Yablokov Commission, despite, or perhaps because of, the association of that organisation with Andrey Zolotkov, who had been the first source of military malpractice on dumping. Only three years later, the institutional framework was to prove far more hostile. In late 1995, the Federal Security Bureau raided the homes and offices of several persons involved in the preparation of a report on the waste management of the Northern Fleet. They later arrested one of them, a Russian citizen formerly with the Northern Fleet but employed at the Moscow office of the Norwegian environmental organisation Bellona; he was accused of espionage and high treason.[103]

Enhancing compliance

Even during the period of military self-regulation, some efforts had been made on the part of Soviet authorities to stimulate *alternatives* to the dumping of radioactive waste. For the Arctic waste problem, as noted, the elaboration of alternatives to marine disposal means the construction of adequate interim storage facilities combined with either on-site treatment facilities and a permanent repository, or a smooth system for transporting parts of the waste out of the region for reprocessing. At an early stage, the Soviet authorities chose the latter option, primarily in order to generate plutonium for weapon use. The first interim storage for spent fuel was ready for operation by the Northern Fleet in 1962; it experienced considerable problems right from the outset.[104] Major leakages from the pools

[101] Canfield, 'Soviet and Russian Nuclear Waste Dumping', p. 429; at that time, this agency was named the Ministry of Nuclear Energy.

[102] R. Vartanov, A. Roginko and V. Kolossov, 'Russian Security Policy 1945–96: The Role of the Arctic, the Environment and the NSR', in W. Østreng (ed.), *National Security and International Environmental Cooperation in the Arctic – The Case of the Northern Sea Route* (Dordrecht: Kluwer Academic Publishers, 1999), pp. 52–102. In Y. I. Zubkov and A. I. Kislov, *Federal Nuclear and Radiation Safety Authority of the Russian Federation (Gosatomnadzor)* (Vienna: International Atomic Energy Agency, 1995), pp. 27–8, the authors note that *Gosatomnadzor* had withdrawn permits from three enterprises engaged in the processing of radioactive waste, including Mayak, due to a 'very complicated situation in radwaste management'; the conference where this article was presented was held one month prior to the decision to reduce the area of competence of *Gosatomnadzor* itself.

[103] *Bellona Press Release* (Oslo: Bellona Foundation, 7 February 1996).

[104] T. Nilsen, I. Kudrik and A. Nikitin, 'Zapadnaja Litsa', *Bellona Arbeidsnotat*, No. 5 (Oslo: Bellona Foundation, 1995), pp. 12–13.

occurred from 1982 to 1983 and resulted in a gradual close-down of this storage, fuel assemblies being transferred to nearby storage tanks meant for low-level liquid waste.[105] Three other main interim storage facilities for fuel assemblies were built as well.[106] In 1973, the Northern Fleet and the Murmansk Shipping Company began transporting spent nuclear fuel by barges to Murmansk and from there to Mayak by rail.[107] There is a particular problem with reprocessing: the separation process also generates considerable volumes of high-level liquid waste that cannot be put back into the fuel cycle and that is more hazardous to store than spent nuclear fuel.[108] In the case of the Mayak complex, this has created one of the gravest environmental disaster areas in the entire Soviet Union; thus the early investments in an infrastructure to permit reprocessing of spent fuel can hardly be seen as indication of a Soviet concern to avoid nuclear contamination.

Just as in the case of assessment and regulation, the internationalisation of the dumping issue since the late 1980s was a turning point for efforts to enhance domestic capacities to avoid dumping of radioactive waste. After having ascribed a 1993 incident of dumping of liquid radioactive waste in the Sea of Japan to irresponsibility on the part of the Navy and the nuclear industry, the Russian Minister of Environment informed the Consultative Meeting of the London Convention that Western technology and financial resources would speed up the process of acquiring the ability to manage without such dumping in the future.[109] In response, an international Technical Advisory Assistance Team was set up to develop projects on treatment and storage facilities.[110] The subsequent year, this team could report to the Consultative Meeting that Japan and Russia had signed an agreement to build a treatment facility in the Far East for low-level liquid waste; and also that there was progress regarding a project to enhance liquid processing capacity at Atomflot, the base of the Murmansk Shipping Company.[111] Furthermore, Norway and Russia had reached agreement on a two-year assessment programme on the nuclear waste

[105] See V. N. Lystsov, 'The Yablokov Commission Report on Soviet Radioactive Waste Dumping at Sea: Additional Comments', *Arctic Research of the United States*, Vol. 8, 1994, pp. 271–2; and Nilsen *et al.*, 'Zapadnaja Litsa', pp. 16–17.

[106] These other main storage facilities for spent fuel assemblies are the naval base at Gremikha, the naval shipyard at Severodvinsk and at the Atomflot base of the Murmansk Shipping Company; see T. Nilsen and N. Bøhmer, 'Sources of Radioactive Contamination in Murmansk and Arkangelsk Counties', *Bellona Report*, Vol. 1 (Oslo: Bellona Foundation, 1994), p. 46. Spent fuel is also stored on several other bases and on barges operated by the Murmansk Shipping Company.

[107] However, the Mayak plant cannot reprocess spent fuel in defective assemblies or from reactors which are liquid-metal cooled or have damaged fuel assemblies; see Baklanov *et al.*, *Radioactive Sources in the Kola Region*, p. 52.

[108] See NATO, 'Cross-Border Environmental Problems', pp. 266–7; in 1987, a pilot vitrification facility was opened at the Mayak complex which by 1993 had solidified 5,000 m^3 of high-level liquid waste; see Y. K. Bibilashvili and F. G. Reshetnikov, 'Russia's Nuclear Fuel Cycle: An Industrial Perspective', *IAEA Bulletin*, Vol. 35, 1993, pp. 31–2.

[109] IMO doc. LC 16/14, Annex 6.

[110] IMO doc. LC 16/14, p. 25.

[111] IMO doc. LC 17/14, Annex 5; the project contributes to the trilateral (Russian–Norwegian–US) Arctic Military Environmental Cooperation mentioned earlier in this chapter. A facility to concentrate and solidify low-level liquid waste had been built in 1991–2; see Castberg and Stokke, 'Environmental Problems in Northwest Russia', p. 35.

challenges in the Mayak plant.[112] In 1995, experts from six NATO countries were invited to an international scientific symposium in Moscow on the decommissioning of nuclear submarines, involving leading figures in the State Committee for Defence Branches of Industry (*Goskomoboronprom*), the Ministry of Atomic Power and the Northern Fleet.[113] And in the same year, thirty-two representatives of a dozen ministries and other agencies in Russia responsible for radioactive waste participated in an IAEA meeting on international cooperation on nuclear waste management.[114]

What seems to be happening today is that the former main components of the coalition resisting openness on nuclear matters, including the Navy and the Ministry of Atomic Power, have consolidated their control over domestic decision-making and are themselves becoming increasingly involved in cooperative programmes generated under the London Convention and other fora. Thus, while the level of domestic participation is on its way down, international contacts are still thriving. With the international focus shifting from regulating and mapping radioactive contamination to the development of practical measures to avoid it, the resisters of yesterday are turning up as today's supporters of international coordination in the nuclear waste area. However, the causal significance of the London Convention in this context should not be overstated, since this development has been supported by a range of other cooperative vehicles, including bilateral and regional ones as well as the International Atomic Energy Agency. The role of the London Convention has been partly to coordinate and partly to encourage and legitimise programmatic activities initiated or financed within other such processes.

CONCLUSIONS

The regime set up by the London Convention on dumping has served to lower domestic access barriers in the Soviet Union and, later, Russia to decisions on disposal of nuclear waste and promoted a step-wise broadening of actual participation of regulative agencies and societal intervenor groups. After two decades of military self-regulation, Soviet implementing legislation of the London Convention in 1979 elevated the nuclear waste issue to the cabinet level and added a civilian regulative agency, *Goskomgidromet*, in the management of low- and medium-level waste. This helped to reduce an access bias which had clearly favoured the target groups, primarily the Northern Fleet. The role of the London Convention regime was the decisive one of generating a set of routine-like bureaucratic responses to uncontroversial but explicit responsibilities defined internationally. While secrecy continued to shield military dumping from broader public

112 IMO doc. LC 17/14, p. 30.

113 *NATO Science and Society Newsletter*, No. 45, 1995.

114 For proceedings of the conference, see IAEA, *International Co-operation on Nuclear Waste in the Russian Federation* (Vienna: International Atomic Energy Agency, 1995).

scrutiny, this change brought about a cautious regulative competition which in the mid-1980s was further nurtured by the political turnabouts of *glasnost* and *perestroika*. Institutional upshots of particular significance in this stage of cautious expansion were an environmental bureaucracy, *Goskompriroda*; a more active legislative body, the Congress of People's Deputies; and a radiotoxically attentive environmentalist movement independent of the state apparatus.

When the handling of radioactive waste became politicised in the early 1990s, the international dumping regime was helpful to the successful efforts of critics of dumping, both among regulative agencies and intervenor groups, to enhance transparency on nuclear activities. Access to information on nuclear safety in the military sector, as well as participation in the associated policy-making processes, reached a high point with the publication of the 1993 governmental Yablokov Report, which also responded to demands articulated by the Consultative Meeting of the London Convention. The various prescriptions set forth in this Convention, moreover, appear to have enhanced the political clout of those critical of dumping, as these prescriptions figure prominently in the unequivocal argument made in the Yablokov Report on the severity of past dumping and the need to invest more in storage and decontamination facilities to avoid future dumping.

Since then, access to military information, including nuclear waste practices, has been tightened at a time when public attention to environmental problems is ebbing. Moreover, the limits of the funds, personnel and experience of the environmental bureaucracy are becoming apparent as the nuclear-industrial complex has begun regaining much of its previous political strength and prestige. Importantly, the civilian regulative apparatus does not have physical access to military bases or shipyards. This contraction in terms of domestic access and participation is the upshot of internal Russian developments, but it is to some extent balanced by steadily wider international participation in programmes designed to monitor radioactivity in Arctic seas and, subsequently, to alleviate the operational needs of the Northern Fleet to continue dumping. These international programmes have required the consent, and increasingly the active participation, of the Navy itself. This support has been secured primarily by the belief that such programmes will be conducive to the transfer of technology and financial resources to Russia from the West.

The consequences of these changes in access and participation for the effectiveness of the international dumping regime have been measured along three dimensions: monitoring, regulation and compliance stimulation, including enhancement of target-group capacity to avoid dumping. The entry of *Goskomgidromet*, and later *Goskompriroda*, into the arena meant a somewhat enhanced *monitoring* of the environmental situation; but, until their internationalisation during the politicisation stage, these activities remained remarkably non-intrusive. Until 1993, measurements were not taken near the actual dumping sites. The same is true for behavioural monitoring of compliance: as noted, inspection of nuclear waste management in military facilities was, and remains, largely left to the Northern Fleet itself.

As to *regulations*, the entry of civilian agencies in the radioactive waste area during the period of cautious expansion after 1979 prepared the ground for controversy. *Goskomgidromet* became increasingly critical of Navy practices and in 1987 withdrew its permit to continue the dumping of low- and medium-level waste. However, this regulative discord did not force the Northern Fleet to halt dumping. On the contrary, unlike the civilian Murmansk Shipping Company, which had comparatively better treatment capacity and had been able to terminate dumping in the mid-1980s, the Navy continued to dump nuclear waste well into the 1990s. The contestedness of regulations was partly shaped by the guidelines set forth under the London Convention. However, the articulation of those guidelines in the regulative process remained fairly meek until the nuclear complex, including its military part, was thoroughly, if temporarily, enfeebled by the ecological disaster of Chernobyl and the political reshuffles of *perestroika*. Thus, the international dumping regime provided the direction but not the energy for this change in regulative implementation.

Regarding *compliance stimulation*, the entry of foreign participants into the implementation in the late 1980s, and especially after the turn of the decade, has been significant – not so much in its control mode of improving compliance by verification, as in the supportive form of helping to enhance the ability of the Northern Fleet to avoid dumping. This has been achieved through cooperative international programmes designed to estimate the hazards involved and, subsequently, to elaborate practical ways to enhance treatment and storage facilities for liquid and solid radioactive waste. Such foreign contribution is probably decisive to the realisation of Russian capacity to provide adequate treatment and storage for radioactive waste: the domestic political impetus available appears to have been largely spent as the Navy and Russian authorities are again shrouding waste management in secrecy.

10 Regulation of navigation and vessel-source pollution in the Northern Sea Route: Article 234 and state practice

R. DOUGLAS BRUBAKER

The seaway known as the Northern Sea Route (NSR), which passes through the ice-infested waters of the Russian Arctic, can potentially halve the distance between Europe and northeast Asia. A voyage between Hamburg and Yokohama would be only 6,600 miles[1] if the NSR is chosen, in comparison to 11,400 miles through the Suez Canal. The NSR itself covers between 2,200 and 2,900 miles of often ice-covered waters.[2]

Russia has put considerable efforts into developing the infrastructure for marine transport along the NSR. In the changing geo-political picture of the Arctic in the post-Cold War era,[3] Russia officially opened the NSR for foreign vessels in 1991. However, various factors, not least the difficult ice conditions, have so far prevented it from becoming widely used by international shipping.

Unlike most other sea routes, there is no single, set channel: ice conditions at any one place decide the further course. The NSR crosses a series of individual seas – the Kara, Laptev, East Siberian and Chukchi Seas – which in turn are linked by almost sixty straits running through archipelagos including Novaya Zemlya, Severnaya Zemlya and the New Siberian Islands.

Surface vessels may encounter along the NSR natural obstacles of various kinds. The continental shelf north of Russia is very shallow, in some straits only 8–13 metres. This places absolute limits on the draught of vessels that can navigate. It is frequently in just those areas with the shallowest depths that the most difficult ice conditions prevail.

Vessels would navigate the NSR in convoys. The hazards involved in convoying have been succinctly characterised: 'Being at sea is risky; being at sea in ice is twice the risk; being at sea in ice in convoy with an ice-breaker is three

[1] All references to miles indicate nautical miles.

[2] In Russian legal regulation and political perception, the NSR stretches from the northern point of Novaya Zemlya as well as its straits in the west to the Bering Strait in the east. The actual length of the NSR in each case depends on ice conditions; see Figure 10.1.

[3] See D. Brubaker and W. Østreng, 'The Military Impact on Regime Formation for the Northern Sea Route', in D. Vidas and W. Østreng (eds.), *Order for the Oceans at the Turn of the Century* (The Hague: Kluwer Law International, 1999), pp. 261–90.

Figure 10.1 The Northern Sea Route

times the risk.'[4] Consequently, the international shipping community has thus far never seriously contemplated acquiring the necessary capability for using the NSR. Neither has operation of this route seemed feasible, given the economic preconditions applying in the world market. However, modern technology may render commercial use of the NSR a real option, with the attendant possible reductions in commercial shipping expenses that could mean increased revenues.[5]

This chapter will analyse the *legal regime of navigation* applying to the NSR, which is yet another controversial aspect of any future international use of this route. We will examine Russian regulations governing navigation and vessel-source pollution, and the consistency of domestic provisions with current international law. The latter includes Article 234 of the 1982 United Nations Convention on the Law of the Sea[6] (the LOS Convention), as well as other relevant treaty law provisions and customary law. Russian legislative practice will also be compared with that of the USA and Canada, which together govern most of the ice-covered exclusive economic zones in the Arctic. The aim is to demonstrate how the state practice, including legislation, of relevant Arctic states has been shaping the contours of Article 234.

THE NORTHERN SEA ROUTE: LEGAL CONTROVERSY

It has been observed that:

> For decades, the NSR has been one of the most contentious legal and political issues in US–Soviet/Russian Arctic relations. A major concern relates to the effects the NSR may have on the Arctic environment.[7]

Article 234 of the LOS Convention states:[8]

> Coastal States have the right to adopt and enforce non-discriminatory laws and regulations for the prevention, reduction and control of marine pollution from vessels in ice-covered areas within the limits of the exclusive economic zone, where particularly severe climatic conditions and the presence of ice covering such areas for most of the year create obstructions or exceptional hazards to navigation, and pollution of the marine environment could cause major harm to or irreversible disturbance of the ecological balance. Such laws and regulations shall have due regard to navigation and the protection and preservation of the marine environment based on the best available scientific evidence.

[4] G. Watson, 'Technical Aspects of Ice Navigation and Port Construction in Soviet Arctic', in L. Brigham (ed.), *The Soviet Maritime Arctic* (London: Belhaven Press, 1991), p. 159.

[5] T. Ramsland and S. Hedels, 'The NSR Transit Study (Part IV): The Economics of the NSR. A Feasibility Study of the NSR as an Alternative to the International Shipping Market', *INSROP Working Paper*, No. 59 (Lysaker: Fridtjof Nansen Institute, 1996).

[6] Text reprinted in ILM, Vol. 21, 1982, pp. 1,261ff.

[7] A. Roginko, 'Environmental Protection in the Soviet Arctic Seas', in Brigham (ed.), *The Soviet Maritime Arctic*, pp. 63–82.

[8] On Art. 234, see also Vukas, Chapter 2 in this book, as well as Rothwell and Joyner, Chapter 7 in this book.

Canada, the USSR and the USA negotiated Article 234 at the Third UN Conference on the Law of the Sea.[9] Of these three, to date (8 June 1999) only Russia has ratified the LOS Convention. This adds to the complexity of the legal picture concerning the scope of Article 234 regarding the NSR. Russia is bound by Article 234 as treaty law. The USA has expressly acknowledged Part XII of the LOS Convention, which also includes Article 234, as customary law.[10] Canada is neither a party to the LOS Convention nor has it expressly acknowledged its Part XII as customary law. All three states have, however, adopted relevant domestic legislation and made various declarations, which will be analysed below.

While it may be uncertain whether the entire Part XII is already customary law,[11] it is important to realise that the Convention's regime for prevention of vessel-source pollution[12] incorporates the rules found in MARPOL 73/78,[13] and as such may be applied by all three states in the EEZ.

Russia claims its entire Arctic EEZ, and possibly the high seas, to be subject to special coastal state rights for ice-covered areas. It further claims that ice-covered straits of the NSR are part of its internal waters, supporting this by several theories, including that of historic waters enclosed by straight baselines.[14]

The USA, through its declarations and submerged navigation, has remained opposed to features of the Russian regime for ice-covered areas governing the NSR. While the USA does accept an extensive coastal state prescriptive and enforcement jurisdiction for ice-covered areas, it has reserved its position with regard to ice-covered *straits*, equally so regarding those in the Russian and Canadian Arctic.[15] In the US view, the Russian Arctic straits are international and thus subject to transit passage.[16] In addition, US declarations have included protests against Russian Arctic baselines and objections to the application of the Russian legislation for the NSR to state vessels.[17]

Both the USA and Russia invoke national security when substantiating and warranting their respective stands. The policy of adducing environmental arguments to restrict foreign shipping seems rather presumptuous, considering the poor Soviet/Russian environmental record in the area.[18] Russian experts themselves have acknowledged the doubtful compliance of the Soviet fleet with strict

[9] See D. McRae, 'The Negotiation of Article 234', in F. Griffiths (ed.), *Politics of the Northwest Passage* (Kingston, Ontario: McGill-Queens University Press, 1987), pp. 98–114.

[10] Presidential Proclamation No. 5030, *Federal Register*, Vol. 48, 1983, p. 605 (codified at *Code of Federal Rules*, Vol. 3, section 5030).

[11] On Part XII and customary law, see the discussion by Vukas, Chapter 2 in this book.

[12] Arts. 211 and 217–220 of the LOS Convention.

[13] See P. W. Birnie and A. E. Boyle, *International Law and the Environment* (Oxford: Clarendon Press, 1992), pp. 271 and 298–9. On MARPOL 73/78 and polar waters see Rothwell, Chapter 3 in this book.

[14] A. Kolodkin and M. Volosov, 'The Legal Regime of the Soviet Arctic – Major Issues', *Marine Policy*, Vol. 14, 1990, pp. 162–7.

[15] On the Canadian Arctic, see also Rothwell and Joyner, Chapter 7 in this book.

[16] J. A. Roach and R. W. Smith, *Excessive Maritime Claims* (Newport, RI: Naval War College, 1994), pp. 48, 58, 200–15 and 227; and interview with J. A. Roach, The Hague, 13 July 1995.

[17] *Ibid.*

[18] E. Franckx, *Maritime Arctic Claims – Canadian and Russian Perspectives* (Dordrecht: Martinus Nijhoff, 1993), pp. 33–4 and 192–3. On the Soviet/Russian environmental record regarding dumping of radioactive waste in the Arctic, see Stokke, Chapter 9 in this book.

environmental regulations, as well as the fact that about one-half of the vessel-source pollution in the Arctic in 1990 was of Soviet origin.[19]

RUSSIAN REGULATION: NON-COMPLIANCE WITH, OR INTERPRETATION OF, ARTICLE 234?

Under Article 234 of the LOS Convention, the conditions for coastal states' domestic laws and regulations for ice-covered areas fall within the following confines. First, their area of application must be 'within the limits of the EEZ', where there are 'particularly severe climatic conditions and the presence of ice covering such areas for most of the year'. Secondly, these must be 'non-discriminatory' and aim at preventing, reducing and controlling marine pollution from vessels. Thirdly, they should have 'due regard to navigation' as well as 'protection and preservation of the marine environment based on the best available scientific evidence'. Article 234 probably enjoys a broad scope of application due to its vague formulation.[20] In implementing Article 234, the Soviet Union adopted various legislation. As successor state to the Soviet Union, Russia has incorporated the entire comprehensive Soviet legislation, except when contradictory to the Russian Constitution, and has itself more recently adopted new Arctic legislation.

The 1984 Edict on the Economic Zone of the USSR[21] (Economic Zone Edict) in Article 14 authorised competent Soviet agencies to:

> establish rules for the prevention, reduction, and control of pollution of the marine environment, as well as for safety of shipping, and ensure the observance of such rules in ice-covered areas or areas possessing special natural characteristics where pollution of the marine environment could inflict grave harm to the ecological balance or irreversibly disturb it.

The 1984 Edict on Intensifying Nature Protection in Areas of the Far North and Marine Areas Adjacent to the Northern Coast of the USSR[22] (Environmental Edict) in its Article 3 specified:

> In the marine areas adjacent to the northern coast of the USSR where specially severe climatic conditions and ice hindrances or increased danger for shipping and pollution of the marine environment could cause grave harm to the ecological balance or irreversibly disturb it, special navigation rules for vessels and other floating means shall be established by the competent Soviet agencies. These rules shall provide for higher construction requirements for vessels and other floating means, for equipment and supplies, for the complement

[19] See R. Vartanov, A. Roginko and V. Kolosov, 'Russian Security Policy 1945–96: The Role of the Arctic, the Environment and the NSR', in W. Østreng (ed.), *National Security and International Environmental Cooperation in the Arctic – The Case of the Northern Sea Route* (Dordrecht: Kluwer Academic Publishers, 1999), pp. 53–102.

[20] D. McRae and D. Goundrey, 'Environmental Jurisdiction in Arctic Waters: The Extent of Article 234', *University of British Columbia Law Review*, Vol. 16, 1982, pp. 215–22.

[21] English translation reproduced in W. Butler, *The USSR, Eastern Europe and the Development of the Law of the Sea* (London: Oceana, 1987), F. 2, pp. 1ff.

[22] English translation reproduced in *ibid.*, J. 4, pp. 1ff.

> and skills of the crew, shall prohibit navigation without pilotage or other escort, shall establish periods and areas closed for navigation, and also other measures ensuring the safety of shipping and the prevention, reduction, and control of pollution of the marine environment. The said rules shall be published in *Notices to Mariners.*

Accordingly, in 1990 the Regulations for Navigation on the Seaways of the Northern Sea Route (NSR Regulations) were adopted.[23] The objectives are stated in Article 2:

> on the basis of non-discrimination for vessels of all States, [to] regulate navigation through the Northern Sea Route for the purpose of ensuring safe navigation and preventing, reducing, and keeping under control marine environment pollution from vessels, since the specifically severe climatic conditions that exist in the Arctic Regions and the presence of ice during the larger part of the year bring about obstacles, or increased danger, to navigation while pollution of sea, or the northern coast of the USSR might cause great harm to the ecological balance, or upset it irreparably, as well as inflict damage on the interest and well-being of the peoples of the Extreme North.

The NSR Regulations contain provisions governing mandatory notification and authorisation to foreign vessels for navigating the NSR (Articles 3 and 8); different forms of leading along the NSR (Articles 3 and 7(4));[24] special requirements for vessels and command personnel (Article 4); civil liability (Article 5); inspection (Article 6); order of navigation (Article 7); control of navigation (Article 8); temporary suspension of navigation in particular areas of the NSR (Article 9); removal of vessels off the NSR (Article 10); limitation of liability for the Northern Sea Route Administration (Article 11); and notification of polluting discharge (Article 12). The 'special requirements' refer to technical and operational rates and standards set forth in publications issued by the Northern Sea Route Administration (NSRA), including the Guide to Navigating through the Northern Sea Route (NSR Navigation Guide) and the Requirements for the Design, Equipment and Supplies of Vessels Navigating the Northern Sea Route (Design Requirements).[25] Important features of the NSR Regulations include the possibility for application on the high seas, application to state vessels, mandatory payment of fees, and the introduction of specially protected areas within ice-covered areas. These features of Russian legislation need to be examined not only in the light of Article 234 but also in the light of the relevant practice of other states.

[23] Regulations were approved by the USSR Minister of Merchant Marine, 14 September 1990. Russian text published in *Izveshcheniya Moreplavatelyam* (Notices to Mariners), No. 29, 18 June 1991; English translation, as used in this chapter, published in *Guide to Navigating Through the Northern Sea Route* (St Petersburg: Head Department of Navigation and Oceanography, Russian Ministry of Defence, 1996), pp. 81–4. An English translation is also available, with minor deviations, in *International Challenges*, Vol. 12, 1992, pp. 121–6. Russia has adopted a plethora of legislation; however only those provisions considered most central will be discussed.

[24] Art. 7(4) lists: (1) leading along recommended routes by shore-based pilotage; (2) aircraft-assisted leading; (3) conventional pilotage; (4) icebreaker leading; and (5) icebreaker-assisted pilotage.

[25] Arts. 1(5) and 4 of the NSR Regulations. English translation of the *Guide to Navigating Through the Northern Sea Route*. An English translation of the Design Requirements is in *ibid.*, pp. 317–23.

Legislation governing vessel-source pollution within the Arctic EEZ: 'due regard to navigation'

'Due regard to navigation and the protection and preservation of the marine environment based on the best available scientific evidence'[26] is an express condition with which coastal states must comply when they unilaterally adopt and enforce their laws and regulations governing vessel-source pollution in ice-covered areas within the limits of the EEZ. These laws and regulations may provide for design, construction equipment, crewing, discharge and safety standards for navigation.

Several views exist regarding the meaning of 'due regard to navigation [etc.]' in the context of Article 234, ranging from traditional freedom of navigation and innocent passage, passage rights in the territorial sea applicable to the EEZ, passage rights varying according to the status and circumstances of the waters in question, to extensive coastal state jurisdiction in the EEZ limited only by navigation upon permit.[27]

The first view may likely be the most technically-legally sound. Article 234 expressly restricts its own scope to the EEZ, which under the LOS Convention is defined as 'an area beyond and adjacent to the territorial sea',[28] not extending 'beyond 200 miles from the baselines',[29] where all states enjoy, *inter alia*, the freedom of navigation.[30]

However, it would be absurd to be able to exercise greater rights in the EEZ than in the territorial sea. Russian legislation, but also that of Canada and the USA, as will be seen, support the view on extensive coastal state jurisdiction. Most of the provisions contained in the Russian Arctic legislation far exceed limitations to innocent passage, and encompass both the territorial sea and the EEZ.

Article 234 aims at preventing, reducing and controlling vessel-source pollution which can cause 'major harm to or irreversible disturbance of the ecological balance'. A main objective of the Russian provisions is to prevent, reduce and control vessel-source pollution of its Arctic marine environment, since it might cause 'great harm to the ecological balance, or upset it irreparably' as well as 'inflict damage on the interests and well-being of the peoples' of the Russian Arctic.[31] As we have seen, Russia has included 'safety of shipping' as a goal to environmental considerations under Article 14 of the Economic Zone Edict and Article 3 of the Environmental Edict. However, where as a coastal state it has special rights based on Article 234, the requirement of that Article to have 'due regard to navigation [etc.]' may be questioned.

The NSR Regulations require that the owner or master of a vessel intending to navigate the NSR submits to the NSRA a notification and a request for

[26] This expression will be referred to as 'due regard to navigation [etc.]' hereinafter.

[27] See A. E. Boyle, 'Legal Regimes of the Arctic – Remarks', *American Society of International Law Proceedings*, Vol. 82, 1988, pp. 327–8.

[28] Art. 55 of the LOS Convention.

[29] *Ibid.*, Art. 57.

[30] *Ibid.*, Art. 58(1).

[31] Art. 2 of the NSR Regulations.

leading,[32] and that the vessel satisfy the special requirements.[33] The NSRA shall then consider this application and inform of the possibility of leading through the NSR,[34] which implies authorisation of navigation. These requirements are clearly contrary to Articles 17 and 58 of the LOS Convention, guaranteeing innocent passage in the territorial sea and freedom of navigation in the EEZ, respectively. They may, however, arguably be permitted under a broad interpretation of 'due regard to navigation [etc.]', due to the dangers of navigation in ice-infested waters.

Leading, especially in ice-bound straits, as required under Article 7(4) of the NSR Regulations, seems reasonable and gives substance to preventing, reducing and controlling vessel-source pollution.[35] However, the NSR Regulations also seem to allow for the possibility of complete prohibition of navigation. This may be deduced from the vaguely phrased passage in Article 7(4), which allows the mandatory prescription of leading in 'other regions', as well as from Article 7(1) which states that the NSRA will determine the beginning and end of the navigational period taking into account, *inter alia*, 'other conditions'. Article 9, which will be discussed below, additionally allows for suspension of navigation in cases where 'an obvious necessity of environment protection or safe navigation' so demands. Thus, due to the open-ended prohibition of navigation without pilotage or other escort under the NSR Regulations,[36] it seems doubtful whether 'due regard to navigation [etc.]' is actually taken.

Furthermore, the NSR Regulations require mandatory 'payment for the services rendered to vessels',[37] which also may raise a question regarding 'due regard to navigation [etc.]'. Under international law, innocent passage may not be hampered by the coastal state, and neither can charges be levied unless 'specific services' are rendered.[38] Such specific services probably include pilotage or rescue services, but charges shall be levied without discrimination.[39] For navigation in the EEZ, charging fees is contrary to freedom of navigation, and may be subject only to pollution prevention provisions giving effect to and conforming to 'generally accepted international rules and standards'.

It may, however, be argued that the fees required under the NSR Regulations are a necessity for the protection of the marine environment. Pilotage and rescue services, though applicable to navigation through the EEZ, may be required by the natural conditions of the NSR. However, for innocent passage under Article 26 of the LOS Convention fees are allowed *only* on a case-by-case basis for 'specific services rendered to the ship', and therefore not by reason of mere passage as such through the territorial sea. It would thus probably be exaggerated

[32] *Ibid.*, Art. 3(1). This should be in compliance with the form and time stated in the NSR Navigation Guide. [33] Art. 4 of the NSR Regulations. [34] *Ibid.*, Art. 3(2).

[35] Such *compulsory* ice-breaker assisted pilotage is established under Art. 7(4) of the NSR Regulations in four straits, due to the adverse navigational situation and ice conditions.

[36] See Art. 3 of the Environmental Edict. [37] Art. 8(4) of the NSR Regulations.

[38] See Arts. 15(1) and 18(1) of the 1958 Geneva Convention on the Territorial Sea and the Contiguous Zone (UNTS, Vol. 516, pp. 205ff); and Arts. 24(1) and 26 of the LOS Convention. See the discussion in R. R. Churchill and A. V. Lowe, The Law of the Sea, 2nd edn (Manchester University Press, 1988), p. 79. [39] For a discussion concerning discrimination see below in this chapter.

to allow expanded coastal state rights regarding obligatory fees for passage through an ice-covered EEZ. Services surrounding icebreakers and ice-strengthened vessels, for which fees could be charged under the NSR Regulations, would likely not be needed if a relatively ice-free season occurred, as was the case during the summer of 1995. Since any mandatory blanket fees may in fact serve to prohibit navigation, their introduction is likely to exceed even a liberal interpretation of 'due regard to navigation [etc.]'.

Enforcement measures – including inspections if deemed necessary, stopping, detention and arrest, suspension if deemed necessary and removal for violations – apply under the Russian legislation.[40] It may be questioned whether this strict enforcement of provisions, permitting navigation only in special instances and largely dependent upon official discretion, does not provide a near prohibition of navigation.

Casting some doubt upon how genuine is the objective of Arctic environmental protection, upon which the Russian legislation is purportedly based, Article 6 of the Environmental Edict requires a 'positive ecologically substantiated opinion' of authorised agencies for various activities, but *not* for vessel traffic. It is only for the construction or renovation of various installations and structures at sea that an environmental impact assessment (EIA) is required. Similarly, an EIA requirement is not explicitly included within the scope of Article 4 of the NSR Regulations, in connection with requirements for vessels, nor is it found in Article 3 of the Environmental Edict, where it conceivably could have been addressed.

Notwithstanding the vague formulations characterising Article 234, the extent to which 'due regard to navigation' is exceeded by the Russian legislation seems beyond even a liberal interpretation. If, however, the Russian provisions are compared to the US legislation as a coastal state, nearly all the Russian rules may be said to fall within similar limits set by the 1990 United States Oil Pollution Act (OPA) for commercial vessels carrying oil.[41] Similarities are found in the Canadian legislation as well: the Arctic Waters Pollution Prevention Act (AWPPA)[42] and the Arctic Shipping Pollution Prevention Regulations (ASPPR).[43] Only a few of the requirements contained in the Russian legislation actually exceed that of the US legislation: blanket fees, icebreaker-assisted pilotage and icebreaker leading. The latter two activities may, however, also be required by Canadian practice in the Northwest Passage. In 1992, a Canadian ice pilot was on board the Russian vessel *Kapitan Khlebnikov*; and a Canadian Coast Guard officer was on board the *USCGC Polar Star* in 1988 with the additional accompaniment of the *CCGS John A.*

[40] See Arts. 6 and 10 of the NSR Regulations; Arts. 14 and 15 of the Economic Zone Edict; Arts. 3 and 15 of the Environmental Edict; and Art. 11 of the Decree of the Council of Ministers of the USSR, 1 June 1990, 'On Measures of Securing the Implementation of the Edict of the Presidium of the USSR Supreme Soviet of 26 November 1984 "On Intensifying Nature Protection in Areas of the Extreme North and Marine Areas Adjacent to the Northern Coast of the USSR"' (1990 Decree); an English translation is available in E. Franckx, 'Nature Protection in the Arctic – Recent Soviet Legislation', *International and Comparative Law Quarterly*, Vol. 41, 1992, pp. 377–83.

[41] Public Law 101-380, *Statutes*, Vol. 104, section 484 (1990) (codified at USC, Vol. 33, section 2701).

[42] *Revised Statutes of Canada*, Chapter 2, section A-12. Text reprinted in ILM, Vol. 9, 1970, pp. 598ff.

[43] *Consolidated Regulations of Canada*, 1978, Chapter 356, as amended.

MacDonald.[44] It is therefore in respect of fees that the legislation and practice of the USA and Canada is consistent, permitting only charges for specific services actually rendered.

Admittedly, the provisions compared are not always directly parallel. The Russian requirement for the notification of an intended passage through the NSR and the approval of a request for leading are more formal and administrative than those under the OPA.[45] However, the USA clearly has procedures for determining authorised passage upon sufficient proof of financial security. Under the OPA, if foreign vessels cannot prove financial security, then denial of clearance, denial of entry into the USA or US navigable waters, detention at the place where the lack of evidence is discovered, and seizure and forfeiture within US navigable waters may result.[46] Canadian enforcement measures under the ASPPR, the AWPPA and the Arctic Waters Pollution Prevention Regulations (AWPPR) appear less comprehensive than those of Russia or the USA, though they are rigorously enforced. They are also based around a certificate, but one which shows compliance with both technical and potential liability standards.[47] Vessels may be inspected and denied navigation if requirements are not met or if there is a danger of discharge.

The US design, equipment and construction standards, including double hulls, though unilaterally adopted, may now comply with MARPOL 73/78.[48] The Russian standards are Arctic-specific and have been established unilaterally, though the Russian ice classes UL, L1, L2 and L3 may resemble the unilaterally established Canadian classes A, B, C and D, respectively.[49]

Discharge standards under the OPA are governed by means of liability for oil damages or threat thereof in navigable waters, the shoreline or the EEZ.[50] For Russia the standards are also governed through liability for damages in the EEZ, but are not limited to oil; moreover, discharges may be totally prohibited in areas 'adjacent to the northern coast'.[51] Canada completely bans discharges in its 100-mile zone.[52]

Lesser forms for Russian leading, including with radio, aircraft and conventional pilot, may have counterparts in the provisions of the OPA[53] and the 1990 Oil Terminal and Oil Tanker Environmental Oversight and Monitoring Act (Monitoring Act).[54] Under the latter, pilotage and some escort for Arctic waters is required as well as oil tanker monitoring, though the functioning is advisory only. Canada may now have year-round surveillance and control of shipping, compulsory pilotage and training.[55]

[44] Franckx, *Maritime Arctic Claims*, pp. 261–4.

[45] See OPA, para. 1016(a) and (b).

[46] *Ibid.* The scope of the term 'navigable waters' will be discussed below.

[47] ASPPR, Schedule VIII and section 15; AWPPA, sections 12 and 15(3); and AWPPR *Consolidated Regulations of Canada* (1978), Chapter 354, as amended, section 12. See E. Gold, *Handbook on Marine Pollution*, p. 82.

[48] OPA, paras. 4109, 4110 and 4115.

[49] See Design Requirements, Appendix; and NSR Navigation Guide, p. 324, referring to the Canadian ASPPR, Schedule VIII.

[50] OPA, para. 1002(a).

[51] Arts. 19(3), (9) and (10) of the Economic Zone Edict; and Art. 11 of the Environmental Edict.

[52] AWPPA, sections 2 and 4.

[53] OPA, para. 4116.

[54] USC, Vol. 33, section 2732.

[55] D. Pharand, *Canada's Arctic Waters in International Law* (Cambridge University Press, 1988), pp. 240–1.

For Russia, ships must satisfy special crewing and training requirements for navigating in ice.[56] For the USA, requirements as to manning, training, qualifications and watchkeeping standards for foreign tankers appear to be generally stricter than international standards.[57]

For Russia, safety considerations are stated to give state agencies the power unilaterally to establish the Russian provisions.[58] For the USA, no limitation of liability is applicable if there was a violation of a federal safety, construction or operation regulation.[59] For Canada, the sixteen Arctic 'safety control zones' are the basis upon which ship construction and navigation standards are governed.[60]

On the whole, despite some inconsistencies and differing approaches, Russia, the USA and Canada seem to have – individually – established remarkably similar standards applicable in their Arctic EEZs, or substantial parts thereof. Thus, while the theoretical limits indicated concerning 'due regard to navigation' may be sound, practice in the Arctic related to surface passage has developed towards favouring extensive coastal state jurisdiction.

Area of application: 'ice-covered' and 'adjacent marine areas'

Article 234 governs ice-covered areas *within* the limits of the EEZ. In addition, 'the presence of ice covering such areas for most of the year' is required. The latter cannot be found either in Article 14 of the Economic Zone Edict or in Articles 2 and 3 of the Environmental Edict. Only broader terms appear: 'ice-covered areas', 'areas possessing special natural characteristics', and 'severe climatic conditions and ice' creating 'hindrances or increased danger for shipping'. 'Ice' is not specifically addressed in the 1990 Decree; and in the NSR Regulations, reference is made only to 'presence of ice', 'severe climatic conditions', 'ice', 'ice conditions' and 'ice-breaking'.[61] The term 'ice-covered' thus remains unclear in Russian legislation.

One view distinguishes 'ice-covered areas' as being those Arctic seas that are covered by ice 'for most of the year', with an average ice cover for six months or more.[62] This may be contrasted with 0.5 ice concentration for more than eight months a year.[63] However, the USA apparently considers such definitional problems surrounding 'ice-covered areas' to play a minor role.[64]

Given the vagueness of exactly where Article 234 is to apply within the EEZ, it is perhaps not surprising that Russian legislation is somewhat ambiguous about limiting its own application to 200 miles. The NSR itself may in some cases fall beyond the EEZ (see Figure 10.1), and it has also been claimed that Arctic states have special privileges with respect to the control of Arctic waters which may

[56] Art. 4 of the Navigation Rules; and Design Requirements, p. 323. [57] OPA, para. 4106.
[58] Art. 14 of the Economic Zone Edict. [59] OPA, para. 1004(c)(1).
[60] Pharand, *Canada's Arctic Waters*, pp. 240–1. [61] Arts. 2, 4, 6, 7 and 11 of the NSR Regulations.
[62] N. Koroleva, V. Markov and A. Ushakov, *Legal Regime of Navigation in the Russian Arctic* (Moscow: Russian Association of International Maritime Law, 1995), p. 75.
[63] Franckx, *Maritime Arctic Claims*, pp. 192 and 225.
[64] Interview with J. A. Roach, US Department of State, Washington DC, 27 June 1994.

include the high seas.[65] Articles 3 and 17 of the Environmental Edict, as well as the title of the Edict itself, indicate that 'marine areas adjacent to the northern coast' are to be included, perhaps also encompassing the high seas.[66] All articles of the 1990 Decree, except for Article 13, contain the term 'marine areas adjacent'. Article 1 does not clarify whether the provisions are also applicable beyond the EEZ; however, Article 12 specifies that Article 3 of the Environmental Edict is to be applied to 'marine areas adjacent' to the Soviet northern coast *within the Soviet economic zone,* and applied to the NSR and 'adjacent areas'. Article 1 of the NSR Regulations specifies that the NSR is situated *within* the inland waters, territorial sea *or* EEZ adjoining the USSR northern coast *and that it includes* seaways suitable for guiding vessels in ice. Due to the vagueness concerning leading in Article 7(4) of the NSR Regulations, not only in the four straits specifically mentioned in that provision but also 'in other regions', leading may be required in any part of the EEZ. The mandatory nature of the measures, including leading, to be taken by the authorities – who 'shall prescribe' rather than 'may' or 'should' prescribe – may seemingly be argued to encompass the high seas, if these are understood as an 'adjacent area' under the legislation and claims indicated.

On the other hand, the scope of application might be seen as limited to the EEZ through use of the terms 'within' and 'or' in Article 1 of the NSR Regulations. Definite east and west limitations for the NSR are specified as 'the extreme points of which in the west are the western entrances to the Novaya Zemlya straits and the meridian running from Mys Zhelaniya northward, and in the east, in the Bering Strait, by the parallel 66°N and the meridian 168°58′37″W'. This may be further supported by Article 18 of the Environmental Edict which, while using the phrase 'marine areas adjacent', refers to other legislation, including the Economic Zone Edict which specifies the EEZ in its title.[67]

The USA is not apparently opposed to a broad interpretation of the ice-covered areas regime – however, this is only *within* the EEZ. Looking at the OPA, application is restricted to the EEZ, and to 'navigable waters' vaguely defined as 'the waters of the United States, including the territorial sea'.[68] Any limitation of scope to that less than the EEZ appears to be regulated mainly on an article-by-article basis, for example as seen regarding the application of US enforcement measures – the USA or US 'navigable waters'. However, the two terms, 'EEZ' and 'navigable waters', also appear concurrently, for instance concerning the scope of application for liability for oil damages. Normally, 'navigable waters' might be restricted to US internal waters and territorial sea, since that appears to be the ordinary meaning. Additionally, the US Coast Guard has apparently so far not actively enforced the OPA in the Arctic EEZ.[69] However, a liberal interpretation of 'navigable waters'

[65] Kolodkin and Volosov, 'The Legal Regime of the Soviet Arctic', pp. 159–60 and 164–5.

[66] Franckx, *Maritime Arctic Claims*, pp. 179 and 219. [67] See *ibid.*

[68] OPA, para. 1001(8) and (21). The EEZ is defined under the former as 'the zone established by Presidential Proclamation Numbered 5030, dated March 10, 1983'.

[69] Interview with E. Gold, Oslo, Norway, 23 June 1994.

might well open the way for full application of the OPA in the EEZ of the USA. The USA claims that the environmental provisions of the LOS Convention are to be regarded as customary law.[70] If the USA recognises the extensive Part XII, including coastal state prescriptive and enforcement jurisdiction over the EEZ of foreign states, it seems likely that the USA as a coastal state would possess implementing legislation of its own with respect to its EEZ. Given the US history of staunchly espousing freedom of navigation, it is noteworthy that the USA then gives its official acceptance to Part XII with the subsequent restrictions on complete freedom of navigation. This means that it accepts that the Part XII environmental benefits are necessary and that the negotiated limits to free navigation are acceptable. Since the OPA is the central US legislation governing vessel-source pollution, it would seem reasonable to argue that, through the term 'navigable waters', the OPA is applicable or may become applicable in the entire EEZ of the USA. In September 1999 it was announced that the US Coast Guard and federal law enforcement agencies will be allowed to enforce US environmental law and to board foreign vessels up to twenty-four miles from the baselines.[71] The Canadian AWPPA and supporting legislation are limited in application to 100 miles from the baselines.

Thus, despite special Arctic conditions, the area beyond 200 miles from the baselines remains the high seas. This is the limit under international law, regardless of vague provisions appearing in the Russian legislation or ambiguous statements that have been put forward as claims.[72]

Specially protected areas

The 'specially protected areas' legislation of the Russian Federation allows the establishment of: special discharge norms and navigational practices; design, construction, crewing and equipment standards; sea lanes; reporting requirements; and suspension of navigation. The meaning of unilaterally adopted rules for specially protected areas within the Arctic, itself an ice-covered area and thus already under the Article 234 regime, is less than clear.[73] Russia proposed to the IMO that the Arctic be declared a special area under MARPOL 73/78;[74] however, this initiative was reportedly delivered to the wrong IMO committee, and no further action was taken.[75]

According to the Environmental Edict, 'specially protected territories,

[70] Presidential Proclamation No. 5030.

[71] 'US Doubles Offshore Control Zone', *International Herald Tribune*, 4–5 September 1999, p. 3.

[72] See also Churchill and Lowe, *The Law of the Sea*, pp. 164–5.

[73] On specially protected areas in the context of Antarctic environmental protection, see Joyner, Chapter 5 in this book.

[74] On MARPOL 73/78 and special areas in polar regions, see Rothwell, Chapter 3 in this book.

[75] Interview with V. Michailichenko, Director of the NSRA, at the Conference on Harmonisation of Polar Ship Rules (Harmonisation Conference), Göteborg, Sweden, 24 November 1994; and interview with K. Grensemann, Head of German Delegation to the IMO, at the Harmonisation Conference, Ålesund, Norway, 20 November 1995.

including marine areas, shall be created and, when necessary, protected zones shall be established'.[76] In these areas, 'all activity . . . which disturbs natural complexes or threatens the preservation of the respective natural objects shall be prohibited',[77] and ice-covered areas may be closed to navigation for indefinite periods or be made subject to special navigational rules.[78] Various organs led by the Russian Ministry of Merchant Marine can unilaterally draft navigational rules based upon safety and environmental concerns for specially protected areas,[79] where the discharge of wastes, materials and articles is totally prohibited.[80] Technical requirements for vessels are spelled out under the Design Requirements. These have been completely based upon the provisions governing ice-covered areas. It seems possible that additional design, construction, crewing and equipment rules could be developed for specially protected areas within the Arctic, rules that would *amplify* the effects of Article 234. These, however, would undoubtedly close designated areas to virtually all vessels.

Sea lanes are mentioned only in relation to specially protected areas, and this is where navigation is solely permitted.[81] No procedure for including the IMO, in the sense of Article 211(6) of the LOS Convention, is provided in the above provisions for establishing sea lanes in specially protected areas. For the Arctic they may probably be justified under the Article 234 'due regard to navigation [etc.]'. The extensive Russian enforcement measures, which are also applicable to specially protected areas,[82] are necessarily in excess of Article 234, to the extent noted.

Though the relation between specially protected areas and ice-covered areas is unclear, Canadian legislation, and to a degree that of the USA, are similar to the Russian provisions for specially protected areas. The ASPPR provides for sixteen Arctic zones, thereby restricting navigation for all vessels not complying with Canadian Class 1-10 icebreaker and A-E ice-strengthened vessels classifications throughout the year. The OPA[83] and the Monitoring Act provide for Arctic specially protected areas, the Prince William Sound and the Cook Inlet, which are governed by more comprehensive rules, though navigation may not be suspended completely.

Non-discriminatory laws and regulations

Under Article 234, coastal states have the right to adopt and enforce non-discriminatory laws and regulations regarding vessel-source pollution. The main

[76] Art. 1 of the Environmental Edict; see also *ibid.*, Art. 2. [77] *Ibid.*, Art. 2. [78] *Ibid.*, Art. 3.

[79] Arts. 1 and 2 of the 1990 Decree. [80] Art. 11 of the Environmental Edict.

[81] *Ibid.*, Art. 4. Leading and routing under Arts. 3, 7 and 8 of the NSR Regulations as well as the NSR Navigation Guide seem to make a requirement analogous to sea lanes for the NSR, including ice-covered areas.

[82] Art. 13 of the Economic Zone Edict; Art. 3 of the Environmental Edict; and Art. 11 of the 1990 Decree. [83] OPA, paras. 5001–7.

thrust of the Russian legislation is environmental protection and safety, thereby necessarily implying application to all vessels, including Russian ones. The principles behind the NSR Regulations, as stated in Article 2 (quoted above), include regulation of navigation, on the basis of non-discrimination. These regulations are to apply to all vessels, regardless of nationality,[84] and the implications from supporting legislation are the same.

However, the fees required under Article 8(4) of the NSR Regulations may indicate questionable adherence to non-discrimination requirements. It seems doubtful that the current fee rate of US$4–12 per ton, depending upon the size of vessel, is also required of Russian vessels.[85] What then is meant by 'non-discrimination'? Does it refer to non-discrimination only among foreign, non-Russian vessels of different nationalities? Or does it refer also to non-discrimination between foreign and Russian vessels?

No clarification is given by the text of Article 234. However, arguments may be made surrounding the use of the terms 'against' and 'among' dealing with non-discrimination in other articles of the LOS Convention.[86] In this context Article 234 seems to be situated in the 'against' cluster, since it safeguards the right of passage in relation to the environmental provisions, albeit in special, particularly adverse conditions. Thus the adoption and enforcement of non-discriminatory provisions is probably meant to be interpreted in a non-discriminatory manner 'against' all vessels – the coastal state's flag vessels as well as the vessels of any other state. It might be argued that it is not discriminatory for a coastal state to differentiate between its own vessels and foreign vessels in its internal waters and territorial sea. Nevertheless, as seen, due to the scope of Article 234 extending from the baselines to the EEZ, the non-discrimination requirement governing all vessels, coastal state and foreign, would seem likely to apply within the territorial sea as well.

This interpretation also appears supported by the practice of the states implementing Article 234, including Russia – with the exception of fees for services rendered. That in turn would mean that the Russian fees, if they are to be justified under Article 234, must apply to *all* vessels. It seems probable that the Russian practice on this point is contrary to this. Specific Arctic practice supports the non-discrimination requirements of Article 234, and it is only Russia which has a blanket fee structure. Passage rights under both Canadian and US legislation are not dependent upon the payment of fees.

[84] See Arts. 1(4) and 2 of the NSR Regulations.

[85] Interview with T. Ramsland (Norwegian Coordinator for INSROP Sub-Programme III, 'Economic Aspects'; former Lieutenant-Commander in the Norwegian Navy and Research Fellow at the Norwegian School of Business and Sociology), Bergen, Norway, 20 May 1996.

[86] M. H. Nordquist (editor-in-chief) with S. Rosenne and A. Yankov (eds.), *United Nations Convention on the Law of the Sea 1982, A Commentary*, Vol. II (Dordrecht: Martinus Nijhoff, 1993), pp. 226, 232, 376–7 and 462; and M. H. Nordquist (editor-in-chief) with S. Rosenne and S. N. Nandan (eds.), *United Nations Convention on the Law of the Sea 1982, A Commentary*, Vol. IV (Dordrecht: Martinus Nijhoff, 1991), pp. 345–7 and 396–7.

Application to state vessels

The Russian provisions governing ice-covered areas are specifically stated as applying to *all* vessels.[87] For vessels enjoying sovereign immunity, navigation in the ordinary sense is thus either greatly restricted or prohibited. This scope of application of the Russian regime appears to be contrary to the provisions of the LOS Convention and customary international law related to sovereign immunity,[88] nor has it any counterpart in US legislation.[89] However, there may be some similarities in relevant Canadian legislation.[90]

Features of the Russian legislation which could be viewed as consistent with the LOS Convention include notification and authorisation of navigation of state vessels in the territorial sea. Several states consider the passage of war vessels in a foreign territorial sea *a priori* not innocent.[91] Additionally, Article 9(e) of the 1993 Law on the State Frontier of the Russian Federation[92] specifically requires submarines to navigate in the territorial sea on the surface, which is consistent with Article 20 of the LOS Convention.

The USA may have been tempted to test the Russian regime, especially its application to state vessels as well as the high seas.[93] Application of the Russian regime seems even less plausible in the case of submarines, to which the NSR Regulations regarding leading in ice would also apply.

Civil liability and compensation

Article 235(3) of the LOS Convention requires states to cooperate in implementing and further developing international law relating to liability, compensation and dispute settlement. Where appropriate, compulsory insurance or compensation funds are to be developed.

Russia is a party to the 1969 International Convention on Civil Liability for Oil Pollution Damage (CLC) and the 1971 International Convention on the Establishment of an International Fund for Compensation for Oil Pollution Damage (Fund Convention).[94] It was also party to the now terminated Tanker Owners' Voluntary Agreement Concerning Liability for Oil Pollution (TOVALOP).[95] The owner of a vessel registered with a state party and carrying over 2,000 tons

[87] See Arts. 1(4), 2 and 7(4) of the NSR Regulations; Art. 14 of the Economic Zone Edict; and Art. 3 of the Environmental Edict.

[88] Art. 236 of the LOS Convention. See Churchill and Lowe, *The Law of the Sea*, p. 260.

[89] Nowhere in the OPA or the Monitoring Act is application to state vessels stated or implied.

[90] See AWPPA, section 12. Franckx, *Maritime Arctic Claims*, pp. 87 and 122 notes that exemptions can have been granted under section 12(2).

[91] J. Shao, 'The Question of Innocent Passage of Warships After UNCLOS III', *Marine Policy*, Vol. 13, 1989, p. 66.

[92] On file with author. Obtained from A. Yakovlev (retired Admiral, Russian Navy, INSROP Project Leader), 26 August 1995.

[93] Franckx, *Maritime Arctic Claims*, pp. 193–5.

[94] Text reprinted in, respectively, ILM, Vol. 9, 1970, pp. 45ff and ILM, Vol. 11, 1972, pp. 284ff. For the status of participation see the IMO website at www.imo.org.

[95] Text reprinted in ILM, Vol. 8, 1969, pp. 497ff. See 'Ocean Orbit', *Newsletter* (of the International Tanker Owners' Pollution Federation), August 1997, p. 5.

of oil cargo is required to carry insurance or other security covering maximum liability.[96] A certificate issued by the flag state verifying this financial security must be carried on board ship;[97] the CLC certificate is probably the most important to Western insurers as far as oil is concerned.[98]

Under the Russian provisions a certain confusion governs. Liability for serious damage caused by oil pollution and hazardous and harmful substances is now 'virtually non-existent', due to inflation.[99] Under the 1981 Edict on the Amounts of Compensation by Ship-Owners for Losses Caused by Pollution of the Sea by Oil and Other Substances Harmful for the Life of People and for Living Resources of the Sea,[100] a ship-owner must compensate for pollution damage caused, not exceeding 120 roubles per registered tonne and 12.5 million roubles per single polluting incident. Article 5 of the NSR Regulations requires a mandatory 'certificate of due financial security with respect to the civil liability of the Owner for damage inflicted by polluting marine environment and the northern coast', without which navigation is not permitted. There appears little compliance with the CLC and the Fund Convention.[101] According to Article 4 of the 1990 Decree, conditions for insurance coverage would appear to be unilaterally determined by state officials. In addition, the Russian certificate of insurance or other financial security for oil pollution damage may be 'irrelevant for the purposes of Western insurance for the NSR'.[102] On the other hand, there has been an ongoing process of harmonisation with international regimes, including the CLC and its 1992 Protocol, as well as the 1996 International Convention on Liability and Compensation for Damage in Connection with the Carriage of Hazardous and Noxious Substances by Sea (HNS Convention).[103] Preparations have been made for Russia to ratify the 1992 CLC Protocol and the HNS Convention, although few further details have been provided and the process has, as of 2 April 2000, not yet been accomplished for the former.[104]

The OPA is entirely a unilaterally adopted liability regime and thus exceeds the Russian provisions.[105] The USA has not ratified any of the international liability treaties – including the CLC, the Fund Convention and the Protocols of 1992. The OPA was apparently enacted with the knowledge that it would make US

96 See CLC, Art. VII.

97 Annex to CLC, Certificate of Insurance or Other Financial Security in Respect of Civil Liability for Oil Pollution Damage.

98 D. Torrens, 'Marine Insurance for the NSR – Pilot Study', *INSROP Working Paper*, No. 1 (Lysaker: Fridtjof Nansen Institute, 1994), pp. 32–3.

99 A. Kolodkin, O. Kulistikova and E. Mokhova, 'Matters of Responsibility for Mariner Pollution under the Legislation of the Russian Federation', *INSROP Working Paper*, No. 88 (Lysaker: Fridtjof Nansen Institute, 1997), p. 5.

100 English translation reprinted in Butler, *The USSR, Eastern Europe*, J. 3, p. 1.

101 T. Scovazzi, Review of Kolodkin *et al.*, 'Matters of Responsibility' (Appendix in *ibid.*, p. 1).

102 Torrens, 'Marine Insurance', p. 33.

103 Text reprinted in ILM, Vol. 35, 1996, pp. 1,415ff. See Kolodkin *et al.*, 'Matters of Responsibility', pp. 6–9.

104 For an updated status, see the IMO website at www.imo.org. Protocols of 1992 to Amend the CLC and the Fund Convention, in force 1996; on file with author, obtained from the Oil Pollution Compensation Fund, London. Russia is not a party to either.

105 T. Wagner, 'The Oil Pollution Act of 1990: An Analysis', *Journal of Maritime Law and Commerce*, Vol. 21, 1990, p. 573.

adherence to the CLC and the Fund Convention plus Protocols impossible.[106] Regulation in the OPA requires proof of sufficient guarantee to meet the maximum amount of liability under its provisions, and vessels not carrying such evidence may be subject to the enforcement measures noted.[107] No limitation of liability to the established limits is allowed if the incident was caused through a violation of the federal provisions for safety, construction or operation.[108]

Canadian legislation requires that, prior to navigation, financial responsibility corresponding to potential liability must be established to the satisfaction of government officials.[109] Strict yet limited liability is imposed on vessel and cargo owners for illegal environmental pollution damage and the costs of a government-ordered clean-up.[110] Prior to navigation, a certificate must be obtained, giving evidence that the vessel has met the necessary standards.[111] Although this is optional, in practice insurers require it, so a mandatory requirement of compliance with the specified standards is achieved *de facto*.[112]

Criminal responsibility

Article 230(1) of the LOS Convention allows only monetary damages in cases beyond the territorial sea where foreign vessels violate domestic environmental legislation or international rules and standards for the prevention, reduction and control of marine pollution. Article 230(2) makes the equivalent limitation for the territorial sea, with the exception of cases of 'wilful and serious acts of pollution in the territorial sea'.

Criminal liability may apparently arise for violations of environmental provisions governing the Russian EEZ under Articles 19 and 20 of the Economic Zone Edict. Neither the 1984 Decree on the Procedure for Applying Articles 19 and 21 of the Edict on the Economic Zone of the USSR (Procedure Decree)[113] nor the 1985 Statute on the Protection of the Economic Zone of the USSR (Protection Statute)[114] provide any clarification. However, Article 5(2) of the Protection Statute notes that agencies responsible for protecting the economic zone shall, when necessary, transmit materials for bringing guilty persons to responsibility.[115] In clear support of this, Article 14 of the Environmental Edict asserts that persons guilty of violations bear criminal and other responsibility. While fines appear applicable to environmental violations associated with ice-covered areas, if these violations 'by their character' entail criminal responsibility under Soviet/Russian law, then the latter attaches.

[106] P. Edelman, 'Oil Pollution Act', *Pace Environmental Law Review*, Vol. 8, 1990, p. 12.

[107] OPA, paras. 1016(a) and (b). Para. 1004(a) gives limitations for liability and removal costs.

[108] OPA, para. 1004(c)(1). [109] AWPPR, section 12.

[110] AWPPA, section 6(1); and AWPPR, section 15.

[111] AWPPR, section 12; and ASPPR, Schedule VIII, section 12. [112] Torrens, 'Marine Insurance', p. 56.

[113] English translation reprinted in Butler, *The USSR, Eastern Europe*, F. 2, pp. 17ff.

[114] English translation reprinted in *ibid.*, F. 2, pp. 23ff.

[115] See E. Franckx, 'The New USSR Legislation on Pollution Prevention in the EEZ', *International Journal of Estuarine and Coastal Law*, Vol. 1, 1986, p. 170.

Are there no limitations as to penalties and fines for pollution violations in the Arctic? Article 11(h) of the 1990 Decree may provide some clarification: fines may be imposed by plenipotentiary officials for violations consistent with Article 14(2) of the Environmental Edict, which the 1990 Decree implements. This provision refers solely to *fines,* and not criminal liability, which is set forth by Article 14(4). It seems probable that criminal liability may still attach under the ambiguous Article 14(4), for violations which 'by their character entail criminal responsibility' under prevailing Soviet/Russian legislation. More recent clarification may be provided under Article 252 of the 1996 Penal Code and supporting legislation, whereby criminal liability arises for 'marine pollution owing to violation of the rules regulating . . . discharge of substances and material harmful to human health and marine life from vessels'.[116] Fines may be imposed, and prison terms of two to five years added for such acts causing substantial harm. From the above, it would seem that criminal responsibility would attach, in clear excess of Article 230(1) and (2) of the LOS Convention, except for cases of wilful and serious acts of pollution in the territorial sea.

If, however, we consider the Russian provisions in relation to US legislation, we find that criminal liability may well fall within limits similar to those set by the USA, applicable in the EEZ for commercial vessels carrying oil. US federal legislation indicates generally that penalties for the discharge of oil or hazardous substances into navigable waters, the contiguous zone or adjoining shorelines, harmful to human health or the welfare of the USA, may result in fines of not more than US$10,000 or imprisonment for not more than one year.[117] Further, under Alaskan legislation the criminal penalties may be up to one year's imprisonment and fines of up to US$5,000 for spills less than 10,000 barrels, and up to five years' imprisonment and fines of up to US$50,000 for spills involving more than 10,000 barrels.[118] Also, in the Canadian Arctic zone, criminal liability for violations of environmental provisions may result in fines and imprisonment.[119]

Thus, though the Russian provisions for criminal liability theoretically exceed the limits indicated by Article 230 of the LOS Convention, practice in the Arctic is developing otherwise. The possibility for criminal responsibility under legislation adopted by the USA, the main opponent to the Russian as well as Canadian regimes, would seem to provide unusually strong support.

PRACTICE OF THE USA AND OTHER STATES IN RUSSIAN ARCTIC WATERS

It is only the USA which, through its declarations and submarine navigation, actually opposes the Russian regime. These declarations have been noted. Navigation of submarines, however, has remained largely secret, although it would

116 See Kolodkin *et al.*, 'Matters of Responsibility', pp. 10–13 and 15–16, for various statutes.

117 USC, Vol. 33, sections 1321(3), (4) and (5). See generally E. Gold, *Gard Handbook on Marine Pollution*, 2nd edn (Arendal: Assuranceforeningen Gard, 1998), pp. 157–8 and 175.

118 Gold, *Gard Handbook*, p. 166.

119 *Ibid.*, pp. 130 and 194.

appear that passages of US submarines have taken place in all the Russian maritime zones, including the territorial sea and internal waters.[120] Though it is conceivable that the UK and France have also been navigating their submarines in the Russian Arctic, this is not fully documented.[121]

Although the US claims for transit passage also govern surface traffic, all US traffic by both commercial and state vessels appears to be substantially *in compliance* with the Russian regime. The same may be said for other foreign vessels. So far it appears that the commercial vessels of Finnish, Latvian and German flags have also been navigating in compliance with the Russian provisions.[122]

There have, however, been a few specific exceptions with respect to this compliance. Here we should note the Vil'kitskii Straits incidents of the mid-1960s, which involved US Coast Guard and Navy vessels navigating in the Laptev, East Siberian and Kara Seas; the straits themselves were not entered.[123] In these the Soviet authorities claimed that the NSR, along which authorisation of passage was required, traverses Soviet territorial and internal waters, including all west–east straits in the Kara Sea as well as the Dmitrii Laptev and the Sannikov Straits, which were claimed as historic waters. The USA claimed that the straits were used for international navigation or contained high seas channels through which rights to navigation were unlimited. The US vessels, however, withdrew. An additional exception may be the several passages made by the Norwegian state vessel *Sverdrup II* in the Kara Sea in 1995 and 1996, with both US and Norwegian personnel on board.[124] This military research vessel was on assignment, mapping the extent of Russian dumping of radioactive material,[125] which included taking sediment samples. Samples were collected 'near the sites of the dumped reactors' in Abrosimov, Stepovoy, Tsivolka and Techeniya Bays, which may indicate that internal waters as well as the territorial sea, though not the straits, were entered.[126] These passages may have been regulated under one or more trilateral agreements entered by the Norwegian, Russian and US ministries of defence, including the 1996 Declaration on Arctic Military Environmental Cooperation (AMEC).[127]

[120] V. I. Aleksin, 'We Are Ready When You Are', *US Naval Institute Proceedings*, March 1993, p. 56; S. Sontag, C. Drew and A. Drew, *Blind Man's Bluff – The Untold Story of American Submarine Espionage* (New York, NY: Public Affairs, 1998), pp. 158–83, 198 and 209–58.

[121] See W. Reisman, 'The Regime of Straits and the Third United Nations Conference on the Law of the Sea', *American Journal of International Law*, Vol. 74, 1980, pp. 52–3 and 69.

[122] D. Brubaker, 'The Legal Status of Straits in Russian Arctic Waters – Internationality', *INSROP Working Paper*, No. 57 (Lysaker: Fridtjof Nansen Institute, 1996).

[123] See Roach and Smith, *Excessive Maritime Claims*, pp. 200–7. US vessels have made several voyages in the mid-1980s, opposing the Canadian Arctic regime; see Rothwell and Joyner, Chapter 7 in this book.

[124] See S. E. King, D. R. Johnson and J. Carroll, *EPOCA-95 Cruise Report* (Washington, DC: Naval Research Laboratory of the US Navy, 1996); interviews with A. Ushakov, Deputy Director, NSRA; with T. Ramsland; and with V. Peresypkin, Director, CNIIMF – all at INSROP Joint Research Committee, Oslo, 17 April 1996; also interview with T. Ramsland, Oslo, 28 August 1996.

[125] On Soviet/Russian dumping of radioactive waste in the Barents and Kara Seas, see Stokke, Chapter 9 in this book.

[126] *EPOCA-95 Cruise Report*, p. E-1.

[127] Text available at www.denix.osd.mil/denix/Public/Intl/AMEC/declar.html. See an analysis by S. G. Sawhill, 'Cleaning-Up the Arctic's Cold War Legacy: Nuclear Waste and Arctic Military Environmental Cooperation', *Cooperation and Conflict*, Vol. 35, 2000, pp. 5–36.

However, it is understood that a formal protest was later delivered by the Russian Foreign Ministry to the Norwegian Ministry of Foreign Affairs.

The *Sverdrup II* passages would seem to strike the Russian regime at one of its weakest points – its application to state vessels. They are in effect a substantive protest by Norway to the Russian regime, although no protest issued by Norway has been officially published.[128] Russian legislation in this respect is similar to that of Canada; however, no other state has attempted to regulate the navigation of state vessels in ice-covered areas, this being regarded as contrary to Article 236 of the LOS Convention.

Except for the Vil'kitskii Straits incidents, US protests to the Russian regime have been essentially declarative in nature. These declarations, lacking enforcement, would seem only to be discredited by the US domestic legislation as a coastal state, the OPA.[129] It is expected that the USA will require its commercial vessels to comply with all the Russian provisions, including fees, except for application on the high seas. This we may conclude from the US practice in the Canadian Arctic, which has been characterised by compliance with domestic legislation,[130] as well as from the USA's own domestic legislation, the OPA. Further, this would seem plausible from the current US role in the Arctic Council, as well as its participation in the preparation of the draft Polar Navigation Code at the Harmonisation Conferences, where the coast guards and maritime directorates of the Arctic littoral states and other interested states also took part.[131] As to the Arctic Council, among its objectives is the promotion of cooperation and the coordination of action on common Arctic issues, particularly on sustainable development and environmental protection.[132] The USA ostensibly does not consider that such disparity in its Arctic coastal and navigational practice sets a detrimental precedent elsewhere. However, should the NSR become economically feasible, the issue of Russian fees may well come under US scrutiny.

Thus we may conclude that a broad interpretation of Article 234 is being practised through substantial compliance with the Russian provisions discussed, relating to surface traffic of both commercial and state vessels. This is apparently the case despite US declarations to the contrary, and includes all known passages of foreign vessels along the NSR, except for those associated with the Vil'kitskii Straits incidents and *Sverdrup II*. Should such compliance continue, it would seem difficult to argue that customary international law is *not* being formed for the Arctic. The legal consequences of occasional passages by foreign submarines, generally held secret by all states, would seem indecisive as long as the coastal state has no official knowledge of this activity. Lacking official knowledge, it has no opportunity to lodge an effective protest.[133] The passage of submarines has been

[128] Interview with D. Mjaaland, Norwegian Ministry of Foreign Affairs, Oslo, 13 September 1994.

[129] See also Roach and Smith, *Excessive Maritime Claims*, pp. 262 and 264.

[130] *Ibid.*, pp. 207 and 227.

[131] On the Polar Navigation Code, see Brigham, Chapter 11 in this book; see also L. Brigham, 'Commentary – An International Polar Navigation Code for the Twenty-First Century', *Polar Record*, Vol. 33, 1997, pp. 283–4.

[132] On the Arctic Council, see Vidas, Chapter 4 in this book.

[133] Churchill and Lowe, *The Law of the Sea*, p. 94.

an aberration in law of the sea earlier as well,[134] and as such may not hinder the formation of customary law.

CONCLUSIONS

Article 234 of the LOS Convention is formulated in very vague terms. This makes it difficult to show whether the Russian Arctic legislation analysed in this chapter is in many respects in excess of Article 234. There are, however, several contentious features of that Russian legislation that do appear to be in excess of current international law.

First, any application of the Russian legislation to the *high seas* finds no counterpart in US or Canadian legislation, nor has it any basis in international conventional and customary law. Secondly, its application to *state vessels* has no basis in international conventional and customary law either; it finds no counterpart in US legislation, though the relevant Canadian legislation shows some similarities to that of Russia. This requirement of the Russian legislation appears, however, to be substantially complied with in practice. Though US submarines may have made submerged passages contrary to the Russian provisions, this has not been the case with surface-vessel navigation. Thirdly, mandatory *fees* for passage along the NSR are acceptable under Article 26 of the LOS Convention, but only as payment for 'specific services' rendered for passage through the territorial sea. Fees might be acceptable under Article 234, as scientifically sound for environmental protection. In that case, however, they must be applied without discrimination, as well as levied as payment for specific services rendered. The Russian fees appear to be blanket, and probably discriminate in fact. Finally, the status of several other requirements is somewhat unclear. Icebreaker-assisted pilotage, ice-breaker leading and the introduction of specially protected areas, though generally exceeding the requirements of US legislation, have parallels in the legislation adopted by Canada. The USA has established specially protected areas but permits navigation, including without the use of ice-breakers. Canada has its Arctic 'safety control zones' upon the basis of which vessel navigational access through ice-construction standards are governed. The Russian and Canadian requirements could probably also be argued to be justified under Article 234, with its provisions concerning 'due regard to navigation [etc.]'.

The substance of Article 234 is still under formation. While Russia may be seen as 'straining' existing international law with its Arctic legislation, most of the requirements have counterparts in the legislation of both the USA and Canada. Furthermore, navigation of foreign vessels through the NSR appears to be substantially in compliance with the requirements of the Russian regime. It is only the USA, as a maritime power, which has most consistently opposed the Russian Arctic

[134] M. Leifer, *International Straits of the World – Malacca, Singapore, and Indonesia* (Alphen an den Rijn: Sijthoff & Noordhoff, 1978), pp. 162–3 and 168–73.

regime, through declarations made by the US State Department and the US Navy, and through passages undertaken by US Navy submarines. However, the surface navigation of US flag vessels has thus far apparently largely adhered to the Russian regime. To the extent that practice among the Arctic states is both consistent and norm-setting, this probably indicates that a process of formation of customary international law for the Arctic is underway, defining and interpreting the contours of Article 234 through relevant state practice.

11 The emerging International Polar Navigation Code: bi-polar relevance?

LAWSON W. BRIGHAM

Throughout history, ships have been the principal means for reaching the remotest regions of the world ocean. Such is the case at the end of the twentieth century as ships carry scientists, explorers, commercial fishermen and tourists, among others, to remote areas in the Southern Ocean and the Arctic Ocean. Since 1977, polar icebreakers have made an impressive total of twenty-eight voyages to the North Pole for science and tourism.[1] Such access to the central Arctic Ocean by surface ship would have been unthinkable only thirty years ago, when the Arctic Ocean was considered the domain of the nuclear submarine. In the Antarctic ships carrying tourists have circumnavigated the continent,[2] and thousands of people have visited the Antarctic Peninsula and the Ross Sea region by tourist ship. These voyages of discovery and adventure reflect unprecedented ship access to nearly all polar marine waters. Thus, it should not be surprising that national and international regulatory bodies have initiated the development of construction and navigation standards for polar ships.

NAVIGATING IN POLAR WATERS

What basic factors make ship navigation in polar waters unique compared to all other ship voyages across the global ocean? Obviously, polar waters are located at the extremities of the world ocean, generally remote from the centres of human civilisation and from the normal availability of port services and rescue capability. Polar waters are also among the least charted marine regions on earth. Further, the Arctic and Southern Oceans are perennially cold, with water temperatures near freezing, and human survival within such waters is exceedingly limited.

[1] Remarkably, twenty-one of these North Pole voyages have been with tourists sailing aboard Russian icebreakers. During 1977–99 polar icebreakers from Russia, Sweden, Germany, Canada and the United States have reached the North Pole.

[2] J. Splettstoesser, R. Headland and F. Todd, 'First Circumnavigation of Antarctica by Tourist Ship', *Polar Record*, Vol. 33, 1997, pp. 244–5.

Above all, the principal difference between polar waters and the remaining oceans is the presence of sea ice formed through the freezing of sea water. On a global scale, 7 per cent of the earth's surface is covered by sea ice at some time each year.[3]

As shown in Figure 11.1, in the Arctic Ocean the average winter maximum extent of sea ice extends well beyond the central Arctic Ocean as such – covering the Canadian Archipelago, the Russian Arctic including regions of the Barents Sea, Baffin Bay, the Greenland Sea, the Sea of Okhotsk and much of the Bering Sea. The Arctic sea ice area fluctuates between 9 million km^2 (summer) and 15 million km^2 (winter).[4] The summer minimum represents the approximate area of multi-year ice that remains in the central Arctic Ocean.[5] In the Southern Ocean, the maximum extent of sea ice occurs in September and surrounds the Antarctic continent north of 55° to 65° South. The fluctuation of sea ice area in the Southern Ocean is considerably greater than in the Arctic Ocean – 3 million km^2 (summer) to 20 million km^2 (winter) – and varies from 1.5 to 10 per cent of the Southern Ocean surface.[6] This variability is due primarily to the open boundary north of the Antarctic coast that allows most of the sea ice to melt during the austral summer.

Ice in all its forms – sea ice and floating glacial ice (icebergs) – presents a formidable obstacle to surface ships operating anywhere in polar waters. Usually such ships require specialised design, added hull strengthening and increased propulsive power. Even ships that are not attempting to break ice, or do not intend passage through ice, should have adequate hull strength, since they may become trapped in drift ice (and be subject to potential damage by the surrounding ice under pressure of the winds and currents). As indicated in Figure 11.1, there are vast regions of open water during the summer minimum extent of sea ice surrounding Antarctica. Most Antarctic research and support ships as well as tourist ships operate near the coast during January through March, so as to maximise their access to the continent. However, low visibility, frequent icebergs and substantial fast ice are predominant conditions in most Antarctic coastal regions. There are very few areas where a ship will not encounter ice in some form. Highly variable and rapidly changing weather and sea ice conditions make Antarctic ship operations challenging and always demanding of a tenacious emphasis on safety. The remoteness of Antarctica and the normally independent operation of ships in the region require that special consideration be given to ship and crew survivability in such an extreme environment.

The land-locked nature of the Arctic Ocean, with only one major opening for sea ice export (between Svalbard and Greenland), results in substantial ice remaining in many Arctic coastal seas for long periods each year. Sea ice and icebergs are also confined by the large island groups that surround the Arctic basin

[3] W. Weeks, 'Sea Ice', *Encyclopedia of Earth System Science*, Vol. 4 (New York: Academic Press, 1992), p. 41. [4] *Ibid.*

[5] Multi-year ice is also found along the northern Greenland coast and around the northern islands of the Canadian Arctic. [6] Weeks, 'Sea Ice', p. 41.

(see Figure 11.1). Many of the passages between the islands, particularly in the Russian and Canadian Arctic regions, remain partially ice-covered throughout much of the year. Many traditional navigation routes can be ice-covered even during the Arctic summer. In several respects, ice navigation in the central Arctic Ocean can be more demanding than Antarctic operations because of the greater extent of summer sea ice. The presence of thick, multi-year ice in the central Arctic Ocean that on occasion intrudes into the coastal seas also influences the need for higher-class polar ships and for additional design requirements. Due to the seasonal melting of sea ice, a short navigation season, mostly in open water, is possible in several Arctic coastal regions – among them Hudson Bay, around Greenland, Svalbard, along the western Alaskan coast and in the southwest Kara Sea.

Navigating in polar waters requires the availability of reliable and timely information on the location of the surrounding sea ice and forecasts of any near-term changes. Ice (and weather) information can be obtained from national ice centres, local observations, and real-time satellite imagery received directly aboard ship. In many polar regions, ships will sail around and avoid areas of more difficult ice conditions by using all available environmental information. During most Antarctic operations, this is a safe and effective strategy of route planning. In the Arctic, however, many navigation straits and coastal seas are constrained by geography; polar ships are either escorted by icebreakers or are forced to navigate independently through areas of considerable ice. Undoubtedly, future polar ships will have the capability of acquiring better ice information, which in turn will improve the effectiveness and safety of ice navigation.

HISTORICAL PERSPECTIVES

Significantly, strategies to enhance marine safety by the establishment of navigation and construction standards for polar ships were pursued by several circumpolar nations prior to the era of the LOS Convention and the formulation of its Article 234. For more than three decades, Russia and Canada developed extensive regulatory and control systems for Arctic shipping. Russia's system for navigation along the Northern Sea Route includes the issuance of ice passports (certificates) to commercial ship classes.[7] Close operational control of all ships, with or without icebreaker escort, remains a hallmark of the Russian system.[8] Canada's ice-regimes shipping system designates zones for seasonal ice navigation by ships of varying ice class. Specific structural standards for polar ship classes are integral components of both the Russian and the Canadian approaches. Detailed and refined rules for ice-going ships in the Baltic Sea have also been developed,

[7] *Guide to Navigating Through the Northern Sea Route* (St Petersburg: Head Department of Navigation and Oceanography of the Ministry of Defence of the Russian Federation, 1996), p. 319 (hereinafter NSR Guide). On Russian legislation and practice relating to navigation along the NSR, see Brubaker, Chapter 10 in this book.

[8] NSR Guide, pp. 84–9.

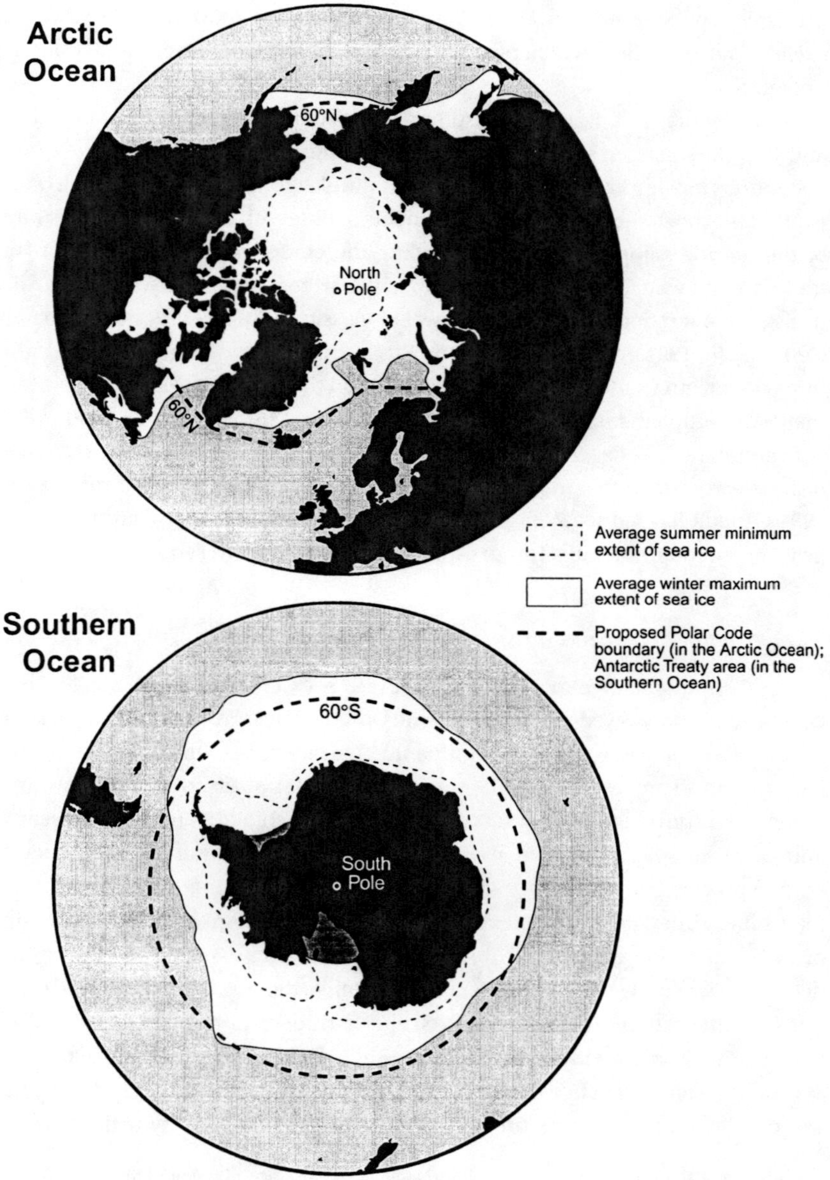

Figure 11.1 Extent of sea ice in the polar oceans

principally by Sweden and Finland. In recent decades, most ship classification societies have developed their own ship construction rules for Arctic and Antarctic operations.[9]

By the early 1990s, the resulting efforts of national maritime authorities and ship classification societies had left a non-uniform, patchwork set of rules and regulations for ships navigating in ice. The multiple domestic systems with a range of different ship ice-classes were clearly incompatible with the internationalisation of the marine shipping industry, including the emerging international tourism trade in polar waters. Most critically, ship and mariner certificates were not readily transferable among nations or between classification societies. Fortunately Germany (in 1991, for safety issues) and Russia (in 1993, for discharge issues) proposed to the International Maritime Organisation (IMO) initiatives for the development of supplemental polar rules for SOLAS, MARPOL 73/78 and other conventions and codes. The IMO agreed that an Outside Working Group (OWG) of technical experts should explore the development of specialised polar rules. Since 1993 Canada has led what has become known as the polar ship 'harmonisation process', an international effort to establish unified standards and rules.[10]

THE HARMONISATION PROCESS AND DEVELOPMENTS

Bi-annual meetings of the OWG in 1993–7 were held in Canada, Finland, Germany, Norway, Sweden, Russia and the United States. From the outset, various key stakeholders were active and influential participants in the open fora of the harmonisation process. Included were members of national and regional maritime authorities, ship classification societies, commercial ship operators, and research and academic specialists.[11] Since the primary emphasis of the effort was on developing a practical instrument for safety and environmental protection, most participants had either technical expertise (ship design, engineering and construction) or polar operation experience. Both backgrounds were essential to crafting the details of a workable body of polar ship rules. Unlike the LOS Convention and other notable international maritime efforts, the harmonisation process was led by marine practitioners with polar expertise: while diplomats and legal experts were at times present and active during the proceedings, the drafting of the polar rules was conducted by maritime professionals. Sixteen nations, many with active bi-

[9] Examples of the ship classification societies include the American Bureau of Shipping, Bureau Veritas (France), Det Norske Veritas, Germanischer Lloyd, Lloyd's Register of Shipping and the Russian Register of Shipping.

[10] Canada's leadership coming from within Transport Canada.

[11] Examples of national maritime authorities participating in the harmonisation process include the Canadian Coast Guard, Transport Canada, the Government of the Northwest Territories, the Icelandic Maritime Administration, the Finnish Maritime Administration, the Danish Maritime Authority, the Swedish Maritime Administration, the Australian Maritime Safety Authority, the US Coast Guard, the Ministry of Transport of the Russian Federation (Northern Sea Route Administration), and the Norwegian Maritime Directorate.

polar interests, participated by sending experts or providing material during the deliberations of the OWG.[12]

The main objective of the harmonisation process was to create a comprehensive, unified code of rules for ships navigating in Arctic and Antarctic waters – the 'International Code of Safety for Ships in Polar Waters' (Polar Code) as submitted to the IMO in the spring of 1998.[13] A key strategy of the OWG was to build upon existing IMO ship rules.[14] The Polar Code was *never intended to duplicate or replace* existing standards for international safety, pollution prevention and training. The *additional* measures of the Polar Code focus specifically and equally on the safety of human life and the protection of the marine environment. From the early deliberations of the OWG, several guiding principles, endorsed by the IMO, have shaped development of the draft Code:

1. ships are to have suitable ice strengthening for their intended voyages;
2. no oil shall be carried against the outer shell;
3. all crew members are to be properly trained in the operation of polar vessels;
4. appropriate navigational equipment shall be carried;
5. suitable survival equipment shall be carried for each person;
6. a unified method of classifying ice conditions is to be used; and
7. consideration of vessel installed power and endurance must also be made.

A second, significant strategy of the OWG was to separate development of the Polar Code – the generalised description of the harmonised polar ship rules – from parallel work by the International Association of Classification Societies (IACS) on the detailed specifics of hull and machinery requirements.[15] The IACS unified set of rules would eventually include a single set of international ice classes for ships, detailed hull structural requirements and specific engineering systems requirements for polar ships. Thus, as technology evolves, the IACS uniform rules can

[12] The nations involved in the harmonisation process are Argentina, Australia, Canada, Chile, Denmark, Finland, France, Germany, Iceland, Italy, Japan, Norway, Russia, Sweden, the United Kingdom and the United States.

[13] The draft Polar Code was submitted by Canada, on behalf of the OWG, to the IMO Sub-Committee on Ship Design and Equipment, 41st Session in London, March 1998.

[14] Notably, the 1973 International Convention for the Prevention of Pollution from Ships as modified by the Protocol of 1978 (MARPOL 73/78); the 1974 International Convention for the Safety of Life at Sea (SOLAS); and the 1978 International Convention on Standards of Training, Certification and Watchkeeping for Seafarers (STCW).

[15] The IACS, representing the major independent ship classification societies, includes 90 per cent of the world fleet by tonnage. IACS members conduct more than 500,000 ship surveys each year (by 6,000 surveyors and 3,700 technical staff). IACS roles include a classification role, i.e. independent assessment of the integrity and mechanical ability of the ship for the intended purpose; and a statutory certification role, i.e. as certifying agent under delegated authority from a maritime administration for requirements of international and national instruments. IACS clients include shipowners, states, designers, shipyards, manufacturers, underwriters, financiers, charterers and cargo interests.

adjust and adapt, leaving the broad rules of the IMO Polar Code essentially unchanged. This fundamental separation agreed to by the national maritime authorities, the IMO and the IACS governing body allowed the bi-annual harmonisation meetings to proceed smoothly with drafting the Polar Code. An IACS *ad hoc* group organised work and discussions on detailed technical issues.

Developments related to navigation and polar mariner certification have also had an important influence on the evolution of the Polar Code. A harmonisation working sub-group focused on the navigation and training elements of the proposed Code. Key topics included a certification process for ice pilots, an international ice navigators' course, and future ice simulator training requirements. In addition, a new organisation, the Circumpolar Advisory Group on Ice Operations (CAGIO), was formed during the harmonisation process.[16] CAGIO was established by the national administrations which have responsibility for domestic ice-covered waters and which may also operate polar ships in their respective national interests. As a unique international forum for discussion of bi-polar ice operations, its birth at the end of the century can be attributed partially to the end of the Cold War, which has allowed the full participation of the Russian Federation. In the future, CAGIO can act in a consultative role to the IMO, the Arctic Council and other polar and maritime organisations.[17]

A major theme of this book focuses on the integration of legal and policy approaches at various levels (global, regional, sub-regional, national) to protecting the polar marine environment. In essence, the aforementioned harmonisation process is uniquely illustrative of integrated global cooperation – among the OWG, IMO, CAGIO and IACS. Significant cooperation is evident within the process between public (the IMO and national administrations) and private (the IACS and commercial shipping) institutions. There was also a unity of purpose among the national maritime administrations and the IACS to develop polar ship standards adhering from the beginning to a 'precautionary approach'.[18] The key stakeholders all recognised the potential risks associated with increased bi-polar shipping and were committed to produce a functional and timely draft instrument for IMO consideration.

KEY ELEMENTS OF THE POLAR CODE

Table 11.1 is a brief outline of the structure and components of the draft Polar Code, as submitted to the IMO in March 1998. In the Preamble, linkage of the

[16] The founding members of the CAGIO were Canada, Finland, Norway, Sweden, Russia and the United States. [17] CAGIO Terms of Reference, November 1996.

[18] In the core of the precautionary principle is an anticipatory or preventive approach, a key facet of the Polar Code. While acknowledging that uncertainty is unavoidable, this principle also stipulates that uncertainty (with regard to environmental impact) should not be used as an escape to delay or avoid the development and implementation of protection strategies or measures. For further discussion of the precautionary principle in the context of the theme of this book, see especially VanderZwaag, Chapter 8; also Boyle, Chapter 1, Rothwell, Chapter 3 and Vidas, Chapter 4 in this book.

OWG to the IMO is established through the Sub-Committee on Design and Equipment under the Maritime Safety Committee.[19] The unique risks of sea and glacial ice, remoteness, the effects of cold temperatures and challenging navigational environments for the polar regions are acknowledged. A defining statement ensures that 'all ships operating in Polar Waters meet internationally acceptable standards of ship safety and pollution prevention' – the objective of the Code.[20] Importantly, the Polar Code is not intended to infringe on national systems: domestic navigation rules and regulations may be retained.[21] As articulated by the CAGIO, national maritime administrations are to be responsible for compliance with the Polar Code.[22] The Code is also to be applied in its entirety, and not piecemeal to suit a special region or sub-region, or particular nation.[23]

One of the most critical definitions in the Code is for 'Polar Waters'. For the Antarctic those waters south of 60° South are considered 'polar'.[24] In the Arctic the definition takes into account the open waters (with no seasonal ice) of the North Atlantic. From Labrador to Greenland, polar waters are north of 60° North; the boundary then proceeds from the southern tip of Greenland to Keflavik, Iceland, and then from the northern shores of Iceland to Bjørnøya Island (south of Svalbard); the boundary continues onwards to Cape Kanin Nos in the Russian Arctic (on the Barents Sea). In the North Pacific Ocean the 60° North parallel marks the polar boundary cutting across the Bering Sea from Alaska to the Russian Far Northeast.[25]

The provisions of the Polar Code are meant to be used in addition to any other applicable code or convention (such as SOLAS, MARPOL 73/78 and the STCW).[26] While most ships operating in polar waters and engaged on international voyages are subject to the Code, specific exemption is granted for several ship types, such as warships.[27] Seven polar ship classes (PC1 through to PC7) are defined on the basis of requirements related to environmental conditions: PC1 is the most capable ship which can operate year-round in all polar waters, whereas PC7 is the least capable, operating in summer/autumn in thin first-year ice (with old ice inclusions).[28] Part A of the Code outlines the general construction requirements for polar ships. The technical details are left to the IACS unified requirements undergoing parallel development. Notable requirements and rules include: hull structures designed to resist global and local ice loads; enhanced stability standards; double bottoms for all ships; no pollutants to be carried next to the outer hull (a pollutant must be separated from the outer shell by a cofferdam); escape measures adapted to cold environments, particularly ice accretion; anchoring, towing and

[19] Polar Code, Preamble, para. 1(1). [20] *Ibid.*, para. 1(4). [21] *Ibid.*, para. 2(6).

[22] *Ibid.*, para. 2(7). [23] *Ibid.*, para. 2(8). [24] Polar Code, Guide, para. 3(19).

[25] *Ibid.* See Figure 11.1 above. [26] Polar Code, Chapter 1, para. 1(1)(2).

[27] *Ibid.*, para. 1(1)(1) and 1(1)(4). Exempted vessels are: warships and troopships; ships operating independently, powered solely by oars, sails or other non-mechanical means; wooden ships of primitive build; and stationary ships permanently anchored or moored in a single location.

[28] *Ibid.*, Chapter 1, Table 1.1. The ice types in the polar class descriptions follow the sea ice nomenclature established by the World Meteorological Organisation (WMO).

Table 11.1 *Structure and components of the draft Polar Code as submitted to the IMO, March 1998*

Preamble		
Guide to the Code		
Chapter	1 –	General
Part A – Construction requirements		
Chapters	2 –	Structures
	3 –	Subdivision and stability
	4 –	Accommodation and escape measures
	5 –	Directional control systems
	6 –	Anchoring and towing arrangements
	7 –	Main machinery
	8 –	Auxiliary machinery
	9 –	Electrical installations
Part B – Equipment		
Chapters	10 –	Fire safety
	11 –	Life-saving appliances and survival
	12 –	Navigational equipment
	13 –	Communications
Part C – Operational		
Chapters	14 –	Operational requirements
	15 –	Crewing
	16 –	Emergency equipment
	17 –	Environmental protection and damage control
Annexes		
	I –	Polar ship safety certificate
	II –	Permit to operate in polar waters
	III –	Breathing apparatus
	IV –	Life-saving appliances and survival equipment
Appendices		
	I –	Draft SOLAS amendment
	II –	Draft amendment to STCW Convention
	III –	Draft IMO resolution: equivalencies for existing ships
	IV –	Draft IMO resolution: equivalencies for existing ice navigators
	V –	Life-saving equipment comparisons

steering systems designed for operating in ice; and machinery and electrical systems specially adapted to function in low temperatures and for the loads and vibrations anticipated when in ice.[29]

Part B is concerned with the unique equipment requirements of polar ships for lifesaving, firefighting, navigation and communication. Specific guidance is provided for the operation of fire safety equipment in view of the cold, remote regions of operation. Ship systems, such as those for ventilation and pumping,

[29] *Ibid.*, Chapters 2–9.

must be protected from low temperatures, freezing and ice accretion.[30] The polar environment imposes significant demands on lifesaving and survival equipment; two (of many) rules reflect this fact: all lifeboats for polar class ships must be enclosed, and ice accretion on lifeboats and liferafts must be adequately dealt with.[31] A certain redundancy of navigation equipment is required: two speed/distance measuring devices and two independent echo-sounding devices.[32] For the higher-rated polar ship classes (PC1–5), one radar should be adapted for ice navigation; those ships should also have a voyage data recorder.[33] For PC1–3 ships, hull stress indicators are mandated, to provide continuous information to those piloting the ship.[34] Satellite communication and navigation systems should be used aboard polar ships if operating outside of reliable coastal coverage of land-based systems.[35] However, PC1–5 ships should also be provided with low-frequency radio equipment when satellite communications are difficult.[36]

Special attention in the Code (Part C) is given to human factors in polar operations, including operational procedures and training. All polar-class ships in polar waters should carry a Polar Ship Safety Certificate, a Permit to Operate in Polar Waters, and sufficient personnel trained and certified for ice navigation.[37]

Part C also contains requirements for operating and training manuals (documenting standard emergency procedures), specific medical equipment, and damage control and repair equipment.[38] Chapter 17 outlines pollution prevention and spill mitigation measures with regard to the unique environmental hazards, lack of waste reception and repair facilities, and the limited assistance available in polar regions.[39] Polar ships are to adhere to current MARPOL 73/78 requirements for operational discharges or those rules of a coastal state, whichever are more stringent.[40] Polar ships are also required to process and store all waste for the duration of the voyage.[41]

Annexes I and II of the Code provide amplifying information on the Safety Certificate and Permit to Operate documents. Annexes III and IV provide specific details on breathing apparatus, lifesaving appliances and survival equipment. The Appendices are not formal sections of the Code. They were added by the OWG as a means of submitting the following to the IMO: proposed draft SOLAS and STCW amendments, 'grandfathering' provisions for existing ships and ice navigators, and

[30] *Ibid.*, Chapter 10. [31] *Ibid.*, Chapter 11, para. 11(5)(1) and 11(5)(2).

[32] *Ibid.*, Chapter 12, para. 12(3)(1) and 12(4)(1). [33] *Ibid.*, para. 12(5)(1) and 12(12)(1).

[34] *Ibid.*, para. 12(11)(1). [35] *Ibid.*, para. 12(6)(2); and *ibid.*, Chapter 13, para. 13(2)(1).

[36] *Ibid.*, Chapter 13, para. 13(3)(1).

[37] *Ibid.*, Chapter 14, para. 14(1)(1). Chapter 1 provides for the following: a Polar Ship Safety Certificate is issued by a national administration (or classification society) after a ship survey confirming compliance with the Code's Parts A and B; a Permit to Operate in Polar Waters is issued by a national administration or flag state stipulating the conditions of operation and when satisfied the operators have made adequate provisions for safety and pollution prevention; all ships (other than those in ice-free waters) should carry at least one certified Ice Navigator.

[38] Polar Code, Chapters 14 and 16. [39] *Ibid.*, Chapter 17, para. 17(1)(1).

[40] *Ibid.*, para. 17(1)(2). MARPOL 73/78 Annexes I, II, IV and V for Special Areas are noted in the draft Code. [41] Polar Code, Chapter 17, para. 17(3)(1).

a comparison of lifesaving equipment between the Code and other IMO instruments.

This brief examination of the draft Code highlights several important environmental and safety improvements proposed for ships navigating in the Arctic and Antarctic. Double bottoms, enhanced stability and damage control, cofferdams between pollutants and the outer hull, and hull stress indicators are key features to reduce spills in polar waters. Operational discharges are to be controlled with existing MARPOL 73/78 rules, waste containment (for entire voyages), and the redesign of propellers and shafts to eliminate leakage. Safety certificates, permits to operate, reporting requirements and ice navigator special certification procedures are all regulatory devices to reduce the environmental impact of ships in polar waters.

ARCTIC ISSUES

The harmonisation process began with a northern focus. A majority of the initial participants and stakeholders have substantial Arctic interests and experience.[42] The existing ship traffic (in ice) in the Baltic, along Russia's Northern Sea Route and in Canada's northern waters was influential and highly relevant to the process. To a lesser extent, ice-going ships operate in the Bering and Chukchi Seas off Alaska and around Greenland. Also useful as models are the domestic marine regulatory schemes very much in place in both the Canadian and the Russian Arctic regions.[43] Taken collectively, there is general awareness that these regional shipping experiences (and the related rules developed by national authorities and classification societies) have been important benchmarks for shaping many of the basic safety tenets of the draft Polar Code.

A key factor allowing greater flexibility in drafting a polar code of navigation is that the Arctic is not bound by any existing holistic marine environmental protection regime.[44] More specifically, the Arctic nations have yet to accept the concept of a MARPOL 73/78 Special Area.[45] Without the limitations imposed by such a regulatory regime, or the global visibility of a recent polar ship catastrophe (galvanising public opinion and forcing hasty action), the harmonisation process can be judged in hindsight as relatively unconstrained.

Several polar boundary issues in the Arctic may emerge from the IMO's review of the draft Code. The North Atlantic regions north of 60° North, where there is no seasonal sea ice, have been accommodated into the definition of polar waters; see Figure 11.1 for the proposed boundary in the North Atlantic that is north of 60° North. Three areas also exempted from the Code and in waters north of 60° North are the Baltic Sea, the White Sea and the Sea of Okhotsk. Arguments can

[42] Moreover, at least three of the 'founding' participants – Germany, Russia and the United States – also have extensive Antarctic operational experience.

[43] For Canada, see Rothwell and Joyner, Chapter 7; and for Russia, see Brubaker, Chapter 10 in this book.

[44] See Vidas, Chapter 4 in this book.

[45] See Rothwell, Chapter 3 in this book.

be made for the exemption of similar waters.[46] A critical and related issue is the consideration of ice-free waters during the summer months or limited periods of the year.[47] For established shipping concerns and selected cruise ships that operate entirely in ice-free waters, national maritime authorities may have considerable discretion in deciding how a permit to operate can be issued. To minimise the impact on a proven, safe operation, the question remains whether a permit can be issued on a seasonal basis. Deciding how the Code can be applied on an annual or seasonal basis is a significant point of discussion and reconciliation for the IMO's deliberations on the draft Code. This issue also has significant ramifications for Antarctic ship operations of a similar nature.

The intent of the Code is not to replace completely all domestic Arctic traffic management systems – those used in the internal and coastal waters of a circumpolar state. Local marine systems, such as those established for the Northwest Passage in Canada and the Northern Sea Route in Russia, are likely to remain in place. National systems will continue to provide traffic information, notice to mariners, environmental protection advice, icebreaker support (escort) and pilotage. Special transit regulations for internal waters will remain within the purview of the coastal state. Arctic states will also continue to exercise authority for marine environmental protection matters in their Arctic coastal waters, on the basis of Article 234 of the LOS Convention.[48] The most significant change for the Arctic will be the replacement of existing national polar classes and structural standards with the harmonised rules for polar ships. These uniform rules should not influence the operating procedures traditionally used in different Arctic regions – ranging from ships operating independently in ice to those requiring escort by polar icebreaker. Such navigational control is to remain at the national level and is not a significant consideration in the draft Code.

A continuing issue is the need for adequate impact assessments and cost–benefit analyses of the Code. These are particularly required in the Arctic due to the seasonal presence of commercial shipping. The economic concerns of the Arctic shipping community must also be addressed in implementation provisions of the Code, including the length of the transitional period (non-mandatory to mandatory status), the grandfathering of existing ships into the Code, and the equivalency of certain ships. An acceptable balance between risk and cost should be demonstrated by an economic analysis of the Code. In the future, it is possible that higher polar class ships will be required (forcing increases in capital costs), since fewer government icebreakers will be available for ice escort. One of the clear indicators that the Code is relevant to the Arctic is that the major circumpolar stakeholders merged their expertise to provide credibility and considerable

[46] A prime example is the waters of southern Alaska (north of 60° North) including Cook Inlet (to Anchorage) and Prince William Sound (to Valdez). These waters are entirely within US jurisdiction.

[47] Important examples of seasonally ice-free polar waters include all of the Bering Sea, regions of the Chukchi and Kara Seas, and the waters around Iceland.

[48] See the discussion by Vukas, Chapter 2 in this book.

momentum to the effort. A primary focus of the OWG was to integrate the research and operating experience of the Arctic nations and formulate this collective Arctic experience into a coherent and realistic set of standards. The harmonisation process was blessed with good timing, since several historic Arctic political developments were taking place as the 1990s unfolded. The maturing of the Arctic Environmental Protection Strategy and the establishment of the Arctic Council, together with arguments that sustainable development issues be given centre stage, all served to highlight environmental and safety concerns for the entire Arctic.[49] The importance and timing of a new polar code of navigation to address the shipping component of any evolving Arctic regime appear rational and highly relevant, given the overall context of change in today's circumpolar North.

ANTARCTIC ISSUES

A central issue until recently was whether, and how, to apply the Polar Code to the Southern Ocean, given the markedly different political regime of the Antarctic Treaty and its more recent Environmental Protocol. It must be noted at the outset that, pursuant to decisions of the Antarctic Treaty Consultative Parties at their June 1999 meeting in Lima, Peru, the Polar Code will not apply to Antarctic waters.[50] An opportunity for direct application would, one can argue, appear plausible, since the language of the Environmental Protocol refers specifically to the design, construction, manning and equipment of ships operating in the Antarctic.[51] Rather than have Consultative Parties develop individual (and potentially conflicting) ship standards to meet the overall objectives of the Protocol, the Polar Code has set out to fulfil this complex requirement. And the path remained open for Consultative Parties and COMNAP to recommend to the IMO adjustments to the Code for any unique aspects of Antarctic operations.[52] A potential benefit of having the Polar Code fill the ship requirements of the Antarctic Environmental Protocol is that, as an IMO code, it would eventually apply to all IMO signatories, who have under their jurisdiction most of the world's shipping tonnage. The Polar Code would thus apply to ships whose flag states are not parties to the Antarctic Treaty but who nevertheless might operate in the Southern Ocean.

Another specific difference posed by the Polar Code involves the exemption of certain ships. While the Polar Code exempts warships (among others), it

[49] See Vidas, Chapter 4 in this book, for discussions on the AEPS and other Arctic political developments.

[50] See *Final Report of the XXIII Antarctic Treaty Consultative Meeting, Lima, Peru, 24 May–4 June 1999*, as well as Decision 2 (1999); Annex B in *ibid.*; all available at www.rree.gob.pe/conaan/meeting1.htm; see also further below, in this chapter.

[51] Environmental Protocol, Annex IV, Art. 10: 'In the design, construction, manning and equipment of ships engaged in or supporting Antarctic operations, each party shall take into account the objectives of this Annex.'

[52] COMNAP (the Council of Managers of National Antarctic Programs) could as an alternative to the Code, develop its own set of rules and standards for Antarctic ships.

does not – in contrast to both MARPOL 73/78 and the Environmental Protocol – offer a sovereign immunity clause for ships operating in non-commercial, governmental roles.[53] As many government ships routinely operate in Antarctic waters for most Consultative Parties, this issue would certainly require discussion and reconciliation with the IMO. Closing the gap for sovereign immunity to include only warships would be a controversial move.[54] Conceivably, Permits to Operate, Polar Ship Safety Certificates and ice navigation certification would not be requirements of state-operated ships in the Antarctic if they were exempt under the Code.

Delineating a Polar Code boundary for Antarctic waters as all waters south of 60° South would seem a logical, effective solution, since this matches the Antarctic Treaty area. But at least two issues have arisen: sea ice can extend north of this boundary in winter; and the 1980 Convention on the Conservation of Antarctic Marine Living Resources (CCAMLR) boundary which extends north of 60° South in several sectors. However, nearly all ship traffic currently in the Antarctic, the vast majority being fishing vessels and tourist ships, operates during the summer months when sea ice has retreated to the Antarctic coast in several regions. Also, the MARPOL 73/78 Special Area designation for the Southern Ocean has remained linked to the Antarctic Treaty area, although attempts have been made to match it with the CCAMLR boundary.[55] Other sometimes onerous challenges to Antarctic ship operations remain, such as the remoteness of the waters (to emergency response, repair yards and discharge facilities), the constant cold water, a periodic mix of sea ice and floating glacial ice in many regions, uncharted waters, and rapidly changing weather. The Polar Code could have provided baseline protection for many of these unusual conditions through requirements over and above SOLAS and STCW rules – for safety and survival equipment, fire safety, communications and ice navigation training. A problem was indeed whether the Polar Code would provide a flexible enough framework to Consultative Parties so as to assure appropriate and adequate regulation of specialised ships that operate in highly variable ice conditions.

Under the Polar Code, flag states will be responsible for issuing Permits to Operate. For Antarctic shipping operations, COMNAP and the recently established Committee for Environmental Protection could provide substantial advice and review of such permits with regard to any restrictions (waste discharges, for example) or limitations appropriate to the anticipated ice conditions.[56] Also in the future, Antarctic shipping – for example, potential increases in tourist ship traffic – may be subject to review by the Protocol's extensive environmental evaluation process.[57] The existence of clearly defined polar ship classes and major construction and operational standards (in the Code and unified technical rules of the

[53] See Environmental Protocol, Annex IV, Art. II.

[54] For more discussion on Annex IV of the Environmental Protocol, see Joyner, Chapter 5 in this book.

[55] For further discussion, see Vidas, Chapter 4 in this book.

[56] See *ibid.*, for a brief account of the role of the Committee for Environmental Protection.

[57] Environment Protocol, Annex I, Arts. 1–3; the environmental impact process includes Initial and Comprehensive Environmental Evaluations.

classification societies) would have certain advantages for and be yardsticks for the impact review of known and planned Antarctic marine operations.

The role of the flag state, however, should not be underestimated when considering application of the Code. Under the Code, each flag state remains responsible for issuing a prospective polar ship with a Permit to Operate in polar waters. The flag state, particularly if a party to the Antarctic Treaty, will fully understand the intent of any stringent requirements of the Environmental Protocol and its obligation when operating in Antarctic waters designated by the IMO as a MARPOL Special Area. Working closely with a ship classification society, which will have detailed technical expertise, a flag state can designate an appropriate polar class for the ship and approve all certifications. Inspections for compliance, for instance conducted by a flag state's maritime authority in consultation with COMNAP, could take place prior to sailing into Antarctic waters. This self-policing aspect is a central tenet of Antarctic cooperation and consensus under the Treaty. With the Environmental Protocol now in force, there is potential for the development of an effective enforcement and monitoring system under the Treaty. An existing Polar Code with international standards, coupled with the support of the IMO and the ship classification societies, adds considerable legitimacy to any collective enforcement efforts involving Antarctic ships. Since there are no 'port or coastal states' considered under the Antarctic Treaty, an alternative approach might be to conduct Treaty inspections from a port of departure, outside of Antarctica. Even without a Permit to Operate requirement, under an initially non-mandatory Code, inspection procedures for compliance (concerning both ship and crew standards) with Treaty and non-Treaty nations will be necessary.

One continuing issue is whether the Antarctic marine environment can be considered similar to the Arctic with regard to the risk (from ice damage) and safety of ship operations. As already noted in this chapter, there is considerably more multi-year ice in the Arctic Ocean compared with the Southern Ocean, and most specialised Antarctic ships operate primarily in the summer (in open water or first-year ice). From another perspective, Antarctic waters can be viewed as significantly more 'hazardous' due to the isolation of the Antarctic continent and the very long transits that would be faced by damaged ships. However, lack of spill response and few adequate ports and fuelling depots are common and serious problems for both regions. While it is believed the polar ship construction standards proposed will satisfy all Antarctic ice conditions, adjustments could have been made to the polar ship classes to include Antarctic-specific considerations. The Code also provides essential protection in polar waters over and above SOLAS and STCW requirements for lifesaving and survival equipment, fire safety, communications and crew training. Each of these requirements, in particular special crew training and qualifications and unique communications standards, could have incorporated Antarctic-specific criteria.

Flexibility and a strategy of inclusion by the IMO were viewed as keys to the Polar Code's application to the Southern Ocean. Expanded reference to the Antarctic within the Code seemed necessary to gain further acceptance from the Antarctic community. The Environmental Protocol has brought new obligations to all Antarctic marine operators – environmental impact assessment; design, construction, manning and equipment concerns; and requirements for emergency preparedness and contingency plans. The draft Polar Code encompassed standards to address each of these more stringent requirements. Perhaps additional issues requiring broad consideration should have included monitoring and inspection measures (for compliance), future liability rules to be developed under the Environmental Protocol, and Antarctic-specific training requirements.

Now, however, it has become clear that the Code will, after all, not apply to the Southern Ocean. Once the draft Polar Code reached the IMO community at large, a debate ensued as to the applicability of the Code to Antarctic waters and whether or not the Code should be a mandatory instrument. At the IMO Marine Safety Committee meeting of 19–28 May 1999, several significant decisions were taken for further work:

1. only recommended guidelines should be developed for SOLAS ships operating in ice-covered waters;
2. guidelines applicable in areas north of 60° North that are ice-free should be resolved; and
3. Antarctic waters are to be excluded from the application of the guidelines, unless the Antarctic Treaty members decide otherwise.[58]

In a follow up, the Antarctic Treaty Consultative Parties at their Twenty-Third Consultative Meeting in Lima, Peru, decided to develop a separate set of guidelines for Antarctic shipping. A meeting of experts is to be held in London in April 2000 to begin the process.[59]

ASSESSING THE POTENTIALS OF THE POLAR CODE

From technical as well as practical navigation perspectives, a polar code of navigation could be implemented today, and it would provide quite effective standards for ships operating in *both* polar regions. However, from a political perspective, several important obstacles remain. Most significantly, the draft Code required the inclusion of more Antarctic-specific language, including direct reference to the Antarctic Treaty and the Protocol on Environmental Protection. Also, a key conceptual difference remained between the two polar oceans – the Southern

[58] See IMO, 'Outcome of Discussion at the 71st Session of the Maritime Safety Committee', doc. XIII ATCM/IP 111, of May 1999, submitted by IMO at the XXIII Antarctic Treaty Consultative Meeting, Lima, Peru, 24 May–4 June 1999; available at www.rree.gob.pe/conaan/meeting1.htm.

[59] See para. 105 of the *Final Report of the XXIII Antarctic Treaty Consultative Meeting*, as well as Decision 2 (1999); Annex B in *ibid.*

Ocean has been designated a MARPOL Special Area; for the Arctic Ocean, no such concept has been accepted by the circumpolar nations. From an environmental perspective, adjustment of the polar ship classes (PC1 through to PC7) may have been necessary to address adequately Antarctic summer operations. One class could have recognised that many ships, particularly those in Antarctic waters, spend nearly all their voyage in open water, albeit with the presence of glacial ice in the form of tabular or other icebergs.

Despite these obstacles and potential adjustments, the obvious usefulness of a new, bi-polar navigation code would have been to permit polar ships to operate in both polar oceans under uniform standards. Polar research ships, logistics or support vessels, and tourist ships would thus be subject to a consistent set of international standards. Issues involving crew qualifications (including ice piloting), structural certification, navigation equipment, discharges, insurance and other safety requirements would be more effectively addressed under the framework of an IMO-sanctioned Polar Code. This approach contrasts with the development of distinct Antarctic ship standards and operating procedures by the Antarctic Treaty Consultative Parties.

In all likelihood, a near-term, mandatory Polar Code would not be implemented by the IMO following the initial reviews of Canada's draft document. A more plausible and perhaps acceptable strategy would be for the IMO to implement the Polar Code in a non-mandatory form for a specified period (for example, ten to fifteen years). The IMO's long-term goal would nevertheless be eventually to make the Code a mandatory instrument, complementary to the mandatory provisions of SOLAS, MARPOL 73/78 and the STCW. During what might prove to be an effective evaluation period, the Code could have been adjusted to include future Antarctic requirements as the Environmental Protocol is implemented. Future deliberations and agreed measures of the Arctic Council, for example on issues related to sustainable development, might also result in appropriate changes to the Polar Code. Moreover, a non-mandatory phase would allow sufficient planning for new polar ships to be designed under the future Polar Code, and allow adequate preparation of certification processes for mariner ice qualifications.

A lengthy period of non-mandatory IMO status of the Polar Code would serve a useful purpose for the harmonisation work of the IACS. One of the most important benefits of the Code's development was the close working relationships established between polar ship operators and technical specialists in ship design and engineering. This merger of technical expertise with broad operational experience was essential to the IACS and its quest for unified design standards. This process can continue with IMO encouragement and support. A non-mandatory status maintains IMO sanction and provides the IACS with an incentive to promulgate polar ship classes and other safety and structural design rules. Without future IMO involvement, the IACS would find it difficult to proceed with the development of the specific technical details required of all polar ship rule-making.

CONCLUSIONS AND THE FUTURE

This chapter has aimed at showing that the emerging Polar Code can be a practical marine safety and environmental protection instrument – a timely, precautionary response to the entering into force of the Antarctic Environmental Protection Protocol and the ongoing Arctic environmental cooperation, now also within the Arctic Council. It is the first international code that deals with the design, construction and operation of polar ships. Although previous standards have been separately developed by circumpolar states and individual ship classification societies, the draft Code provides the IMO with a framework of provisions to complement already established international standards. The vision was that ships built to a new Polar Code would be much safer and would have access to both Arctic and Antarctic waters. The key advantage of such a Code is that it would apply to all polar ships irrespective of their flag.

It is clear that the draft IMO Polar Code and the IACS unified rules are very much works in progress. Significant bi-polar, political attention has also been drawn to the harmonisation process. This visibility and the needed political support could have assisted the further refinement and adjustment of the Code to specifically Antarctic issues, as well as to specifically Arctic ones. Eight Arctic governments at the 1997 AEPS ministerial meetings in Alta were of the view that the Arctic countries should take joint action to promote completion of the Polar Code.[60] A report of the Twenty-Second Antarctic Treaty Consultative Meeting in Tromsø, Norway, reveals significant interest and discussion of the draft Polar Code; it contains key language emphasising that Consultative Parties should provide the IMO with relevant information of importance to the Code.[61] COMNAP was requested to develop training guidelines for Antarctic navigation and compile current Antarctic shipping standards.[62] However, the Antarctic Treaty Consultative Parties, following up on the IMO's decision of May 1999, decided at their June 1999 Consulatative Meeting in Lima to develop a separate set of guidelines for Antarctic shipping.

The evolution of an international polar code of navigation embodies one of the major themes of this book: it is the result of complex interrelationships among global, regional and national regimes for protection of the marine environment. Each level of regulation, coordination and legal basis has influenced the broad features of the Polar Code. International cooperation has been focused appropriately on the IMO and a host of stakeholders with polar marine expertise. Approaching the twenty-first century, the Code embodied an opportunity to

[60] See para. 15 of the Declaration on the Arctic Environmental Protection Strategy, signed by the Arctic ministers in Alta, Norway, 13 June 1997.

[61] See paras. 85–96 of the *Final Report of the Twenty-Second Antarctic Treaty Consultative Meeting, Tromsø, Norway, 25 May–5 June 1998* (Oslo: Norwegian Ministry of Foreign Affairs, 1998).

[62] *Ibid.*, para. 93.

become a unique and highly relevant bridge between ocean activities in the Arctic and Antarctic.

At the end of the twentieth century, however, the initial development of a 'bi-polar' code of navigation has met an uncertain future. Nevertheless, one key impact of the Polar Code initiative has been to mobilise the polar maritime community, particularly those involved in the Antarctic, to action regarding enhanced standards for polar ships. The further evolution, now of an 'Arctic code' and 'Antarctic guidelines', will certainly continue to reflect a central theme of this book – the examination of complex international instruments to protect the polar marine environment and enhance the safety of all human activities in both Arctic and Antarctic waters.

Index of international instruments and national legislation

Treaties and other international instruments

National legislation

Australia

Canada

Russia/Soviet Union

United States

Subject index

Printed in the United States
120374LV00005B/46-66/A

9 780521 032964